Refining Petroleum for Chemicals

A symposium co-sponsored by the Division of Petroleum Chemistry and the Division of Industrial and Engineering Chemistry at the 158th Meeting of the American Chemical Society, New York, N.Y., Sept. 10–12, 1969.

L. J. Spillane and H. P. Leftin,
Symposium Co-chairmen

ADVANCES IN CHEMISTRY SERIES **97**

AMERICAN CHEMICAL SOCIETY

WASHINGTON, D. C.   1970

Coden: ADCSHA

Copyright © 1970

American Chemical Society

All Rights Reserved

Library of Congress Catalog Card 76-140417

ISBN 8412-0109-9

PRINTED IN THE UNITED STATES OF AMERICA

Advances in Chemistry Series
Robert F. Gould, *Editor*

Advisory Board

Frank G. Ciapetta

Gunther Eichhorn

Frederick M. Fowkes

F. Leo Kauffman

Stanley Kirschner

John L. Lundberg

Fred W. McLafferty

William E. Parham

Jack Weiner

AMERICAN CHEMICAL SOCIETY PUBLICATIONS

FOREWORD

ADVANCES IN CHEMISTRY SERIES was founded in 1949 by the American Chemical Society as an outlet for symposia and collections of data in special areas of topical interest that could not be accommodated in the Society's journals. It provides a medium for symposia that would otherwise be fragmented, their papers distributed among several journals or not published at all. Papers are refereed critically according to ACS editorial standards and receive the careful attention and processing characteristic of ACS publications. Papers published in ADVANCES IN CHEMISTRY SERIES are original contributions not published elsewhere in whole or major part and include reports of research as well as reviews since symposia may embrace both types of presentation.

CONTENTS

Preface vii

1. Some Recent Developments in the Gas-Phase Pyrolysis of Hydrocarbons 1
   Sidney W. Benson, Stanford Research Institute, Menlo Park, Calif.

2. The Chemistry of Aromatics Production *via* Catalytic Reforming .. 20
   E. L. Pollitzer, J. C. Hayes, and Vladimir Haensel, Universal Oil Products Co., Des Plaines, Ill.

3. Chemistry of Hydrocracking 38
   G. E. Langlois and R. F. Sullivan, Chevron Research Co., Richmond, Calif.

4. Secondary Reactions of Olefins in Pyrolysis of Petroleum Hydrocarbons 68
   Tomoya Sakai, Kazuhiko Soma, Yoichi Sasaki, Hiroo Tominaga, and Taiseki Kunugi, University of Tokyo, Tokyo, Japan

5. Rate Modeling for the Butane–Butenes System 92
   John Happel, Miguel A. Hnatow, and Reiji Mezaki, New York University, Bronx, N. Y.

6. Some Problems of the Cracking Mechanism of Hydrocarbons 110
   R. S. Magaril, Tyumen Industrial Institute, Tyumen, U.S.S.R.

7. The Chemical Refinery in Perspective 123
   Matthew F. Stewart and James T. Jensen, Arthur D. Little, Inc., Cambridge, Mass.

8. Product Optimization in the Petrochemical Refinery 137
   Harry M. Walker, Monsanto Co., Alvin, Tex.

9. The Manufacture of Propylene 153
   Alvin H. Weiss, Worcester Polytechnic Institute, Worcester, Mass.

10. Kinetics of the Demethylation of Methylcyclohexane 179
    M. M. Johnson and H. J. Hepp, Phillips Petroleum Co., Bartlesville, Okla.

11. Catalytic Dehydrogenation of Higher Normal Paraffins to Linear Olefins 193
    James F. Roth, Joseph B. Abell, Loyd W. Fannin, and Andrew R. Schaefer, Monsanto Co., St. Louis, Mo.

12. **The Octafining Process for Isomerization of C_8 Aromatics Containing Ethylbenzene** 204
    H. F. Uhlig and W. C. Pfefferle, Engelhard Minerals and Chemicals Corp., Newark, N. J.

13. **Process Evaluation of Improved Solvents for Butadiene Recovery** .. 215
    G. D. Davis, E. C. Makin, Jr., and C. H. Middlebrooks, Monsanto Co., St. Louis, Mo.

14. **Tetraethylene Glycol—A Superior Solvent for Aromatics Extraction** 228
    G. S. Somekh and B. O. Friedlander, Union Carbide Corp., Tarrytown, N. Y.

15. **A New Solvent for Aromatics Separation** 242
    Julian Feldman, U. S. Industrial Chemicals Co., Cincinnati, Ohio

16. **Carbon Black from Petroleum Oil** 264
    T. A. Ruble, Continental Carbon Co., Houston, Tex.

17. **Petrochemical Feedstocks from Residuals** 271
    A. R. Johnson, S. Alpert, and L. M. Lehman, Hydrocarbon Research, Inc., New York, N. Y.

Index 289

PREFACE

This volume reviews the processing by which liquid petroleum may be converted to a maximum yield of chemicals, not fuels. Petroleum is considered to include natural gasoline, condensate, crude oil, or fractions of these materials obtained by physical methods. Excluded are liquefied petroleum gas (LPG), ethane, and natural gas. Processing is restricted for the most part to production of the first materials saleable as chemicals. In these contexts an alternate name for the symposium might well be the Technology of the Petrochemical Refinery.

A considerable volume of literature has been published on the production of petrochemicals and even a few prophetic papers on the petrochemical refinery (1, 3, 4, 5). To date, however, the petrochemical refinery concept has been used in two slightly differing contexts: one wherein the plant output (using crude oil as feed) is exclusively chemicals (4); the other wherein the production of petrochemicals is integrated into an existing fuels refinery (1, 5). The former has been reduced successfully to practice on a commercial scale only in one acknowledged instance (6), and this operation has been restricted to condensate as feedstock. The second context does not differ much from present petrochemical operations of oil companies except in degree. From a technological viewpoint the former, operating on heavy crude oil, would represent perfection of the petrochemical refinery concept and would demand the most of new technology. Hence, in the present symposium the chemistry and new processing techniques will be directed exclusively to the production of chemicals, not fuels.

Whether or not the production of chemicals from crude oil can be independent of the production of fuels has still not been determined. It will indeed be a new industry, but its future depends largely on the development of new technology. Technology employed so far, however, has largely been an adaptation of processes designed for ethylene production from gaseous hydrocarbons or for maximum gasoline production by petroleum refining.

Since the introduction of hydrodealkylation processes some 11 years ago (2), the pace has quickened in the development of processes specifically designed to manufacture chemicals from liquid petroleum feedstocks. This volume is designed to unify and stimulate further development of new concepts, ultimately leading to establishment of the petrochemical refinery as an independent and economically viable enterprise.

Any refinery is based upon taking a crude mixture of raw materials and producing a spectrum of desired products. It differs from a chemical plant primarily in that several products are desired—*i.e.*, the coproduct concept. In common with the fuels refinery, the technology of the petrochemical refinery is influenced by four fundamental considerations:

(a) Selection of raw material
(b) Definition of product mix
(c) Selection of basic processes
(d) Integration of processes and products

All but the last, which depends so much on individual circumstances, are discussed to some degree in this book. Emphasis is on basic processing, the main point of impact for chemical science.

With crude oil as feed the production of chemicals by known methods rests upon four types of processing:

(1) Thermal pyrolytic cracking
(2) Catalytic reforming
(3) Catalytic or thermal hydrocracking
(4) Separation and purification

The papers in this volume have been organized around four main topics which combine both the broad considerations of the petrochemical refinery and the specific types of processing:

(1) Review of the basic chemical reactions employed in the present-day petrochemical refinery.

(2) Description of new chemical developments in dehydrogenation and cracking reactions.

(3) Discussion of the products and economics of the petrochemical refinery.

(4) Presentation of typical new processing concepts—chemical as well as physical—applicable to the petrochemical refinery of the future.

Although the papers in this volume do not represent an exhaustive review of the petrochemical refinery concept, they do offer for the first time a "status report" on the general subject. Former publications have directed themselves to some specific phase, usually economic or engineering, with little attention directed to their interface with chemistry and process development. Hopefully, the papers presented here will offer a starting point for continuing development of a fascinating and most important subject on an integrated basis.

Literature Cited

(1) Beavon, D. K., *Hydrocarbon Processing* **1969**, 48 (1), 110.
(2) Bethea, S. R., Heinrich, R. L., Souby, A. M., Yule, L. T., *Ind. Eng. Chem.* **1958**, 50, 1245.

(3) Gambro, A. J., Caspers, J., *World Petrol.* **1967,** 38 (4), 88–94.
(4) Gambro, A. J., Caspers, J., Newman, J., *World Petrol.* **1968,** 39 (13), 36–73.
(5) Struth, B. W., List, H. L., Symposium on New Developments in Petrochemicals, *National Meetg. AIChE, 64th,* March 16-20 (1969).
(6) Walker, H. M., Symposium on Steam Cracking of Heavier Feedstocks, *National Meetg. AIChE, 64th,* March 16-20 (1969).

Leo J. Spillane

Gulf States Asphalt Co., Inc.
Houston, Tex.
May 1970

1

Some Recent Developments in the Gas-Phase Pyrolysis of Hydrocarbons

SIDNEY W. BENSON

Department of Thermochemistry and Chemical Kinetics, Stanford Research Institute, Menlo Park, Calif. 94025

Recent work on elementary steps of importance in hydrocarbon pyrolysis are described and summarized from a quantitative and predictive point of view. This includes the normal H atom metathesis reactions, addition of radicals to unsaturates, and fission reactions responsible for chain propagation, as well as the initiation and termination steps. The large negative activation energies characteristic of kinetic chain lengths put an upper limit of about 1000°K on the temperature at which propagation steps are important. Arrhenius parameters for molecular reactions of unsaturated hydrocarbons are discussed, in particular Cope rearrangements and 1,5-H shift reactions. Some anomalies from shock tube data are described briefly as are the kinetic properties of small biradicals. Some difficulties of kinetic modeling are discussed along with the importance of deviations of unimolecular rate constants from Arrhenius behavior at high temperatures.

From a scientific, or a technological, or an economic point of view, the pyrolysis of hydrocarbons must rank as one of the most important chemical processes of our 20th century. Research activity on the pyrolysis of hydrocarbons has mirrored this intense interest since the early part of the century. These studies have yielded a number of simple conclusions. One is that pyrolytic systems are an extremely complex assembly of chemical reactions taking place mainly through the aegis of a relatively small number of very active free radicals.

The kinetic understanding of the role played by free radical processes stems from two historic papers by F. O. Rice (52, 53) outlining plausible chain sequences to account both for the products of pyrolysis and the over-all concentration dependence of the rate. Although all the important chain steps and the cogent thermochemical arguments for their importance were made in these two papers by Rice, it has become common practice to refer to the Rice-Herzfeld mechanism of chain pyrolysis on the basis of a third paper (54). This third paper differs from the earlier ones only in extending the Rice scheme to include acetone and acetaldehyde and also in assigning A-factors and activation energies to all of the individual chain steps in an effort to reproduce the over-all kinetic parameters.

Two important elementary step reactions were described in these radical chains of Rice. One was the ubiquity of the metathetical reaction of H atom abstraction by free radicals from hydrocarbons. The second was the rapid unimolecular decomposition of free radicals into olefins and secondary atoms or radicals, as a chain step competitive with metathesis.

A few years later, a third type of reaction was added to the scheme, the isomerization of large radicals by internal abstraction of H atoms (9). This was shown (41) to account satisfactorily for the product distribution arising from the pyrolysis of long chain hydrocarbons (e.g., n-$C_{16}H_{34}$). Very little has happened in the approximately 25 years since the last of these contributions to alter our conceptual understanding of the kinetics of hydrocarbon pyrolysis. Instead, the very extensive research done since then has generally been devoted to determining the quantitative kinetic parameters associated with the elementary step reactions of the pyrolysis chain. Much of this work has been summarized in some recent books (62) and reviews (26, 51).

One new field related to hydrocarbon pyrolysis of small ring compounds (3- and 4-membered rings) has been developed in the past decade, and that is the subject of biradicals. These are discussed briefly. In the present article, I consider in some detail the present status of our experimental and theoretical understanding of these step reactions and how they relate to the very practical problems of kinetic modeling of pyrolysis systems.

Molecular Reactions of Hydrocarbons

One of the early controversies in the field of pyrolysis revolved about the question of the importance of concerted, molecular elimination reactions of hydrocarbons to give olefins and alkanes as products. Numerous experiments (51) have given a decisive answer to this problem in favor of the completely radical mechanism for alkane pyrolysis. At the same time, ironically, very clear cut evidence has been elicited to show that both mono- and polyolefins can react quite readily in concerted, molecular processes to form isomeric products or fission products (26). These usually involve the formation of a cyclic transition state with a six-membered ring. The Cope rearrangement provides one of the oldest examples (25) of such an isomerization:

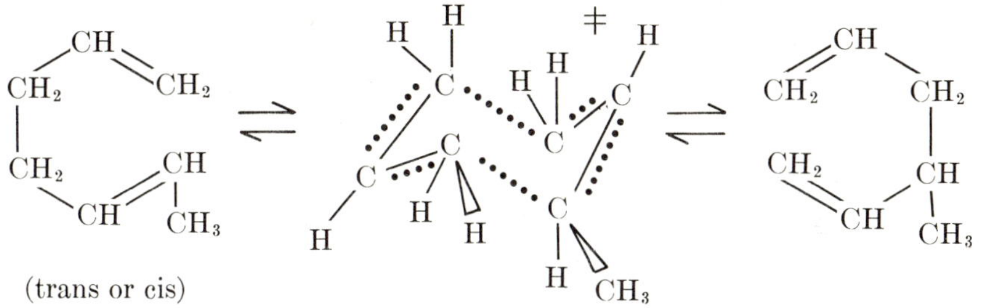

$$\log [k_{\text{cis}}(\text{sec}^{-1})] = 10.66 - 36.72/\theta$$
$$\log [k_{\text{trans}}(\text{sec}^{-1})] = 10.39 - 35.36/\theta$$

where $\theta = 2.303\, RT$ in units of kcal/mole.

In the illustration shown, the transition state involves a six-membered ring, and it is presumed on structural grounds that it has the thermodynamically favored conformation of the chair form of cyclohexane. Since it also has a mobile π-electron system, it seems reasonable to picture it as two allylic radicals interacting with each other *via* their terminal (carbon atoms 1 and 3) π-electrons. For such a structure, the cis diolefin will have its bulky methyl group in the structurally unfavorable quasi-axial position. The higher activation energy of 1.36 kcal/mole observed for the reaction of the cis isomer supports such a structure.

The "ene" reaction, or more prosaically, the 1,5-H shift reaction, provides another more recent example of a molecular reaction:

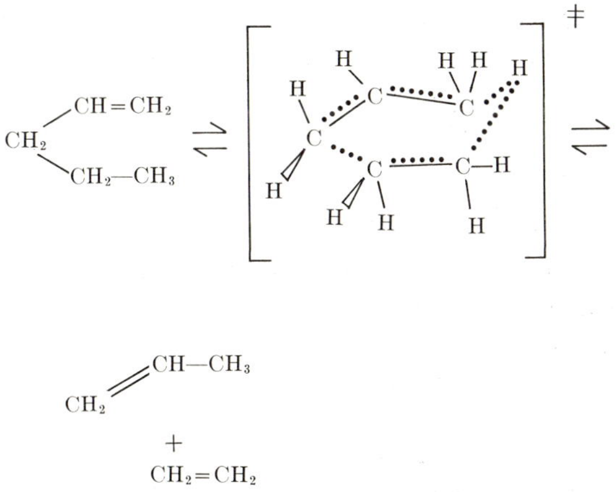

Such reactions were initially studied for the vinyl ethyl ethers (14, 70).

A second example of this is provided by the pyrolysis of *cis*-1-methyl-2-vinylcyclopropane, which decomposes at about 120°C lower than its trans isomer to give *cis*-hexa-1,4-diene (*17, 18, 58*):

$$\text{Log} [k\text{cis (sec}^{-1})] = 11.03 - 31.24/\theta$$

For both examples of 1,5-H shifts, the transition state is probably best regarded as involving a puckered five-membered ring of carbon atoms with the transferring H atom out of the plane. In one of the simplest examples of the 1,5-H shift reaction, however, namely the suprafacial transfer of the secondary H atom of cyclopentadiene around the C-5 ring, the ring itself is forced to be planar (*46, 57*).

The *trans*-1-methyl,2-vinyl compound decomposes (*20*) to give 92% of the same cis diene product and also 8% of 4-methylcyclopentene. The diene rate constant is:

$$\log [k_{\text{trans}} (\text{sec}^{-1})] = 14.74 - 48.64/\theta$$

with very different Arrhenius parameters from the cis isomer. The mechnism is different as well. The trans compound decomposes *via* a biradical intermediate (*11*).

The low activation energies with which these concerted reactions can take place ($\sim 35 \pm 5$ kcal) make them very important in all pyrolysis systems in which olefins are major products. They lead to an enhanced fragmentation of large monoolefins and have probably been responsible for much misidentification of cracking rate parameters.

While it is not possible to make *a priori* estimates of the activation energies of these reactions, such estimates can be made from values observed in homologous compounds. The A-factors for concerted processes can be estimated on the basis of some simple rules proposed by Benson and O'Neal (*10*). These rules suggest that the chief loss of entropy in forming the transition state lies in converting hindered internal rotations in the open chain compounds into more or less "stiff" out-of-plane torsions in the transition state. In its crudest form, one can calculate the A-factors for these cyclic transition states by assuming that one loses about 4.0 gibbs/mole of entropy per rotor in forming the transition state. The A-factor is related to $\Delta S^{\ddagger}$, the entropy of activation, by (*4*):

$$A = \left(\frac{ekT_m}{h}\right) e^{\Delta S^{\ddagger}/R}$$

where T_m is the mean temperature of the range used. At 600°K the factor $ekT_m/h = 10^{13.55}$ sec^{-1}.

Table I lists a few representative examples of estimated A-factors for some hydrocarbon reactions proceeding through six-membered ring, cyclic transition states. We see that although the simple rotor rule is fairly effective in estimating A-factors, it generally errs on the high side. The more complex analysis (*10*) which takes into account all the changes in frequencies in going to the cyclic complex does a somewhat better job of estimating the A-factors.

Metathesis Reactions

While the molecular reactions of olefins are of considerable importance, free radical reactions are still the major route for pyrolysis of both alkenes and alkanes. For metathesis reactions of free radical with molecules, methods have now become available for predicting both lower and upper limits of the A-factors for such reactions (*3, 6*).

The upper limits are provided by the collision frequency of the radical–molecule pair which is about $10^{11.3}$ 1/mole-sec at 400°K. This result can also be arrived at from transition state theory by assuming that the centers of the colliding pair lie on a spherical shell 3.5 A in radius and 0.10 A thick. This corresponds to a tight transition state since the small amplitude of motion of 0.10 A is characteristic of bond vibration amplitudes in molecules. The only bimolecular reactions whose A-factors come close to this upper limit are the methathesis reactions of I atoms (*27*) for which the A-factors equal, or slightly exceed, the collision frequency.

The lower limits are obtained by equating the intrinsic entropy of the transition state to a molecule of similar structure and molecular weight. Thus, for the reaction $H + C_2H_6 \rightleftarrows C_2H_5 + H_2$, we can assume

Table I. Measured Arrhenius Parameters and Estimated A-Factors Ring Cyclic

	Reactant	Product
a	$CD_2=CHCH_2CH_2CH=CD_2$	$CH_2=CHCD_2CD_2CH=CH_2$
b	(1,5-heptadiene skeleton)	1, 5-heptadiene (cis + trans)
c	(diene)	(cyclohexene)
d	(allene-diene)	(methylene cyclohexene)
e	(substituted diene)	(substituted cyclopentane)
f	(diene)	(diene)
g	(diene)	(diene)
h	(vinyl cyclopropane)	(cyclopentene)
i	(cyclohexadiene)	(benzene) $+ H_2$

that the transition state has an intrinsic entropy in excess of C_2H_6. By intrinsic entropy is meant the entropy of the species corrected for symmetry, optical isomerism, and electron spin (4).

Thus, at 400°K

$$\Delta S^{\ddagger}_{400} = S^{0\ddagger}_{400}(C_2H_6 \ldots H) - S^0_{400}(C_2H_6) - S^0_{400}(H)$$
$$\geq R\ln 2 + R\ln 6 - S^0_{400}(H)$$
$$\geq 1.4 + 3.6 - 28.9 = -23.9 \text{ gibbs/mole}$$

for Unimolecular Reactions Proceeding through Six-Membered Transition States

E_{obs}, kcal/mole	Log [A(sec^{-1})], observed	Log [A(sec^{-1})], estimated[a]	Ref.
35.5	11.1	11.1 (10.9)	63
35.0	10.85	11.3 (10.9)	25
29.9	11.85	11.5 (11.7)	42
28.47	9.97	10.8 (11.2)	23
35.2	9.06	9.6 (9.9)	31
32.76	11.24	11.1 (12.3)	19
32.5	11.8	11.9 (12.1)	24
31.2	11.03	11.1 (12.3)	17, 18, 58
43.8	12.36	12.4 (—)	13, 21
42.7	12.02		

[a] Values in parentheses are obtained by making proper corrections for symmetry (reaction path degeneracy) and subtracting from $S^{0\ddagger}$, 4.0 gibbs/mole for every rotor involved.

On correcting to moles/liter standard states, this leads to an A-factor for the metathesis:

$$A = \frac{ekT_m}{h} e^{\Delta S\ddagger/R} \geq 10^{13.4 - 3.2} = 10^{10.2} \text{ l/mole-sec}$$

The reported "best" value is $10^{11.0}$ l/mole-sec (4).

For the similar reaction of $CH_3 + C_2H_6 \rightleftarrows CH_4 + C_2H_5$, we find:

$$\Delta S^{\ddagger}_{400} = S^{0\ddagger}_{400}(C_2H_6 \ldots CH^3) - S^0_{400}(C_2H_6) - S^0_{400}(CH_3)$$
$$\geq R\ln 12 + R\ln 2 + [S^0_{400}(C_3H_8) - S^0(C_2H_6)] - S^0_{400}(CH_3)$$
$$= 6.4 + 11.2 - 49.2 = -31.6 \text{ gibbs/mole}$$

Hence, $A \geq 10^{13.4 - 4.9} = 10^{8.5}$ l/mole-sec

In the above calculation, $R\ln 2 = 1.4$ gibbs/mole has been added to the entropy of the transition state to allow for the increased moment of inertia of the transition state over that of the model, propane. The reported experimental A-factor of $10^{8.5}$ l/mole-sec is sufficiently close to the lower limit to suggest both that the transition states for CH_3 abstractions of H from RH molecules are extremely "tight" and that the experimenal A-factors are probably about a factor of 2 too low. Most of these reactions have been measured only once and that one some time ago.

The activation energies for radical–molecule metathesis reactions are generally low. If we define the "intrinsic activation energy" of a reaction as the activation energy for the reaction when carried out in its exothermic direction, it has been observed that most intrinsic activation energies fall into the range 8 ± 3 kcal/mole. The activation energy for the reverse reaction, which is then endothermic, equals the "intrinsic activation energy" plus the endothermic heat of reaction.

Intrinsic activation energies tend to be higher for atoms and radicals in the first two rows of the periodic table (*e.g.*, F, H, CH_3, O, etc.) and lower for homologous species in the later rows, such as Br, I, HS·, and Cl. They tend to be lower in a homologous series of reactions as the exothermicity increases. There is currently no simple method for predicting them, although the bond-energy–bond-order scheme (BEBO), originally proposed by Johnston and Parr (*32*) and recently extended (*43, 44, 45*) has been very successful. The accuracy is about ± 2 kcal.

Radical Reactions with Unsaturates

Recent measurements of the C–H bond strengths in benzene (*55*) and C_2H_4 (*56*) yield values of 112 kcal/mole for the former and a lower limit of 107 kcal/mole for the latter. These are so much higher than other C–H bond strengths that it suggests that abstraction of H atoms from an aromatic ring, or a double-bonded carbon atom, must be very rare. Instead, a much faster process is usually the addition of atoms or radicals to unsaturated carbon atoms to produce radicals.

The chlorination and bromination of benzene and benzene derivatives have been interpreted successfully and quantitatively in terms of such a mechanism (*55*). The first product is actually a reactive cyclohexadiene which is dehydrohalogenated rapidly by radical attack to the observed product:

$$Cl + \underset{}{\bigcirc} \underset{-1}{\overset{1}{\rightleftarrows}} \underset{}{\bigcirc}\!\!\!\!\overset{H}{\underset{Cl}{\diagup}}$$

$$Cl_2 + \underset{}{\bigcirc}\!\!\!\!\overset{H}{\underset{Cl}{\diagup}} \underset{-2}{\overset{2}{\rightleftarrows}} \underset{}{\bigcirc}\!\!\!\!\overset{H}{\underset{\underset{Cl}{H}}{\diagup Cl}} + Cl\cdot$$

$$Cl + \underset{}{\bigcirc}\!\!\!\!\overset{\overset{H}{\diagup Cl}}{\underset{Cl}{\diagdown H}} \underset{-3}{\overset{3}{\rightleftarrows}} \underset{}{\bigcirc}\!\!\!\!\overset{Cl}{\underset{Cl}{\diagup H}} + HCl$$

$$\underset{}{\bigcirc}\!\!\!\!\overset{Cl}{\underset{Cl}{\diagup H}} \underset{}{\overset{-1'}{\rightleftarrows}} \underset{}{\bigcirc}\!\!\!\!\overset{Cl}{} + Cl\cdot$$

The concerted elimination of HCl from the dichlorocyclohexadiene to form PhCl can be shown to be probably slower than the competing radical path to form PhCl.

Many anomalies in the reactions of CH_3 radicals with benzene and toluene can be explained by similar mechanisms (7, 50, 64). It is very likely that for C_2H_4 reactions with radicals, a similar addition–disproportionation mechanism holds. For higher olefins there is a verified competition between addition to the double bond and abstraction of a weakly bonded allylic H atom.

A-factors for the addition reactions are generally low, of the order of 10^8 l/mole-sec, suggesting a very tight transition state.

Isomerization of Radicals

Although isomerization of hydrocarbon radicals was one of the earliest reactions postulated to account for the products of hydrocarbon pyrolysis, the first direct evidence for such isomerization was relatively recent (22, 29, 30). It appears that internal abstraction of an H atom can take place relatively easily from the carbon atom in the 4th, 5th, or 6th position from the radical center. The transition state must of necessity be cyclic, and the strain-free six-membered ring appears to be favored relative to the five- or seven-membered rings, both of which have about 6 kcal/mole of ring strain.

In the oxidation of hydrocarbons such internal isomerization leads to the formation of unusual products. Thus, in the oxidation of 2,4-dimethylpentane (59), the major product is the 2,4-dihydroperoxide obtained by the route:

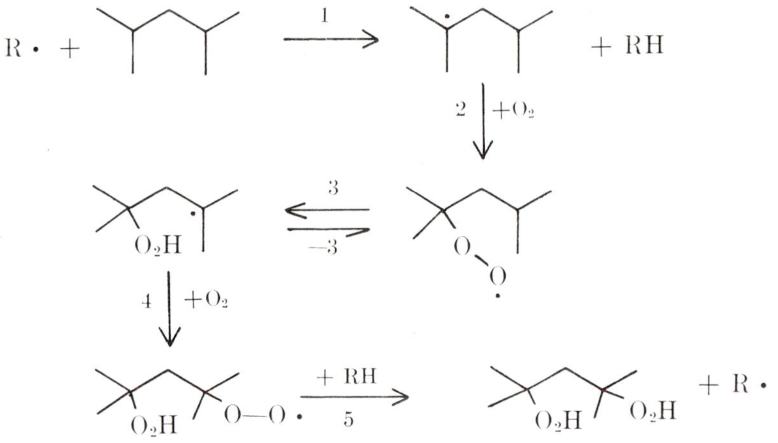

Very elegant studies, using D-isotope labeling (32, 33, 34, 35, 36), have shown that 1,2- and 1,4-H shifts in hydrocarbon radicals do not occur. This implies that they must have extremely high activation energies—for isopropyl radical in excess of 45 kcal/mole. There is no good evidence that larger saturated groups, such as alkyl groups, can migrate in free radical isomerizations.

Despite the potential interest in isomerization reactions, very little quantitative data exist. Only a few scattered and probably not very reliable activation energies have been reported, while two recent studies of 1,5-H atom shifts give unbelievably low values of the Arrhenius A-factors (22, 71).

Initiation Reactions

There are two important rules in selecting initiation reactions. The first is that unimolecular fission reactions, because of their high A-factors ($\sim 10^{16 \pm 1}$ sec^{-1}) will generally be faster than bimolecular reactions for which A-factors run about $10^{9 \pm 1}$ l/mole-sec. At 1 atm of reactant the ratio of rates for initiation is of the order of 10^9 in favor of the first-order processes. A rough rule of thumb is that the bimolecular initiations will be significant only if they have activation energies at least 40 kcal less than the competing first-order initiation. This is very rare. One example occurs in the pyrolysis of 1,3-cyclohexadiene (12). The initiation reaction is:

$$2 \: C_6H_8 \rightleftharpoons C_6H_7\cdot + C_6H_9\cdot$$

and it is only 22 kcal endothermic. The competing first-order process is:

$$C_6H_8 \rightleftharpoons C_6H_7\cdot + H$$

which is 72 kcal endothermic and negligible even at low pressures.

The second rule is that among competing first-order initiation reactions, generally the one having the lowest activation energy will dominate whether it involves initial reactants or products. Thus, in the hydrogenation of $C_2H_4 + H_2 \rightarrow C_2H_6$, the initial stages of the reaction are probably initiated by the bimolecular reaction:

$$H_2 + M \rightarrow 2H + M$$

However, this reaction which is 104 kcal endothermic is almost immediately replaced by the much faster initiation:

$$C_2H_6 \rightarrow 2CH_3$$

This latter is first order and has a lower activation energy of 88 kcal compared with 104 kcal for its competition.

At 800°K as soon as 0.01% of C_2H_6 is formed in the system, it will provide radicals faster than the bimolecular process, even at $H_2 = 1$ atm. Note that the H atoms essential for the propagation of the chain hydrogenation are provided by the rapid and thermoneutral, transfer reaction:

$$CH_3 + H_2 \rightleftharpoons CH_4 + H$$

It is very likely that some of the irreproducible induction periods observed in many pyrolysis reactions can be attributed to the transition from very slow to more rapid initiation as intermediates are formed in the system. Not much kinetic information has come from studies of these induction periods.

Chain Lengths

The kinetic chain length, λ, is defined as the ratio of rate of production of products to the rate of initiation of radicals. If we choose a chain in which H abstraction is irreversible, we can represent the chain length by:

$$\lambda = \frac{\text{rate of propagation}}{\text{rate of initiation}}$$

$$= \frac{k_p (R\cdot)(R'H)}{2k_i (R'H)} = \frac{k_p}{2k_i}(R\cdot)$$

Here we have assumed for simplicity that propagation and initiation arise solely from the parent hydrocarbon R'H. If we can simplify further by assuming uniform termination by R· radicals only, so that the steady-state concentration of R· is given by:

$$(R\cdot) = \left[\frac{k_i}{k_t}(R'H)\right]^{1/2}$$

then, substituting into the expression for λ, we have:

$$\lambda = k_p(k_t/k_i)^{1/2}(RH)^{1/2}/2$$

In general, the activation energy for k_p is small (8 ± 3 kcal), while that for k_t is zero, and E_i is very large, $\sim$ 80 to 100 kcal. In consequence, λ will appear to have a large negative activation of the order of $E_p - \frac{1}{2}E_i$ $\approx$ −37 kcal/mole. This places some very severe restraints on the use and accurate measurement of chain reaction rates. At the very lowest temperatures where chain lengths can be very long ($\sim 10^2$ to 10^3), the rates can be extremely sensitive to adventitious impurities such as O_2 and wall effects. There is a natural floor imposed, however, on the use of low temperatures since at temperatures below about 750°K, initiation rates are so slow that chain rates are too slow to measure.

There is also a natural ceiling temperature for chain studies at about 1000°K. Above 1000°K, λ has become so small that the rates and product distributions are dominated by the initiation and termination steps above.

Although the lower temperatures for chain studies can be reduced by using chain sensitizers (initiators) such as azo compounds or peroxides, it is amazing how few chain studies have been done under such conditions.

Termination Reactions

One of the big surprises of the recent past has been the observation that radical–radical termination processes *via* recombination or disproportionation have no activation energies and very high A-factors. A-factors for recombination reactions of alkyl recombinations vary from $10^{10.4}$ l/mole-sec for CH_3 radicals (28) to a low of about $10^{9.5}$ for *tert-*

butyl radicals (47). For CH_3 radicals at 400°K this corresponds to recombination at every singlet collision with an impact parameter of 3.5 A. One-fourth of all Me–Me collisions have singlet total spin, while three-fourths have triplet spin states.

These high collision efficiencies have some interesting consequences for the theory of unimolecular reactions. In particular, they imply that the preferred path for fission reactions involve appreciable storage of internal energy in two rotational modes (5). For pyrolysis reactions, they suggest that atom–atom recombination will usually contribute very little to over-all recombination relative to atom–radical or radical–radical reactions. The reason is that the atom–atom recombination is third order.

As a rough rule of thumb, because the rate constants show little structural sensitivity, we can usually attribute the bulk of recombination to that radical which is present in highest concentration in the system. This can be determined generally from the propagation steps.

Non-Arrhenius Rate Constants

Almost as old as quantitative, gas-phase kinetics is the appreciation of the fact that not all rate constants are well behaved. From the theory of unimolecular reactions it is possible to conclude that at sufficiently high temperatures, all gas-phase, unimolecular rate constants tend to depart from Arrhenius-like behavior and become pressure dependent. If these reactions correspond to fission into two fragments, there is a corollary of this high temperature behavior for the reverse, association reaction of these two fragments. The second-order association reaction also becomes pressure dependent (*i.e.*, the association tend towards third order) and can have an appreciable negative activation energy.

In the past two decades, advances in technology have produced increased interest in high temperature (above 1000°K) reaction systems where such behavior is observed. The departure from Arrhenius behavior can be quite marked.

In our own laboratories, a new technique called "very low pressure pyrolysis" (VLPP) (60) has permitted us to make direct studies on high temperature rate processes at sufficiently low pressures that secondary processes are negligible. Under such conditions (1200°K, 10^{-4} torr), we have found that the four-center elimination of HI from isopropyl iodide can be as much as 1/1000 as fast as one would estimate from extrapolating the Arrhenius rate constant from lower temperatures.

Another example is provided by the chain hydrogenation of acetylene (8). One step in the chain, $H\cdot + C_2H_2 \rightarrow \dot{C}_2H_3$, will be essentially a third-order reaction and have a negative activation energy of as much as -17 kcal/mole at 1 atm and 1400°K. Since there is no simple way of describing either the concentration or temperature dependence of such non-Arrhenius rate constants, one must have direct recourse to the theory of unimolecular reactions.

Shock Tube Techniques

One of the most important techniques for kinetic studies at high temperatures to emerge during the post war years has been the shock tube (2). Two modes for its operation have been in common use. One is to observe chemical reactions, *via* spectroscopic detectors, directly behind the incident shock wave. A second is to use the shock wave reflected from the end of the shock tube to heat a gas mixture for longer times (1 to 10 msec) and then analyze the products chemically. Because the reflected shock wave has a peculiar time-temperature history, it does not yield very precise or accurate rate constants.

A useful variant of the reflected shock technique has been to add a chemical substance in trace amounts whose unimolecular rate constant is well known. The products of this trace reaction are then used as an internal standard for measuring the mean temperature in the system (66, 67, 68, 69). A number of rate constants for four-center and six-center elimination reactions measured in this way have been found to be in excellent agreement with rate constants obtained more conventionally at lower temperatures. However, almost all of the rate constants for simple fission of branched hydrocarbons obtained in this way have yielded anomalies (48) which have not yet been resolved.

Briefly, the shock tube data give A-factors for radical fission which are about 1 order of magnitude lower than that calculated from the A-factor for the reverse reaction (which is radical recombination) and the estimated entropy of the over-all reaction. In addition, the shock tube activation energies are from 3 to 6 kcal/mole higher than the bond dissociation energies for branched radicals.

It is possible that at higher temperatures ($> 1000°K$) radical recombinations may have small activation energies of the order of 2–6 kcal/mole. However, for this to resolve the problem would also require that the reported recombination rate constants for isopropyl and *tert*-butyl radicals be about 100-fold too high. A related solution is that our current estimates of the entropies of *tert*-butyl and isopropyl radicals may be too high by about 2 gibbs/mole at $1000°K$.

These anomalies may be restricted to radical reactions, but until they are resolved, they will inject considerable uncertainty into the utilization of shock tube data. A parallel, but not necessarily related anomaly, of shock tube data comes from studies of the dissociation of diatomic molecules behind incident shocks. Although the data can be fitted reasonably well with theoretically derived rate constants, there are invariably systematic deviations in such a direction as to suggest that rate constants with activation energies about 25% lower than the known bond strengths provide better fits to the data (37, 39). There is no theoretical basis for such a result, and it casts a considerable pall on the credibility of Arrhenius parameters obtained from shock tube measurements.

A recent indication of some of the problems associated with single pulse shock tube techniques has been reported by two independent groups working on the pyrolysis of cyclopropane (1, 15). Barnard and Seebohm (1) found that their rate constants for $C_2H_5Cl \rightarrow C_2H_4 + HCl$ (800°–1250°K) were in excellent agreement with conventional work in the range 650°–800°K. Cyclopropane isomerization to propylene also fitted conventional Arrhenius parameters from 800° to 1200°K. However, above 1200°K there was a change in products and almost no increase in apparent first-order rate constant from 1200° to 1600°K. Bradley and Frend (15) reported similar results. Their apparent rate constant in the range 1350°–1870°K was $\log [k(\sec^{-1})] = 4.75 - 11.6/\theta$. They also reported similar anomalies above 1350°K for C_2H_6 and above 1400°K for n-butane.

Biradical Reactions

One of the more fascinating developments currently enjoying vogue in hydrocarbon chemistry deals with the formation and reactivity of biradicals (16, 40). While they do not appear to have much relevance to normal hydrocarbon pyrolysis, they are too much of a key to our understanding of hydrocarbon chemistry not to be mentioned.

The simplest biradicals are the carbenes, such as $\ddot{C}H_2$, $\ddot{C}HCH_3$, etc. $\ddot{C}H_2$ is isoelectronic with O atoms and shows many analogous properties to O. In particular, it appears as though the $\ddot{C}H_2$ ground state is a triplet, while a slightly higher energy state is a singlet. In common with ^1O oxygen atoms, the singlet state is extremely reactive in abstraction addition and so-called insertion reactions:

$$\ddot{C}H_2 + >C=C< \rightarrow >\overset{CH_2}{\overset{\wedge}{C-C}}< \quad \text{(addition)}$$

$$\ddot{C}H_2 + RH \begin{cases} \rightarrow \dot{C}H_3 + \dot{R} & \text{(abstraction)} \\ \rightarrow R-CH_2-H & \text{(insertion)} \end{cases}$$

The triplet species does not insert, and mechanistically it appears as though the singlet insertion is simply a two-stage process of abstraction followed by recombination of the 2 adjacent radicals. The addition reaction is of great synthetic interest since it is stereospecific. Both insertion and addition lead to the formation of hot molecules with about 99 kcal of excess energy, and in the gas phase this is sufficient to produce secondary reactions before thermalization can occur.

Other biradicals are formed as intermediates in the pyrolysis of small ring compounds. If we consider olefins as two-membered rings, the thermal cis-trans isomerization of olefins can be considered as proceeding through a transition state which is a twisted (90° twist) 1,2-biradical:

$$\begin{array}{c} Y \\ \end{array} C=C \begin{array}{c} \\ X \end{array} \rightleftarrows \left[\begin{array}{c} Y \\ \end{array} \dot{C}-\dot{C} \begin{array}{c} \\ X \end{array} \right]^{\ddagger} \rightleftarrows \begin{array}{c} Y \\ \end{array} C=C \begin{array}{c} X \\ \end{array}$$

Cyclopropanes yield 1,3-biradicals, and cyclobutanes yield 1,4-biradicals. These are presumably all singlet biradicals, and currently there is no information on the rate of spontaneous singlet–triplet conversion. At the moment some preliminary communications give a very approximate lifetime for singlet–triplet spontaneous conversion for substituted tetramethylene type radicals of about 10^{-8} sec. This is much longer at pyrolysis temperature than the lifetimes of the singlet radicals with respect to ring closure or product formation. The lifetimes of all of these biradicals are estimated at between 10^{-10} to 10^{-12} sec. However, at room temperatures the lifetimes of trimethylene type radicals with respect to ring closure or olefin formation are about 10^{-4} to 10^{-6} sec (*10*), and one should expect to see triplets adding to olefins or unsaturates to form singlet products but in a nonstereospecific manner which is, in fact, observed.

For cyclobutanes and polycyclic compounds, the formation of biradicals competes with concerted splits into two or more unsaturated compounds:

$$\square \longrightarrow \parallel + \parallel$$
$$\longrightarrow C_2H_4 + \diagup\!\!=\!\!\diagdown \text{ (cis and trans)}$$

$$\bigtriangleup\!\!\!\!\bigtriangledown \longrightarrow \bigcirc + C_2H_4$$

The now much-discussed Hoffman-Woodward rules provide a guide to which concerted processes may proceed *via* available low lying electronic states thermally, but they do not provide quantitative data on activation energies. Thermochemical data and methods permit evaluation of the energetics of the biradical paths (*49*). Where experimental data is available this is usually enough to decide between concerted and biradical paths.

Modeling of Pyrolytic Systems

There is now sufficient kinetic understanding and both kinetic and thermochemical data available to describe the behavior of even the most complex pyrolysis in terms of a finite number of elementary step reactions. Even more important, the available data will permit us to reject most of the unimportant reactions. On this basis, it might be expected that hydrocarbon pyrolyses are ripe candidates for quantitative modeling.

If we could restrict our discussion to pyrolysis of pure hydrocarbons (*e.g.*, *n*-butanes, isobutane, *n*-hexane, etc.), at low conversions ($<20\%$), this would be undoubtedly true. However, commercial applications have a number of important imponderables, one of which is the role of reactor walls. We know that some terminations are wall controlled while others are homogeneous. Quantitative rate data are not available or predictable for wall reactions. In a general modeling program we would have to put in unknown first-order terms for radical termination on walls.

Another difficulty has to do with industrial flow reactions. Pyrolysis is an endothermic reaction. In a fast flow system, the reaction can become nearly adiabatic, and consequently temperature drops of 100 or more degrees can occur. This means that heat transfer equations must be introduced into the kinetic scheme. These must be done in oversimplified (*i.e.*, one-dimensional) form, or the model can become intractable even on a large computer.

Finally, most cracking processes do not use pure compounds. Instead, they use complex hydrocarbon mixtures containing appreciable amounts of sulfur compounds which can act as both initiators and inhibitors of radical chain reactions. This can become a serious problem since our thermochemical and kinetic data on sulfur compounds are limited. In addition, the inclusion of too many starting reactants can impose intolerable borders on even large computers. Under the circumstances, the scheme must be simplified by a combination of methods which include careful adjustment of "averaged" kinetic parameters with available data. Many of these can be done reasonably well by analyzing product ratios. Others cannot be done by any simply specifiable procedures but **only** by trial and error in terms of unique systems.

Literature Cited

(1) Barnard, J. A., Seebohm, R. P., *Symp. Gas Kinetics,* Hungarian Chemical Society, Szeged, Hungary, **1969,** 6.
(2) Bauer, S. H., *Ann. Rev. Phys. Chem.* **1965,** 16, 245.
(3) Benson, S. W., "Thermochemical Kinetics," pp. 62, 99, Wiley, New York, 1968.
(4) *Ibid.,* p. 66.
(5) Benson, S. W., *J. Am. Chem. Soc.* **1969,** 91, 2152.
(6) Benson, S. W., DeMore, W. B., *Advan. Phys. Chem.* **1965,** 16, 397, 406–9.
(7) *Ibid.,* pp. 397, 419–20.
(8) Benson, S. W., Haugen, G. R., *J. Phys. Chem.* **1967,** 71, 4404.
(9) Benson, S. W., Kistiakowsky, G. B., *J. Am. Chem. Soc.* **1942,** 64, 80.
(10) Benson, S. W., O'Neal, H. E., *J. Phys. Chem.* **1967,** 71, 2903.
(11) *Ibid.,* **1968,** 72, 1866.
(12) Benson, S. W., Shaw, R., *J. Am. Chem. Soc.* **1967,** 89, 5351.
(13) Benson, S. W., Shaw, R., *Trans. Faraday Soc.* **1967,** 63, 985.
(14) Blades, A. T., Murphy, G. W., *J. Am. Chem. Soc.* **1952,** 74, 1039.
(15) Bradley, J. N., Frend, M. A., *Symp. Gas Kinetics,* Hungarian Chemical Society, Szeged, Hungary, **1969,** 22.
(16) DeMore, W. B., Benson, S. W., *Advan. Photchem.* **1964,** 2, 219.
(17) Ellis, R. J., Frey, H. M., *Proc. Chem. Soc.* **1964,** 221.
(18) Ellis, R. J., Frey, H. M., *J. Chem. Soc.* **1964,** 4184.
(19) *Ibid.,* p. 4770.
(20) *Ibid.,* p. 5578.
(21) *Ibid.,* **1966A,** 553.
(22) Endrenyi, L., LeRoy, D. L., *J. Phys. Chem.* **1966,** 70, 4081.
(23) Frey, H. M., Lister, D. H., *J. Chem. Soc.* **1967A,** 26.
(24) Frey, H. M., Pope, B. M., *J. Chem. Soc.* **1966A,** 1701.
(25) Frey, H. M., Solly, R. K., *Trans. Faraday Soc.* **1968,** 64, 1858.
(26) Frey, H. M., Walsh, R., *Chem. Rev.* **1969,** 69, 103–24.
(27) Golden, D. M., Benson, S. W., *Chem. Rev.* **1969,** 69, 127.
(28) Gomer, R., Kistiakowsky, G. B., *J. Chem. Phys.* **1951,** 19, 85.
(29) Gordon, A. S., McNesby, J. R., *J. Chem. Phys.* **1959,** 31, 853.
(30) *Ibid.,* **1960,** 33, 1882.
(31) Huntsman, W. D., Curry, T. H., *J. Am. Chem. Soc.* **1958,** 80, 2252.
(32) Jackson, W. M., McNesby, J. R., *J. Am. Chem. Soc.* **1961,** 83, 4891.
(33) Jackson, W. M., McNesby, J. R., *J. Chem. Phys.* **1962,** 36, 2272.
(34) *Ibid.,* **1963,** 38, 692.
(35) Jackson, W. M., McNesby, J. R., de B. Darwent, B., *J. Chem. Phys.* **1962,** 37, 1610.
(36) *Ibid.,* p. 2256.
(37) Jacobs, T. A., Cohen, N., Giedt, R. R., *J. Chem. Phys.* **1967,** 46, 1958.
(38) Johnston, H. S., Parr, C., *J. Am. Chem. Soc.* **1963,** 85, 2544.
(39) Kaufmann, F., *Ann. Rev. Phys. Chem.* **1969,** 19, 45.
(40) Kirmse, W., "Carbene Chemistry," Academic, New York, 1964.
(41) Kossiakoff, A., Rice, F. O., *J. Am. Chem. Soc.* **1943,** 65, 590.
(42) Lewis, K. E., Steiner, H., *J. Chem. Soc.* **1964,** 3080.
(43) Mayer, S. W., *J. Phys. Chem.* **1967,** 71, 4159.
(44) Mayer, S. W., Schieler, L., Johnston, H. S., *J. Chem. Phys.* **1966,** 45, 385.
(45) *Ibid.,* **1968,** 72, 2628.
(46) McLean, S., Haynes, P., *Tetrahedron* **1965,** 21, 2329.
(47) Metcalfe, E. L., *J. Chem. Soc.* **1963,** 3560.
(48) O'Neal, H. E., Benson, S. W., *Intern. J. Chem. Kinetics* **1969,** 1, 221.
(49) *Ibid.,* **1970,** 2.
(50) Price, S. J., Trotman-Dickenson, A. F., *J. Chem. Soc.* **1958,** 4205.

(51) Purnell, J. H., Quinn, C. P., "Photochemistry and Reaction Kinetics," P. G. Ashmore, F. S. Dainton, T. M. Sugden, Eds., Cambridge University Press, 1967.
(52) Rice, F. O., *J. Am. Chem. Soc.* **1931,** 53, 1959.
(53) *Ibid.*, **1933,** 55,3035.
(54) Rice, F. O., Herzfeld, K. F., *J. Am. Chem. Soc.* **1934,** 56, 284.
(55) Rodgers, A. S., Golden, D. M., Benson, S. W., *J. Am. Chem. Soc.* **1967,** 89, 4578.
(56) Rodgers, A. S., Golden, D. M., Benson, S. W., Furuyama, S., unpublished data.
(57) Roth, W. R., *Tetrahedron Letters* **1964,** 17, 1009.
(58) Roth, W. R., Konig, J., *Ann.* **1965,** 688, 28.
(59) Rust, F. F., *J. Am. Chem. Soc.* **1957,** 79, 4000.
(60) Spokes, G. N., Benson, S. W., *J. Am. Chem. Soc.* **1967,** 89, 2525.
(61) Spokes, G. N., Benson, S. W., *Proc. Intern. Symp. Combust.*, 11th, **1967,** 95.
(62) Stepukhovitch, A. D., "Kinetics and Mechanism of Alkane Pyrolysis," Saratov University, 1965 (in Russian).
(63) Toscano, V., Doering, W. von E., unpublished work.
(64) Trotman-Dickenson, A. F., Milne, G. S., "Tables of Bimolecular Gas Reactions," U. S. Government Printing Office, Washington, D. C., 1967.
(65) Trotman-Dickenson, A. F., Steacie, E. W. R., *J. Chem. Phys.* **1951,** 19, 329.
(66) Tsang, W., *J. Chem. Phys.* **1964,** 40, 1498.
(67) *Ibid.*, **1965,** 42, 1805.
(68) *Ibid.*, 43, 352.
(69) Tsang, W., *J. Chem. Phys.* **1966,** 44, 4283.
(70) Wang, S., Winkler, C. A., *Can. J. Res.* **1943,** 21B, 97.
(71) Watkins, K. W., Ostreko, L. A., *J. Phys. Chem.* **1966,** 70, 4081.

RECEIVED December 16, 1969. Sponsored in part by grant AP 00353-05 from the Air Pollution Control Administration, Department of Public Health.

2

The Chemistry of Aromatics Production *via* Catalytic Reforming

E. L. POLLITZER, J. C. HAYES, and VLADIMIR HAENSEL

Universal Oil Products Co., Des Plaines, Ill. 60016

> *Catalytic reforming involves the use of dual functional (platinum–alumina) catalysts which promote a variety of reactions, the most important being aromatics formation from cyclohexane, cyclopentane, and paraffin precursors. Data obtained with pure compounds indicate that temperature and pressure control the relative rates of aromatization vs. isomerization and cracking. The boiling range of the charge stock can be used to control the range of products obtained. The final product mix is affected by dealkylation of* gem-*dialkylcyclohexanes during dehydrogenation and by isomerization and dealkylation of the aromatics themselves. Higher molecular weight compounds are aromatized more readily, and unreacted paraffins concentrate in the lower boiling fractions. Dealkylation, transalkylation, and isomerization can be used as secondary processes to modify further reformer aromatics.*

The notion of a "chemical refinery" has been suggested repeatedly over the past 30 years or more, and the possibilities have been explored in the literature (*20, 21, 37, 68, 91, 109, 110, 120*). This subject also occurs frequently in reviews on advances in petroleum technology (*7, 23, 36, 38, 52, 113, 115, 116, 121, 122, 133, 137, 138, 149*).

The large scale use of petroleum products as chemical raw materials really began with the advent of catalytic reforming, particularly catalytic reforming over platinum catalysts. These processes provided a ready and convenient route to a range of aromatic hydrocarbons from readily available and relatively inexpensive starting materials. Some of the earliest patents and publications on catalytic reforming (*49, 54, 75, 86, 140, 141*) point specifically to the use of this process for producing aromatics as chemical intermediates as opposed to the upgrading of naph-

Table I. Reactions in Catalytic Reforming

Isomerization of paraffins
Hydrocracking
Dehydrogenation of cyclohexanes
Isomerization/dehydrogenation of cyclopentanes
Dehydrocyclization of paraffins

thas for fuel use. Many subsequent publications elaborated on this theme (*12, 17, 31, 33, 35, 55, 71, 72, 84, 93, 96, 102, 103, 105, 106, 117, 131, 142, 146, 151, 153*).

Catalytic reforming is a refinery process designed primarily to increase the octane number of a naphtha or a gasoline. Over-all, it is rather complex, involving several different types of reactions which are summarized in Table I. If these reactions are examined in more detail, it will be found that paraffin isomerization (Figure 1) occurs rather easily, and contributes only a limited extent to octane improvement under the reaction conditions employed. Hydrocracking does contribute substantially to octane enhancement (essentially by cracking out low octane components), but it also represents a yield loss because of the formation of gaseous products. The catalyst and operating conditions are therefore adjusted carefully to minimize hydrocracking. It follows, therefore, that a refiner using this process for octane improvement is basically seeking to produce aromatics (Figure 2) from naphthenes and paraffins, and he would, of course, share this as a common goal with a petrochemical manufacturer who is using reforming as a source of aromatic raw materials.

Catalysts

While the emphasis in this discussion is on the chemistry of aromatics production from the point of view of product distribution and mechanisms, it is necessary to describe the catalyst used briefly. Until about 20 years ago a number of different catalysts were used for catalytic reforming, including vanadium (*126*), molybdenum (*47*), cobalt-molybdenum (*159*), and chromia (*3, 41, 85, 100*) catalysts. Since the time when platinum reforming catalysts were introduced to the industry (*2, 24, 56,*

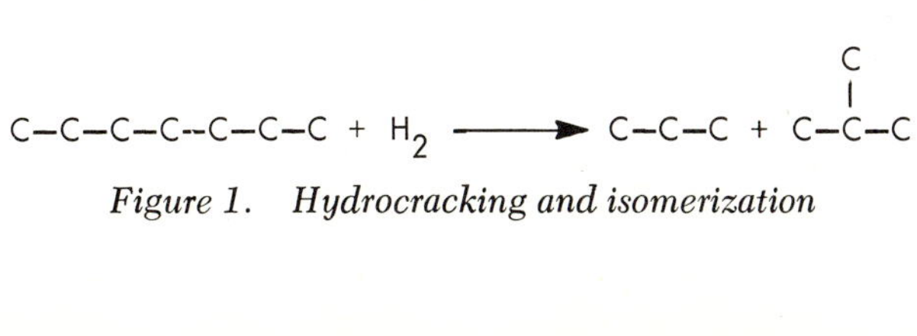

$$C-C-C-C-C-C \longrightarrow C-\underset{|}{\overset{C}{C}}-C-C-C$$

$$C-C-C-C-C-C-C + H_2 \longrightarrow C-C-C + C-\underset{|}{\overset{C}{C}}-C$$

Figure 1. Hydrocracking and isomerization

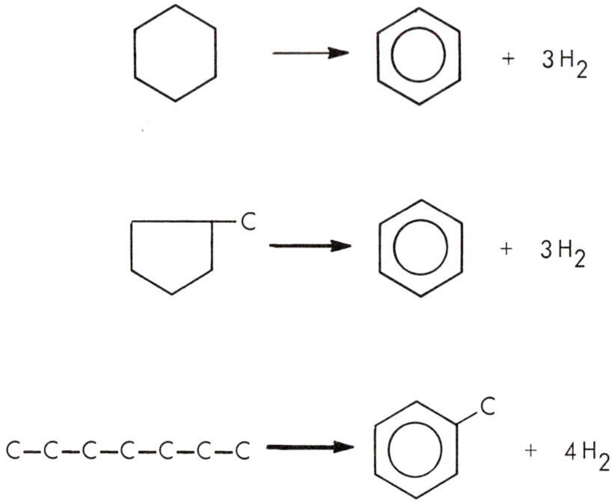

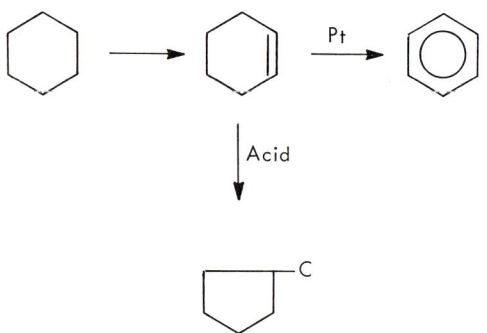

Figure 2. Aromatics formation

Figure 3. Reforming of cyclohexane

57, 58, 83) they have dominated the field. A number of modifications and improvements have been made and continue to be made, but reforming catalysts today are basically still platinum on an acidic alumina support.

Catalysts such as these are referred to as "dual functional" because the platinum and the acid sites are discrete components. They cooperate in promoting the desired over-all reactions, but each appears to be responsible for certain steps. Details of the reaction mechanisms will be discussed later, but the concept of "dual functionality" can be illustrated by the simple reaction shown in Figure 3. Cyclohexane is dehydrogenated to benzene *via* cyclohexene. While the second dehydrogenation step is quite rapid, some of the cyclohexene is also trapped on acid sites and isomerized, appearing in the product largely as methylcyclopentane. The dual functional nature of the catalyst can be demonstrated by adding "poisons" to the system. Ammonia will poison the acid site and essentially eliminate the formation of methylcyclopentane. On the other hand, sulfur will attenuate the platinum activity and slow down the second dehydrogenation step, increasing methylcyclopentane formation relative to benzene.

Naphtha Reforming

The catalytic reforming process in all its aspects has been the subject of many review articles (*5, 9, 10, 11, 25, 29, 42, 50, 59, 60, 61, 62, 67, 69, 97, 114, 123, 127, 132, 143, 148, 154, 158*), and no attempt is made to cover the details here. The aromatic yield levels attainable can be illustrated by an example involving a commercial operation. Data assembled in Table II show the results obtained from a modern high severity reformer operating on a full boiling range Midcontinent naphtha. This unit produces gasoline at over 105 leaded octane, but the results have been presented in terms of aromatics distribution. The aromatics yield is over 60 vol % from a stock originally containing about 10 vol % aromatics. The aromatics distribution is typical of catalytic reforming. Relatively little benzene is produced, the toluene yield is higher than expected, and the C_8 aromatics are not quite at thermodynamic equilibrium. A little less than half of the aromatic product is represented by C_9^+ aromatics which were not identified further.

Table II. Reforming of Full Boiling Range Mid-Continent Naphtha

300 psig, 1.6 LHSV

Products	Yields, Vol % of Charge
C_5^-	24.0
Benzene	3.0
Toluene	12.0
Ethylbenzene	3.1
p-Xylene	3.9
m-Xylene	8.7
o-Xylene	4.5
C_9^+ Aromatics	27.1

Several publications have been concerned specifically with such heavier aromatics (*98, 136, 157*), but in general these derivatives have thus far been less important as chemical starting materials. For petrochemical use, therefore, refiners process lower boiling fractions (*1, 21, 120, 139*) in what is usually described as a "BTX" (benzene–toluene–xylene) operation. Table III shows some results obtained in reforming

Table III. Reforming of Light Kuwait Naphtha
1.5 LHSV

Pressure	300	200	150	100
Temperature, °F	980	973	968	962
Aromatics, vol % of charge	45	48	51	54

a light Kuwait naphtha. The data illustrate several point which recur throughout the subsequent discussion. The conversion to aromatics increases as the pressure is reduced, and the conversions are, in addition, lower than those shown in Table II despite the use of more severe conditions. This is attributable to two factors—the Kuwait naphtha is lower in naphthenes, which makes it more difficult to process, and lower molecular weight materials are generally more resistant to reforming than their higher homologs. Table IV shows the aromatic distribution at two conditions, and the products are mainly toluene and C_8 aromatics. There is still not much benzene produced, and the heavy ends have been eliminated by fractionation of the charge stock. The slight shift towards heavier products at lower pressures is believed to be real and is discussed later.

The effect of pressure on catalytic reforming has been the subject of several specific investigations (*30, 92, 150*). At otherwise equivalent conditions lower pressures not only increase the conversion to aromatics but also reduce the extent of hydrocracking and hence the loss to light gas formation. The general behavior is consistent with the fact that aromatics formation is accompanied by hydrogen production while hydrocracking consumes hydrogen.

While lower pressures thus appear desirable from several points of view, their use unfortunately is accompanied by poorer catalyst stability. Several approaches have been tried to circumvent this limitation. Catalytic reformers have been designed with fluidized beds (*94, 112*), allowing for continuous catalyst regeneration. Some catalytic reformers are designed with two reactor trains, permitting cyclic regeneration (*16*), *125, 129, 147*). Within the past year or two, new catalyst formulations have become available which are reported to be stable enough to permit operation at lower pressures and higher severities.

Table IV. Reforming of Light Kuwait Naphtha
1.5 LHSV

Pressure	Temp., °F	\multicolumn{5}{c}{Aromatics Distribution}

Pressure	Temp., °F	C_6	C_7	C_8	C_9	C_{10}
300	980	9.8	37.9	46.2	5.3	0.2
100	962	8.5	34.7	49.2	7.0	0.2

At this point it is necessary to delve briefly into the mechanism of reforming reactions. This has been the subject of many publications (*13, 15 18, 26, 27, 32, 45, 46, 63, 64, 65, 70, 73, 74, 77, 79, 82, 87, 107, 108, 118, 130, 134, 152*), but the details have not been settled completely. There appears to be general agreement about the mechanism by which cyclohexane or cyclopentane derivatives are converted to aromatics, and the controversy centers on the conversion of paraffins. The most popular concept postulates the primary formation of cyclopentane rings which then expand to cyclohexane rings and dehydrogenate. Other workers, however, propose that cyclohexane rings form directly as long as a six-carbon chain is available. Figure 4 shows the generally accepted scheme for the conversion of paraffins to aromatics *via* cyclopentane intermediates. Examination of the details indicates that some of the steps are acid catalyzed while others are promoted by the platinum function.

In principle, it should be possible to use experimental results on pure compounds coupled with mechanistic considerations to predict the product distribution in a catalytic reformate. In practice the situation is considerably more complex since the final product mix depends not only on the relative rates at which different compounds are converted to aromatics but also on isomerization that may precede aromatization, on

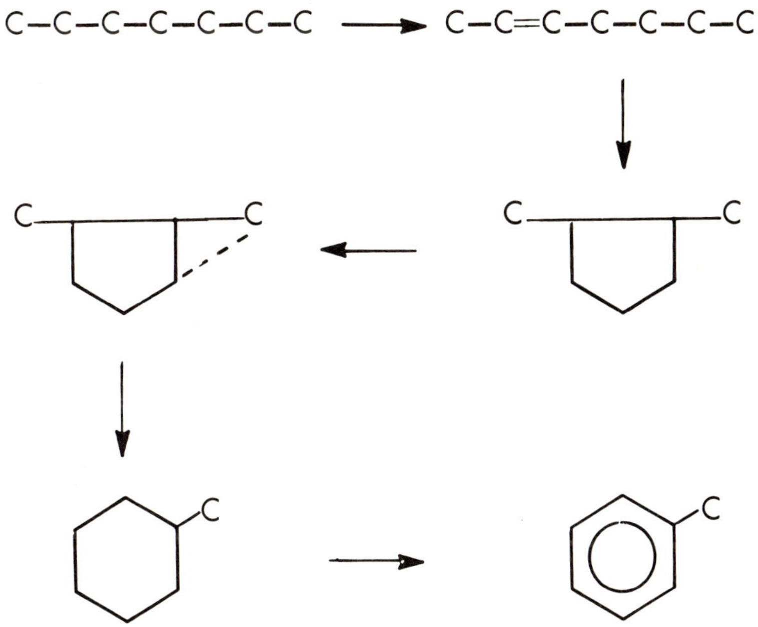

C–C–C–C–C–C–C ⟶ C–C=C–C–C–C–C

Figure 4. Mechanism of paraffin dehydrocyclization in catalytic reforming

side reactions that may produce different aromatics or nonaromatics, etc. Operating conditions will affect the relative rates of all these reactions and will, therefore, have a marked effect on selectivity. It is possible, however, to estimate the contribution of some of these factors on the basis of data obtained with pure compounds.

Reforming of Pure Compounds

It was pointed out that aromatics can be formed from cyclohexanes, cyclopentanes, or paraffins. Within this group of reactions, the dehydrogenation of cyclohexane (65, 79) is the most direct and by far the easiest. Some results obtained in the reforming of cyclohexane at various conditions are shown in Table V. The conversion increases sharply as the pressure is reduced, and the conversion level is very sensitive to temperature, at least at the higher space velocity. The selectivity to benzene is not particularly affected by temperature but is very much influenced by hydrogen pressure. At the lower space velocity a pressure change from 500 to 100 psig represents a selectivity shift from 21 to 95%. The major by-product is methylcyclopentane, which is presumably formed by isomerization of the intermediates as depicted in Figure 3. Pressures cannot be changed readily between reactors in a commercial reforming

Table V. Reforming of Cyclohexane

Pressure, psig	500	100	500	100	100	100
LHSV	2	2	4	4	4	4
Temperature, °F	750	750	750	700	750	800
Conversion of Cyclohexane, %	35	68	18	41	65	85
Selectivity to Benzene, %	21	95	46	96	92	97
Selectivity to MCP, %	54	1.6	48	1.6	0.7	0.6

Table VI. Reforming of Ethyl- and Dimethylcyclohexanes

300 psig, 1.5 LHSV, 920°F

	Conversion, %	Selectivity To C_8 Aromatics
Standard Reforming Catalyst	97	97
High Acidity Catalyst	94	65

unit, and temperature as well as space velocity are used as controls. The conversion of six-membered ring naphthenes is accomplished in the first reactor, and the selectivities are kept high by using high space velocities.

Table VI shows results obtained with a mixture of dimethyl- and ethylcyclohexanes reformed with catalysts of different acidity. The two functions of a reforming catalyst—the acidity and the hydrogenation–dehydrogenation function of the platinum—are balanced carefully for

Table VII. Reforming of Methylcyclopentane

Pressure, psig	100	100	100
LHSV	4	2	2
Temperature, °F	925	860	900
Conversion, %	60.5	66.1	85.4
Selectivity to Benzene, %	61.7	66.6	74

reaction conditions. If the balance is shifted, side reactions will occur. The data in the table show the performance of a high acidity catalyst compared with that of a normal reforming catalyst. The selectivity was low with the high acidity catalyst because of side reactions which lead to cracked products, benzene, toluene, etc.

Table VII shows some results obtained in reforming methylcyclopentane, and it is noteworthy that there appear to be striking differences between these data and the earlier results with cyclohexane. When compared at similar space velocity and pressure conditions, the conversions (and particularly the selectivities) are markedly poorer than those in the cyclohexane case despite the use of higher temperatures. The selectivity would be still poorer at higher pressures.

Higher homologs of cyclopentane are converted more readily, but it is important to note that there is a fundamental difference between cyclohexane reactions and those of all other hydrocarbon types. The aromatization of cyclohexane is largely a function of the platinum component. The acid function will, in fact, promote side reactions unless conditions are controlled carefully. In the case of methylcyclopentanes (or paraffins) the extent of conversion is a function of the catalyst halogen content (*i.e.*, the acidity) and depends on the platinum content only up to a very low critical level of platinum.

Table VIII shows similar data on the conversion of *n*-hexane. Again, even at the favorable low pressure, the selectivity to benzene is relatively poor. As the temperature is increased, the selectivity to aromatization increases somewhat, but hydrocracking increases at an even greater rate. The data in Table VII show better conversion and selectivity for the aromatization of methylcyclopentane to benzene. It appears, therefore, that the rate-limiting step in hexane conversion is the conversion of hexane to methylcyclopentane. This could be explained by reference to

Table VIII. Reforming of *n*-Hexane at 100 psig

LHSV	Temp., °F	Conversion, %	Selectivity to Benzene	Selectivity to iso-Hexane	Selectivity to Hydrocracking
2	885	80.2	16.6	58	25
2	932	86.8	24.1	36.9	38
2	977	90.4	27.4	23.4	49

the mechanistic scheme cited above. Only the normal hexane structure can yield methylcyclopentane through an intermediate such as C=C—C—C—Ċ—C. Isomerization to a methylpentane would result in localizing the carbonium ion on the tertiary position. This structure can no longer ring close to a cyclopentane but only to a highly strained cyclobutane. Since skeletal isomerization occurs readily at reforming conditions, most of the n-hexane would be converted to isohexane or cracked products rather than to cyclopentane.

The low reactivity of hexane and methylcyclopentane accounts for the relatively low benzene content in reformates (*see* Tables II and IV). Since the direct preparation of benzene by catalytic reforming would be economically attractive, considerable effort has been devoted to modifications (*124-131*) designed to achieve this end.

Factors Influencing Aromatics Distribution

While isomerization of n-hexane appears to hinder cyclization of the resultant molecule, higher derivatives should isomerize to branched structures which can still form five-membered rings and ultimately yield aromatics. The conversion of higher paraffin homologs to aromatics should therefore be facilitated. This can be illustrated by comparing the earlier hexane results with data in Table IX on the reforming of heptane. Both the conversion and selectivity are higher for

Table IX. Reforming of *n*-Heptane at 100 psig

LHSV	Temp., °F	Conversion, %	Selectivity to Toluene, %
2	890	88.4	37.8
2	932	94.7	48.5
2	977	99.1	62.0

heptane, and as the temperature is increased, the aromatization reaction rate increases markedly faster than the hydrocracking rate. As a result, the selectivity for toluene nearly doubles.

Toluene is a major aromatic constituent in catalytic reformate and one of the most important chemical intermediates obtained from this reaction. The kinetics of n-heptane reforming have been studied (*39, 40*), and the interest in toluene from this source antedates the use of platinum catalysts (*76, 119, 145*). While heptane is converted with considerably greater ease than hexane, this reaction is, on mass balance calculations, not sufficient to account for the toluene found in reformates.

Another reaction path by which toluene might be formed is illustrated in Figure 5. Whenever a branch (a tertiary carbon) is present, it can

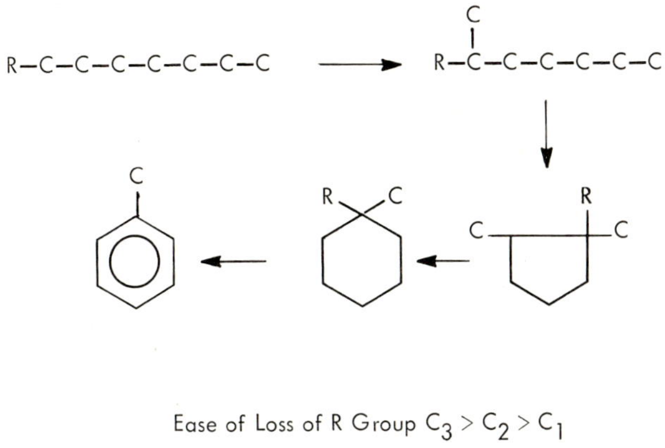

$$R-C-C-C-C-C-C-C \longrightarrow R-\underset{\underset{\downarrow}{|}}{\overset{\overset{C}{|}}{C}}-C-C-C-C-C$$

[Reaction scheme: benzene ring ← cyclohexane with R and C substituents ← cyclopentane with R and two C substituents]

Ease of Loss of R Group $C_3 > C_2 > C_1$

Figure 5. Alkyl group loss in dehydrocyclization

be the site of ring closure, and if it survives through the cyclopentane stage, it will form a *gem*-dialkylcyclohexane. In the process of forming an aromatic ring one of the alkyl groups is lost, and the indications are that higher groups are lost preferentially. Therefore toluene is the preferred product whenever a *gem*-dialkyl group is present.

Eight carbon aromatics (xylenes and ethylbenzene) represent another very important chemical intermediate derived from catalytic reforming (43, 99). These products can be formed by a similar combination of reactions—directly from octanes or from higher hydrocarbons by the loss of alkyl groups. The isomer distribution is the result of secondary reactions which are influenced strongly by operating conditions.

Table X shows the results of reforming a mixture of 8-carbon naphthenes obtained, in turn, from the hydrogenation of mixed xylenes. At low pressures and temperatures relatively little change from the original isomer distribution is found. As the severity is increased, the product aromatics tend more and more towards an equilibrium mixture. These

Table X. Reforming of C_8 Naphthenes

Pressure	100	300	500		
LHSV	1.5	1.5	1.5		
Temp., °F	850	920	920		
Products, % of Charge				Charge Stock	
Non-Aromatics	2.0	2.8	9.4		
Aromatics					
Ethylbenzene	22.6	20.2	15.6	EtCH	24.2
p-Xylene	11.5	14.4	16.1	1,4-Me$_2$CH	15.2
m-Xylene	52.1	46.9	40.6	1,3-Me$_2$CH	52.0
o-Xylene	10.1	12.6	14.4	1,2-Me$_2$CH	6.5

results give no indication as to whether isomerization precedes aromatization, but experimental evidence is available to show that xylenes will indeed isomerize under such conditions.

Thus, Table XI shows some results on the isomerization of m-xylene under reforming conditions. Again, there is little reaction at the low pressure, high space velocity conditions. At higher pressures there is not only isomerization to o- and p-xylene, but also considerable dealkylation to toluene. This confirms that the isomerization of xylenes occurs *via* hydrogenation to naphthenes, at least some of which can isomerize to gem-dimethyl derivatives as shown in Figure 6.

Table XI. Reactions of m-Xylene at Reforming Conditions
4.0 LHSV, 920°F

	Aromatics Distribution, wt %				
	Xylenes				
Pressure, psig	o-	m-	p-	Benzene	Toluene
500	11.2	59.0	11.8	1.2	17.0
100	1.5	91.0	3.0	1.5	3.0

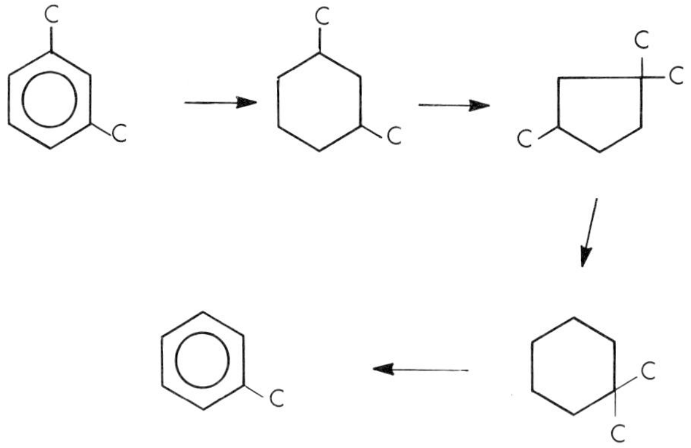

Figure 6. Isomerization of xylenes

Controlling Aromatics Distribution

In the above series of experiments the mixture did not achieve thermodynamic equilibrium despite the use of typical reforming conditions. As a result, the C_8 aromatics in a commercial reformate are rarely at equilibrium except at very high pressures. Furthermore, the reaction of ethylbenzene is markedly slower than the interconversion of xylenes, and the ethylbenzene is therefore typically furthest from its equilibrium value.

Table XII. Reforming of n-Nonane

950°F, 1.5 LHSV

Product Distribution, wt %

Pressure, psig	Nonaromatics		Aromatics			
	C_1–C_4	C_5^+	C_6	C_7	C_8	C_9
100	10.5	19	1.6	3.1	6.2	54.5
300	21.5	20	2.0	5.8	10.1	36.4

Results obtained in reforming n-nonane are shown in Table XII and illustrate the effect of processing conditions on product distribution and yield. The conversion was essentially 100% at both conditions but the percentage of light products (resulting largely from hydrocracking) essentially doubled as the pressure was raised from 100 to 300 psig. There was also a marked shift in the distribution of aromatics—at the low pressure well over 80% of the aromatics produced contained nine carbon atoms, while at the higher pressure there was a definite shift to lower homologs. This can be explained on the basis of the mechanism cited in Figure 5. In a normal hydrocarbon there is no opportunity to form gem-dialkyl derivatives, and the loss of alkyl groups is avoided. At higher pressures the rate of dehydrocyclization is lower, and there is more opportunity for isomerization to occur before cyclization. The result is increased formation of xylenes and toluene. Dealkylation reactions of this type are largely responsible for the formation of methane and ethane in reforming since these are not normal hydrocracking products. Because there is no particular tendency to lose a single side chain, this mechanism would not lead to benzene formation.

Considerable advances have been made in the area of catalytic reforming since the introduction of the process, particularly in recent years. Studies of the thermodynamics of the system (*124, 156, 160*), and the use of mathematical models (*8, 19, 155, 161, 162*) have given us a better understanding of the reactions involved.

A process has been introduced (*88*) based specifically on improved selection and control of operating variables. Potentially the most important new developments have been in the area of reforming catalysts. Up to a few years ago, work in this area centered on metal distribution studies (*95*) and general optimization (*51*). This work has been covered in reviews (*14, 94, 135*). Very little has been published on more recent catalyst development although some papers have been presented (*53, 81, 111, 128, 144*). Available information on the use of such new catalysts points to greatly improved stability, particularly at severe conditions, and also to increased yields by reduction of hydrocracking. There is no indication that the use of such catalysts involves any basic changes in our understanding of the chemistry of reforming. Aromatic production *via* reforming will undoubtedly be fostered by these developments, both by generally improved economics and specifically because the new catalyst systems show promise of allowing the use of lower pressures.

Catalytic reforming permits only limited control over the product distribution. Selection of the proper charge stock will affect the amount and kind of aromatics produced, with the boiling range, molecular weight, naphthene content, and isomer distribution all playing a role. Selection of the proper reaction conditions can play a critical part. Since these controls often do not go far enough, a number of different schemes have been devised for further processing of the aromatics obtained from catalytic reforming (44, 90). Figure 7, for example, shows the scheme for

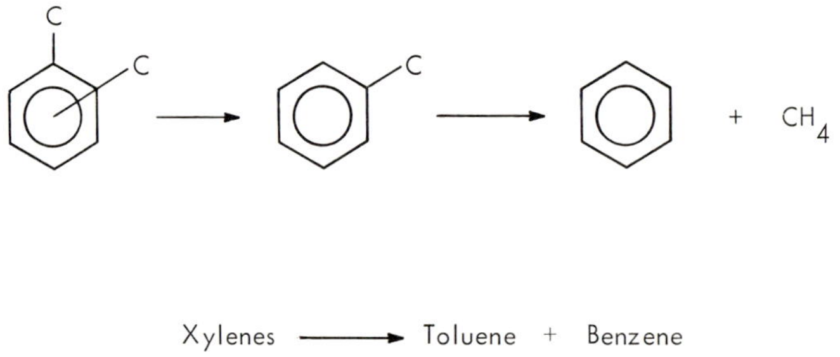

Xylenes ⟶ Toluene + Benzene

Figure 7. Demethylation of aromatics

demethylation. This can be carried out thermally or over catalysts such as chromia–alumina. It is presently used largely to convert toluene to benzene or methylnaphthalene to naphthalene, but it could be used to demethylate higher homologs.

Transalkylation is a more recent entry into this field and uses an acidic catalyst. The scheme is illustrated in Figure 8. As presently designed, it is used to convert toluene to benzene plus xylenes, but C_9 and higher aromatics could be introduced.

In the area of C_8 aromatics, *o*-xylene can, of course, be separated by distillation, and ethylbenzene can be fractionated out, although with

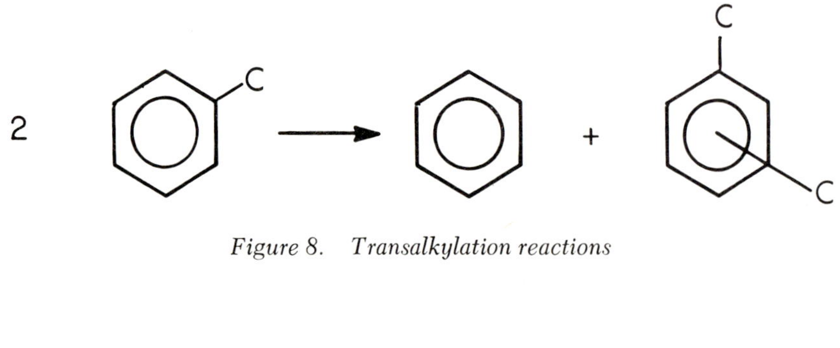

Figure 8. Transalkylation reactions

difficulty. Until recently *p*-xylene has been separated only by fractional crystallization. Last year a process called Parex was announced (*22*), which accomplishes the job by means of a selective adsorbent. In addition, since the remaining C_8 aromatics can be isomerized to an equilibrium mixture, it is at least theoretically possible to convert a mixture of C_8 aromatics to any desired pure isomer.

Advances in such ancillary processes, as well as in the basic catalytic reforming process itself, continue to be made and continue to provide greater flexibility. This, in turn, tends to make catalytic reforming even more attractive as a source of raw materials for the chemical industry.

Literature Cited

(1) Aliev, V. S., Indyukov, N. M., Sidorchuk, I. I., Goncharova, M. A., Gasanova, R. I., *Proizv. Benzola, Vses. Nauch.-Issled. Inst. Neftekhim. Protessov* **1962**, 34.
(2) Annable, W. G., U. S. Patent **2,510,673** (June 6, 1950).
(3) Archibald, R. C., Greensfelder, B. S., *Ind. Eng. Chem.* **1945**, 37, 356.
(4) Barkan, S. A., Maslyanskii, G. N., *Neftekhimiya* **1967**, 7, 746; transl. *Intern. Chem. Eng.* **1968**, 8, 218.
(5) Bart, J. C., Oblad, A. G., Heinemann, H., *Bull. Assoc. Franc. Techniciens Petrole* **1954**, 449.
(6) Beckberger, LaVern H., Woerner, R. C., U. S. Patent **2,551,383** (Dec. 1, 1953).
(7) Belchetz, L., *Rev. Petrol. Technol.* **1955**, 14, 57.
(8) Bel'tsov, B. A., *Khim i Tekhnol. Topliv i Masel* **1966**, 11 (8), 18.
(9) Bland, W. F., *Petrol. Process.* **1950**, 5, 351; *Petrol. Refiner* **1950**, 29 (4), 128.
(10) Bland, W. F., Davidson, R. L., "Petroleum Processing Handbook," pp. 3–25 to 3–28, McGraw-Hill, New York, 1967.
(11) Bogen, J. S., Haensel, V., *Proc. Am. Petrol. Inst., Midyear Meetg.* **1950**, 30M, III, 319.
(12) Borisovich, G. F., Kalashinikova, Z. S., *Khim. Prom.* **1964**, 161.
(13) Bragin, O. V., Kazanskii, B. A., Liberman, A. L., Preobrazhenskii, A. V., *Izv. Akad. Nauk. SSSR* **1967**, 2260.
(14) Bridger, G. W., *Chem. Process Eng.* **1966**, 47, 39.
(15) Brill, E. J., *Ind. Eng. Chem.* **1951**, 43, 2102.
(16) Brooks, J. A., *Oil Gas J.* **1962**, 60 (37), 145.
(17) Brunshtein, B. A., Klimenko, V. L., Tyabukhova, S. F., *Khim. Prom.* **1966**, 42, 889.
(18) Budec, R., Seler, M., Sutic, B., *Nafta (Zagreb)* **1963**, 14, 209; abstr. *J. Appl. Chem.* **1964**, 14 (7), II.
(19) Burnett, R. L., Steinmetz, H. L., Blue, E. M., Novel, J. J., *Am. Chem. Soc., Div. Petrol. Chem., Preprints* **1965**, 10 (1), 17–24.
(20) Caspers, J., Gambro, A. J., *World Petrol.* **1967**, 38 (4), 88.
(21) *Chem. Eng. News* **1968**, 46 (26), 32.
(22) *Chem. Eng. News* **1969**, 47 (10), 47.
(23) Ciapetta, F. G., Pichler, H., *Erdoel Kohle* **1968**, 21 (2), 66.
(24) Ciapetta, F. G., U. S. Patent **2,550,531** (April 24, 1951).
(25) Ciapetta, F. G., *Petro/Chem. Engr.* **1961**, 33 (5), C19–C31.
(26) Ciapetta, F. G., *Am. Chem. Soc., Div. Petrol. Chem., General Papers* **1955**, 33, 167.

(27) Ciapetta, F. G., Dobres, R. M., Baker, R. W., "Catalysis," Vol. 6, pp. 495–692, Reinhold, New York, 1958.
(28) Connor, J. E., Jr., Ciapetta, F. G., Leum, L. N., Fowle, M. J., *Ind. Eng. Chem.* **195**, 47, 152.
(29) Danilov, B., *Petrol si Gaze (Bucharest)* **1956**, 7, 492 (German summary).
(30) Dart, J. C., Oblad, A. G., Schall, J. W., *Proc. Am. Petrol. Inst., 17th Midyear Meetg.* **1952**, 32M, Sect. 3, 208; discussion p. 235 (1953).
(31) Davis, W. H., U. S. Patent **2,653,175** (Sept. 22, 1953).
(32) Donaldson, G. R., Pasik, L. F., Haensel, V., *Ind. Eng. Chem.* **1955**, 47, 731.
(33) Earhart, H. W., Heinrich, R. L., Lewis, E. W., Newson, T. W., Wadley, E. F., *Proc. World Petrol. Congr., 5th, New York*, **1960**, 4, 1.
(34) Ebeid, F. M., Habib, R. M., *J. Chem. U.A.R.* **1963**, 6 (1), 79.
(35) Edgar, J. L., "Introduction to Petroleum Chemicals," H. Steiner, Ed., p. 101, Pergamon Press, Elmsford, N. Y., 1961.
(36) Edminster, W. C., *Petrol. Engr.* **1946**, 18 (2), 51.
(37) Egloff, G., *Petrol. Engr.* **1944**, 15 (8), 98.
(38) Egloff, G., *Riv. Combust.* **1951**, 5, 269.
(39) Feizkhanov, F. A., Panchenkov, G. M., Kolesnikov, I. M., *Neftekhimiya* **1962**, 2, 716.
(40) *Ibid.*, **1964**, 4, 722.
(41) Fisher, F. E., Watts, H. C., Harris, G. E., Hollenbeck, C. M., *Ind. Eng. Chem.* **1946**, 38, 61.
(42) Fowle, M. J., Bent, R. D., Ciapetta, F. G., Pitts, P. M., Leum, L. N., "Progress in Petroleum Technology," pp. 76–82, American Chemical Society, Washington, D. C., 1951.
(43) Fuller, D. L., Greensfelder, B. S., U. S. Patent **2,378,210** (June 12, 1945).
(44) Girelli, A., *Petrol. Mat. Prima Ind. Chim. Mod. Relaz. Commun., Giornate Chim. (Milan)* **1961**, 117.
(45) Green, S. J., Nash, A. W., *Nature* **1941**, 148, 53.
(46) Green, S. J., *J. Inst. Petrol.* **1942**, 28, 179.
(47) Greensfelder, B. S., Archibald, R. C., Fuller, D. L., *Chem. Eng. Progr.* **1947**, 43, 561.
(48) Greensfelder, B. S., Fuller, D. L., *J. Am. Chem. Soc.* **1945**, 67, 2171.
(49) Grosse, A. V., Morrell, J. C., U. S. Patent **2,172,535** (Sept. 12, 1940).
(50) Gruse, W. A., Stevens, D. R., "Chemical Technology of Petroleum," p. 395, McGraw-Hill, New York, 1960.
(51) Gunn, D. J., *Chem. Eng. Sci.* **1967**, 22, 963.
(52) Haensel, V., *Chem. Eng.* **1961**, 68 (24), 105.
(53) Haensel, V., Addison, G. E., *Proc. World Petrol. Congr., 7th, Mexico City* **1967**, 4, 113.
(54) Haensel, V., Berger, C. V., *Petrol. Process.* **1951**, 6, 264.
(55) Haensel, V., U. S. Patent **2,799,627** (July 16, 1957).
(56) Haensel, V., U. S. Patent **2,479,110** (Aug. 16, 1949).
(57) Haensel, V., U. S. Patent **2,479,109** (Aug. 16, 1949).
(58) Haensel, V., Sterba, M. J., "Progress in Petroleum Technology," p. 60, American Chemical Society, Washington, D. C., 1951.
(59) Haensel, V., *Petrol. Refiner* **1950**, 29 (4), 131; *Oil Gas J.* **1950**, 48, 47, 82, 114–9.
(60) Haensel, V., "The Chemistry of Petroleum Hydrocarbons," B. T. Brooks *et al.*, Eds., Vol. 2, p. 189, Reinhold, New York, 1955.
(61) Haensel, V., Berger, C. V., "Advances in Petroleum Chemistry and Refining," K. A. Kobe and J. J. McKetta, Jr., Eds., Vol. 1, pp. 386–427, Interscience, New York, 1958.
(62) Haensel, V., *Am. Chem. Soc., Div. Petrol. Chem., Preprints* **1961**, 6 (3A), 93.

(63) Haensel, V., Donaldson, G. R., *Proc. World Petrol. Congr., 3rd, Hague* **1951**, 225.
(64) Haensel, V., Donaldson, G. R., *Petrol. Process.* **1953**, 8, 236.
(65) Haensel, V., Donaldson, G. R., Riedl, F. J., *Proc. Intern. Congr. Catalysis, 3rd, Amsterdam* **1964**, 1, 308; discussion, p. 317 (1965).
(66) Haensel, V., U. S. Patent **2,911,451** (Nov. 3, 1959).
(67) "Handbook on Catalytic Reforming," The Petroleum Publishing Co., Oklahoma, 1967.
(68) Hatch, L. F., *Hydrocarbon Process.* **1969**, 48 (2), 77.
(69) Heinemann, H., Walser, F. R., Schall, J. W., Oblad, A. G., *Erdol. Kohle* **1955**, 8, 621.
(70) Heinemann, H., Mills, G. A., Hattman, J. B., Kirsch, F. W., *Ind. Eng. Chem.* **1953**, 45, 130.
(71) Hemminger, C. E., Taff, W. O., U. S. Patent **2,697,684** (Dec. 21, 1954).
(72) Henningsen, J., *Dan. Kemi* **1967**, 48 (8), 124.
(73) Herington, E. F. G., Rideal, E. K., *Proc. Roy. Soc. (London)* **1945**, A184, 434.
(74) Hettinger, W. P., Jr., Keith, C. D., Gring, J. L., Teter, J. W., *Ind. Eng. Chem.* **1955**, 47, 719.
(75) Hoog, H., Canadian Patent **391,278** (Sept. 10, 1940).
(76) Howes, D. A., *Ind. Chemist* **1946**, 22, 403.
(77) Iijima, K., Isume, Y., Yamada, S., Mita, K., *Sekiyu Gakkai Shi* **1962**, 243; Engl. transl. *Intern. Chem. Eng.* **1964**, 4, 267.
(78) Indyukov, N. M., Goncharova, M. A., Gasanova, R. I., *Khim. Tekhnol. Topliv i Masel* **1965**, 10 (5), 15.
(79) Isagulyants, G. V., Ryashentseva, M. A., Derbentsey, Yu. I., Minachev, Kh. M., Balandin, A. A., *Neftekhimiya* **1965**, 5, 507.
(80) Izmailov, R. I., Ikruzhnov, A. M., Virobyants, R. A., *Khim. i Tekhnol. Topliv i Masel* **1962**, 7 (11), 29.
(81) Jacobson, R. L., Davis, R. W. et al., *Am. Petrol. Inst., Div. Refining, 34th Midyear Meetg., Chicago*, **1969**, Preprint 37–69.
(82) Kalechits, I. V., *Kinetika i Kataliz* **1967**, 8, 1114.
(83) Kassel, L. S., U. S. Patent **2,573,149** (Oct. 30, 1951).
(84) Kaufmann, H., von Sahr, E., *Chem. Tech. (Berlin)* **1959**, 8, 403.
(85) Kazanskii, B. A., Liberman, A. L., Plate, A. F., Sergienko, S. R., Zelinskii, N. D., *Compt. Rend. Acad. Sci. U.S.S.R.* **1940**, 27 446.
(86) Komarewsky, V., U. S. Patent **2,212,026** (Aug. 20, 1941).
(87) Komarewsky, V. I., Shand, W. C., *J. Am. Chem. Soc.* **1944**, 66, 1118.
(88) Kopf, F. W., Pfefferle, W. C. et al., *Am. Petrol. Inst., Div. Refining, 34th Midyear Meetg., Chicago* **1969**, Preprint 23–69.
(89) Kuvshinova, N. I., Nasledskova, G. G., Skvortsova, E. V., Usov, Yu. N., *Neftekhimiya* **1966**, 6, 872.
(90) Lavergne, J. C., *Rev. Inst. France Petrol Ann. Combust. Liquides* **1966**, 21, 1707.
(91) Liberman, A. L., *Russian Chem. Rev.* **1961**, 30, 237.
(92) Lifland, P., U. S. Patent **3,111,481** (Nov. 19, 1963).
(93) Love, R. M., Pfennig, R. F., "Progress in Petroleum Technology," p. 299, American Chemical Society, Washington, D. C., 1951.
(94) Lyster, W. N., Prengle, H. W., Jr., *Hydrocarbon Process. Petrol. Refiner* **1965**, 44 (9), 161.
(95) Maat, H. J., Moscou, L., *Proc. Intern. Congr. Catalysis, 3rd, Amsterdam* **1964**, 2, 1277; discussion p. 1285.
(96) Manci, C., *Riv. Combust.* **1961**, 15, 801.
(97) Maretic, M., *Nafta (Yugoslavia)* **1958**, 9, 198.
(98) Maslyanskii, G. N., Potapova, A. A., Skornyakova, V. F., *Neftekhimiya* **1967**, 7, 392.

(99) Maslyanskii, G. N., Pannikova, R. F., Barkan, S. A., *Neftekhimiya* **1966**, 6, 40.
(100) Mattox, W. J., *J. Am. Chem. Soc.* **1944**, 66, 2059.
(101) Mattox, W. J., U. S. Patent **2,316,271** (April 13, 1943).
(102) Mayor, Y., *Genie Civil* **1951**, 128, 448.
(103) McGrath, H. G., *Chim. Ind. (Paris)* **1968**, 80, 561.
(104) McGrath, H. G., Hill, L. R., "Progress in Petroleum Technology," p. 39, American Chemical Society, Washington, D. C., 1951.
(105) Meerbott, W. K., Cherry, A. H., Chernoff, B., Crocoll, J., Heldman, J. D., Kaemmerlen, C. J., *Ind. Eng. Chem.* **1954**, 46, 2026.
(106) Metcalf, E. L., *Can. Chem. Process.* **1952**, 36, 50.
(107) Mills, G. A., Heinemann, H., Milliken, T. H., Oblad, A. G., *Ind. Eng. Chem.* **1953**, 45, 134.
(108) Minachev, Kh. M., Isagulyants, G. V., *Proc. Intern. Congr. Catalysis, 3rd, Amsterdam* **1964**, 1, 308; discussion p. 317.
(109) Moos, J., Schultze, G. R., *VDI (Ver. Deut. Ingr.) Z.* **1963**, 105, 693.
(110) Moy, J. A. E., *Petroleum (London)* **1962**, 25 (5), 150.
(111) Munari, S., Rossi, C., *Intern. Ind. Chem. Congr., 35th, Warsaw*, **1964**; abstr. *Brit. Chem. Eng.* **1965**, 10, 712.
(112) Murphree, E. V., "Progress in Petroleum Technology," pp. 58–59, American Chemical Society, Washington, D. C., 1951.
(113) Nelson, W. L., "Encyclopedia of Chemical Technology," 2nd ed., Vol. 15, p. 1, Wiley, New York, 1968.
(114) Notari, B., *Riv. Combust.* **1962**, 16 (2), 51.
(115) Oblad, A. G., *Oil Gas J.* **1955**, 53 (46), 184.
(116) Oblad, A. G., Shalit, H., Tadd, H. T., *Advan. Catalysis* **1957**, 9, 510.
(117) Oblad, A. G., Mills, G. A., U. S. Patent **2,636,909** (April 28, 1953).
(118) Oblad, A. G., Marschner, R. F., Heard, L., *J. Am. Chem.* **1940**, 62, 2066.
(119) Oblad, A. G., U. S. Patent **2,383,072** (Aug. 21, 1945).
(120) *Oil Gas J.* **1968**, 66 (25), 55.
(121) *Oil Gas J.* **1952**, 50 (45), 187.
(122) *Oil Gas J.* **1955**, 53 (46), 140.
(123) *Oil Gas J.* **1967**, 65 (8), 70.
(124) Panchenkov, G. M., Zhorov, Yu. M., Pershin, A. D., *Tr. Mosk. Inst. Neftekhim. Gazov. Prom.* **1967** (74), 26.
(125) *Petrol. Process.* **1957**, 12 (6), 117.
(126) Plate, A. F., Tarasova, G. A., *J. Gen. Chem. (U.S.S.R.)* **1943**, 13, 21, 36.
(127) Plate, A. F., *Usp. Khim.* **1940**, 9, 1301.
(128) Pollitzer, E. L., Sterba, M. J., Lickus, A. G., Hayes, J. C., *Oil Gas J.* **1968**, 63 (53), 140.
(129) Radford, H. D. *et al.*, *Am. Petrol. Inst., Div. Refining, 34th Midyear Meetg., Chicago* **1969**, Preprint 24–69.
(130) Raik, S. E., *Probl. Kinetiki i Kataliza Akad. Naus SSSR* **1949**, 6, 262.
(131) Read, D., *Petrol. Refiner* **1952**, 31 (5), 97.
(132) Riediger, B., *VDI-Z.* **1958**, 100, 763.
(133) Schmitz, E., *Ann. Mines Carburants* **1943**, 2 (14), 155.
(134) Sergienko, S. R., *Bull. Acad. Sci. U.S.S.R., Classe Sci. Chim.* **1941**, 177.
(135) Sinfelt, J. H., *Advan. Chem. Eng.* **1964**, 5, 37.
(136) Smeykal, K., Hauthal, H. G., Engler, W., *Chem. Tech. (Berlin)* **1962**, 14, 732.
(137) Smoley, E. R., Bowles, V. O., *Petrol. Engr.* **1946**, 17 (7), 59.
(138) Smoley, E. R., Whipple, T. G., *Mech. Eng.* **1947**, 69, 293.
(139) Smolnik, Y. E., Sobol, E. P., Yampolskii, N. G., Zhurba, A. S., Bryanskaya, E. K., Cherednichenko, G. I., *Khim. i Tekhnol. Topl. i Masel* **1968**, 13 (6), 10.
(140) Standard Oil Development Co., British Patent **537,532** (June 25, 1941).

(141) Standard Oil Development Co., British Patent **557,291** (Nov. 15, 1943).
(142) Steiner, H., *J. Inst. Petrol.* **1947**, 33, 410.
(143) Sulimov, A. D., *Khimiya* **1964**.
(144) Sutherland, R. E., Haensel, V., *Oil Gas J.* **1967**, 65 (15), 68.
(145) Taylor, H. S., Fehrer, H., U. S. Patent **2,336,900** (Dec. 14, 1943).
(146) Taylor, H. S., Fehrer, H., Turkevich, J., U. S. Patent **2,357,271** (Aug. 29, 1944).
(147) Taylor, W. F., Welty, A. B., Jr., *Oil Gas J.* **1963**, 61 (48), 142.
(148) Teter, J. W., Borgerson, B. T., Beckberger, L. H., *Oil Gas J.* **1953**, 52 (23), 118.
(149) Thomas, C. L., *Chim. Ind. (Paris)* **1962**, 87, 496.
(150) Tomasik, Z., Wrzyszcz, J., Kulak, S., *Nafta (Katowice)* **1964**, 20, 300.
(151) Verti, V., *Riv. Combust.* **1952**, 6, 225.
(152) Verworner, M., Faatz, G., Gelbin, D., *Chem. Tech. (Berlin)* **1962**, 14, 328.
(153) Voss, G., *Erdol. Kohle* **1962**, 15, 387.
(154) Weinert, P. C., Sterba, M. J., Haensel, V., Grote, H. W., *Proc. Am. Petrol. Inst., 17th Midyear Meetg.* **1952**, 32M, III, 187; discussion p. 235.
(155) Welter, W. G., *Chem. Eng. Progr.* **1963**, 59, 78.
(156) Wermann, J., Lucas, K., *Chem. Tech. (Berlin)* **1964**, 16, 342.
(157) Wingerter, K. H., *Chem. Tech. (Berlin)* **1962**, 14, 333.
(158) Wright, J. F., *Petrol. Refiner* **1950**, 29 (9), 163.
(159) Zakhra, Yu., Paushkin, Ya. M., *Izv. Vysshikh Uchebn. Zavedenii, Neft i Gaz* **1962**, 5 (7), 57.
(160) Zhorov, Yu. M., Panchenkov, G. M., Zel'tser, S. P., Tirak'yan, Yu. A., *Khim. i Tekhnol. Topliv. i Masel* **1965**, 10, 12.
(161) Zhorov, Yu. M., Panchenkov, G. M., Zel'tzer, S. P., Tirak'yan, Yu. A., *Kinetika i Kataliz* **1965**, 6, 1092.
(162) Zhorov, Yu. M., Panchenkov, G. M., Tirak'yan, Y. A., Zel'tser, S. P., Fradkin, F. R., *Kinetika i Kataliz* **1967**, 8, 658.

RECEIVED January 12, 1970.

3

Chemistry of Hydrocracking

G. E. LANGLOIS and R. F. SULLIVAN

Chevron Research Co., Richmond, Calif. 94802

The chemistry of catalytic hydrocracking of hydrocarbons is reviewed. Hydrocracking catalyzed by dual-functional, acid catalysts and hydrogenolysis on nonacidic metal-type catalysts are contrasted. Reactions on acidic catalysts are characterized by extensive and deep-seated rearrangements. However, the over-all reactions are surprisingly specific and selective, resulting in a high degree of conservation of ring structures and in the production of high yields of branched paraffins and certain cyclic species. Data on the hydrocracking of typical paraffins, cycloparaffins, aromatics, and polycyclic aromatics are presented; reaction mechanisms to account for the observed products are discussed. Shape-selective catalysis with crystalline aluminosilicate catalysts is briefly reviewed.

Catalytic hydrocracking is practiced extensively commercially in petroleum refining to produce high quality gasoline, jet fuel, and lubricants. Its use in manufacturing chemicals is more limited but still important. Examples are the production of light paraffins and aromatic hydrocarbons by hydrocracking (38, 44, 45, 61). Because of the specificity of many of the reactions, hydrocracking is potentially a low cost method for producing a number of important hydrocarbon chemicals. This chapter reviews the chemistry of catalytic hydrocracking of hydrocarbons. Examples of the hydrocracking of typical pure compounds are presented to illustrate the types of reactions and the principal products. Reaction mechanisms to account for the observed products are discussed.

Types of Hydrocracking Catalysts

Many hydrocracking catalysts of commercial importance are dual-functional catalysts containing both a hydrogenation-dehydrogenation component and an acidic component. The reactions catalyzed by the two catalytic components are quite different. In specific catalysts the relative strengths of these two components can be varied, and, hence, the reactions occurring and the products formed depend critically on the balance between the contributions of each of the components (56).

Here we discuss and contrast the chemistry of hydrocracking with two major categories of hydrocracking catalysts: (1) dual-functional catalysts in which the acidic component is strong and the hydrogenation component is relatively weak, and (2) nonacidic catalysts with an active hydrogenation-dehydrogenation component.

The catalysts used in most major commercial petroleum hydrocracking processes fall in the first category. Reactions of these dual-functional catalysts are characterized by extensive isomerization and skeletal rearrangement typical of acid-catalyzed reactions. With this type of catalyst, the isomerization and cracking reactions occur primarily on the acidic component of the catalyst, and in the mechanisms discussed later, carbonium ion-type intermediates are postulated. Both intermolecular and intramolecular rearrangements may occur before cracking so that the product structures often bear no simple relationship to the reactant structure. Bond cleavage is quite selective, and products tend to be highly branched. Very little cracking (hydrogenolysis) occurs on the hydrogenation sites. The hydrogenation component dehydrogenates saturated reactants to produce reactive olefin intermediates, hydrogenates the unsaturated products from cracking, and prevents catalyst deactivation by hydrogenating coke precursors.

With nonacidic or weakly acidic catalysts the reactions occur primarily on the hydrogenation component and are generally less complex. With this type of catalyst relatively little isomerization or structural rearrangement occurs. Bond cleavage occurs by direct hydrogenolysis, and products are generally simple fragments of the original reactant.

With catalysts intermediate between these two extremes, contributions of both acid-catalyzed cracking and hydrogenolysis are important, and the reaction products are intermediate in character.

In a special category are the so-called shape-selective catalysts. These are derived from crystalline aluminosilicates in which the catalyst pore structure is such that access of reactants and egress of products are limited to molecules of a particular size or shape. With these catalysts the molecules reacting and the products formed are primarily determined by the geometrical characteristics of the catalyst pores.

Reactions of Paraffins

The reaction mechanisms of paraffin hydrocracking have been studied extensively (2, 7, 15, 16, 22, 30, 34, 40, 44). A number of reactions may occur with varying degrees of importance: dehydrogenation and hydrogenation, isomerization, scission of carbon–carbon bonds, hydrogen transfer, disproportionation, and cyclization. Although there is not complete agreement as to all the details of the reaction mechanisms, most of the differences in product distributions reported by various workers can be attributed to relative differences in catalyst acidity or hydrogenation activity, the nature of the catalytic sites, or to differences in reaction conditions (6).

Strongly Acidic Catalysts with Mild Hydrogenation Activity. REACTION MECHANISM. Typical catalysts with strongly acidic properties and relatively mild hydrogenation activity are metal sulfides such as nickel, tungsten, or cobalt sulfide on supports such as silica-alumina.

Important characteristics of the product from hydrocracking of normal paraffins with this type of catalyst are:

(1) The product consists mainly of isoparaffins.

(2) Little or no methane or ethane is formed.

(3) Only a small amount of the normal paraffin reactant isomerizes without cracking.

(4) An important side reaction is an apparent disproportionation process.

Product distributions from the hydrocracking of typical normal paraffins with nickel sulfide on silica alumina catalyst (34, 59) are shown in Figure 1 (n-hexadecane) and Table I (n-decane). Table I also includes results with silica-alumina and with nickel on silica-alumina (discussed later).

The mechanism for hydrocracking of paraffins with a strongly acidic catalyst containing a mild hydrogenation component is generally believed

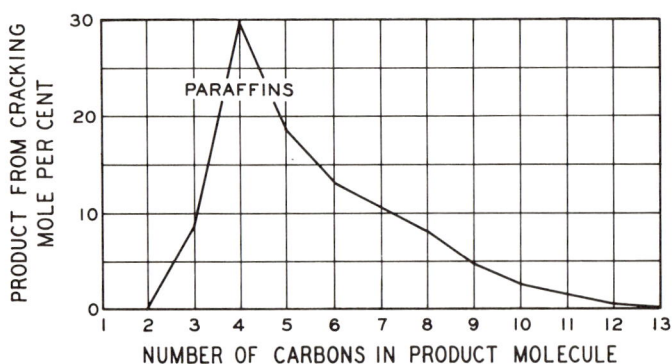

Figure 1. Products from hydrocracking n-hexadecane over nickel sulfide on silica-alumina at 290°C, 82 atm, and 51% conversion

Iso-to-normal ratios: iso-C_4/n-C_4 = 4; iso-C_5/n-C_5 = 13; iso-C_6/n-C_6 = 11.

Table I. Hydrocracking of n-Decane

Catalyst	Silica-Alumina	Ni on Silica-Alumina	Ni_3S_2 on Silica-Alumina
Ni in catalyst, %	0	6.6	6.6
Temp, °C	←---------	288	---------→
Pressure, atm	←---------	82	---------→
Feed rate, LHSV	←---------	8	---------→
H_2/hydrocarbon mole ratio	←---------	10	---------→
Total conversion, %	2.5	7.8	52.8
isomerization, %	0.3	4.5	5.5
cracking, %	2.2	3.3	47.3
Product, moles/100 moles of feed			
Methane	0.01	0.28	0.02
Ethane	—	0.05	0.05
Propane	0.3	0.33	7.9
Isobutane	1.2	0.07	31.1
n-Butane	0.2	0.90	6.8
Isopentane	1.3	0.07	21.5
n-Pentane	0.1	1.20	2.6
Isohexanes	0.9	0.12	18.7
n-Hexane	0.06	1.0	2.0
Isoheptanes	0.3	0.07	4.6
n-Heptane	0.01	0.81	0.2
Octanes	0.03	0.47	1.1
Nonanes	0.01	0.52	1.0
Isodecanes	0.3	4.5	5.5
n-Decane	97.5	92.1	47.2
Olefins	0.05	—	—
Iso-to-normal ratio, C_4–C_7	10	0.08	6.6
Excess butanes, moles/100 moles of decane cracked	20	—	33

to be similar to that for catalytic cracking with features of hydrogenation and hydroisomerization superimposed. The mechanism discussed below is essentially that proposed by Flinn, Beuther, and Larson (6, 22) and Archibald, Greensfelder, Holzman, and Rowe (2).

The following reaction steps are proposed for the hydrocracking of a normal paraffin.

(1) A normal olefin is formed by dehydrogenation of the normal paraffin reactant.

(2) The normal olefin is adsorbed on the acidic sites of the catalyst to form an ionic species. (Ionic intermediates can also be formed by hydride ion exchange as described below for catalytic cracking.)

(3) The carbonium ion thus formed can isomerize to a more stable species (a tertiary carbonium ion), crack to form a smaller ion and an olefin, or desorb as a normal olefin.

(4) The tertiary carbonium ion formed in the previous step can undergo further isomerization, crack, or desorb.

(5) Olefinic products are saturated by hydrogenation or hydrogen transfer reactions.

The mechanism for hydrocracking of paraffins can be related to mechanisms proposed for catalytic cracking (*27, 28, 51*). In catalytic cracking, the olefins required for Step 2 above can be formed by small amounts of thermal cracking. Then, additional ionic species are formed by hydride ion transfer. In the cracking step, the preferred bond for dissociation is one beta to the carbon atom with the positive charge. Rates of cracking of branched ions are more rapid than those with straight chains, and extensive isomerization to branched intermediates occurs before cracking. Formation of methane or ethane is unfavorable.

The similarity between the product distribution from catalytic cracking and from hydrocracking of a normal paraffin at a given temperature suggests that a similar mechanism occurs in both processes. For example, the product distributions for catalytic cracking of n-decane with silica-alumina and for hydrocracking of n-decane with nickel sulfide on silica-alumina at 288°C are compared in Table I and in Figure 2 (*34*). [It is important that such a comparison be made with both processes at the same conditions. Apparent differences in product distribution between catalytic cracking and hydrocracking can be ascribed largely to altered experimental conditions, especially temperature, between the two reac-

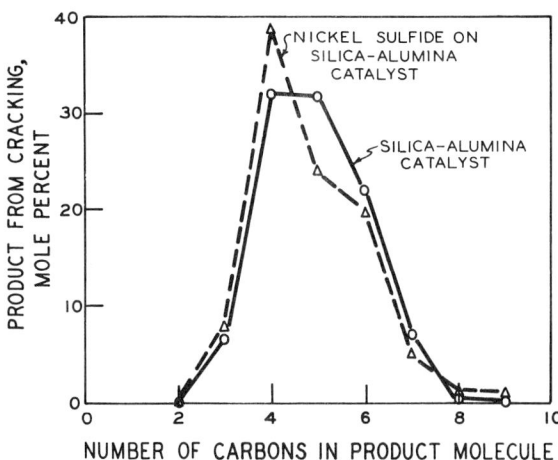

Figure 2. Products from hydrocracking n-*decane at 288°C and 82atm. Distribution by carbon number.*

tions rather than a fundamental difference in the mechanism of the cracking reaction (2).]

Despite a large difference in reaction rates, the products obtained from the two catalysts shown in Figure 2 are strikingly similar. Cracking is the principal reaction in both cases, and only minor quantities of isodecanes are produced. The products from cracking are almost entirely paraffins in the C_3–C_7 range. The ratio of branched to straight chain species in the product (namely, the iso-to-normal ratio) is high, far in excess of thermodynamic equilibrium. Quaternary paraffinic structures are not found in the product.

The reactions of isoparaffins are analogous to those of normal paraffins, but in general the rates of reaction are much more rapid for isoparaffins than for a normal paraffin of the same carbon number. [Exceptions to this are branched paraffin reactants in which all of the branching is on quaternary carbon atoms (2).] The comparatively rapid rates of cracking of isoparaffins explain why only small amounts of branched intermediates with the same molecular weight as the reactant are found in the products from hydrocracking of n-paraffins.

The rate of hydrocracking of n-paraffins increases with increasing molecular weight (2, 22). For example, Flinn, Larson, and Beuther (22) observed that 95% conversion of n-hexadecane was obtained at conditions at which 53% n-octane was converted.

HYDROISOMERIZATION. Another reaction which can contribute to the high ratios of isoparaffins to normal paraffins in product from hydrocracking is hydroisomerization. If n-olefins are formed in the cracking step, they can be converted to isoparaffins by hydroisomerization.

For example, Frye et al. (23) demonstrated that if n-olefins are isomerized over nickel sulfide on silica-alumina in the presence of hydrogen, the iso-to-normal ratios of product paraffins are substantially greater than the thermodynamic equilibrium ratios of either olefins or paraffins. This can be explained by either or both of the following factors: (1) the ratio of tertiary to secondary carbonium ions on the catalyst surface is high, or (2) the rate of hydride ion transfer to the tertiary carbonium ion is exceptionally high. Both factors appear to be important with silica-alumina (50, 54). The similarity between hydroisomerization reactions over a nickel sulfide on silica-alumina catalyst and silica alumina alone was demonstrated by Platteeuw et al. (41). They showed that the same product distribution is obtained with both catalysts. However, the rate of reaction is much lower with silica-alumina owing to catalyst fouling. In the absence of the mild hydrogenation component, diolefin intermediates formed in the hydrogen transfer process act as coke precursors.

DISPROPORTIONATION. Disproportionation reactions of paraffins to form higher and lower molecular weight species are well-known acid-catalyzed reactions (32). Such reactions are believed to occur by a mechanism involving an alkylation process in which an olefinic intermediate adds to an adsorbed carbonium ion. The resulting cation then can crack to form both a fragment of lower carbon number and one of higher carbon number than the reactant.

A reaction resembling such a disproportionation process is an important side reaction in the hydrocracking of paraffins. This can be demonstrated by making a stoichiometric material balance of products from cracking (34). For example, if n-decane hydrocracking involves only simple scission of one or more carbon–carbon bonds, the distribution of products from cracking should have a relatively simple relationship. Methane and ethane formation is negligible, and the product from cracking consists almost entirely of C_3–C_7 species. Therefore, the only possible direct cracking reactions are

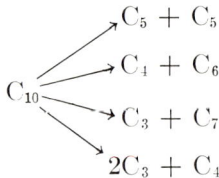

The total butane produced should be related to the other species by the equation:

$$\text{moles of butane} = \text{moles of hexane} + \frac{\text{moles of propane} - \text{moles of heptane}}{2}$$

The actual butane content of the product from the hydrocracking of n-decane can be substantially higher than that calculated on the above simple basis. For example, at 288°C and 82 atm, about 25 moles of "excess butane" are formed from n-decane over a nickel sulfide on silica-alumina catalyst. This result indicates that reactions other than simple scission are involved, such as the apparent disproportionation process discussed above. In the above example, butane would not be the only species formed by such reactions. However, because butane is the most common product from cracking, it is the one to appear in excess in this material balance calculation.

Catalysts with High Hydrogenation Activity. Catalysts of higher hydrogenation activity relative to acidity give a different product distribution than that for strongly acidic catalysts with weak hydrogenation activity. Another set of reactions is dominant. These include isomerization of the reactant (without cracking), hydrogenolysis, and, particularly in the case of noble metal catalysts, cyclization. The extent of each of these reactions varies with the particular catalyst. The most important catalysts in this category are unsulfided metal catalysts. Some important differences are noticed between catalysts in which the hydrogenation component is a noble metal and those with other metals (6).

NONACIDIC SUPPORTS. Hydrogenolysis is the only cracking reaction of paraffins observed with unsulfided metals on nonacidic supports. No appreciable isomerization occurs. However, at controlled conditions, some bonds may be cracked selectively. For example, Haensel and Ipatieff (29) showed that methyl groups attached to secondary carbons could be removed much more easily than those attached to tertiary carbons with nickel or cobalt on nonacidic supports.

WEAKLY ACIDIC SUPPORTS. With unsulfided metals on acidic supports, isomerization is an important reaction. For example, Ciapetta and Hunter (12, 13, 14) demonstrated the effectiveness of nickel on silica-alumina and cobalt on silica-alumina for isomerization of n-paraffins.

Langlois, Sullivan, and Egan (34) report that the primary reaction of n-decane with nonsulfided nickel on silica-alumina is isomerization to produce singly branched isodecanes (Table I). Although a significant amount of cracking occurs, the products are quite different from those found for sulfided nickel on silica-alumina. At relatively low conversions, the product from cracking consists almost entirely of n-paraffins, the distribution by carbon number is more random, and a substantial quantity of methane is formed. Iso-to-normal ratios for the reactions of n-decane with silica-alumina, nickel on silica-alumina, and nickel sulfide on silica-alumina are compared in Table I. In contrast to cracking with more strongly acidic catalysts, isoparaffins do not crack at rates significantly greater than n-paraffins of the same carbon number with nickel on silica-alumina catalyst. There is no evidence of the apparent disproportionation side reaction that is characteristic of cracking with strongly acidic catalysts. All of these features suggest that the cracking is occurring by hydrogenolysis on the nickel metal surface.

It is suggested that during impregnation with a soluble nickel salt, the acidic protons of the catalytic sites on the silica-alumina support are exchanged with nickel to form nickel salts of the silica-alumina acid sites. The resulting salts have sufficient acidity to catalyze isomerization but not acid-catalyzed cracking. One of the effects of sulfiding a nickel on silica-alumina catalyst is that H_2S reacts with these salts and regenerates the original strong acid sites of the silica-alumina support.

NOBLE METAL CATALYSTS. The most frequently mentioned noble metal hydrocracking catalysts are platinum and palladium on silica-alumina. The unsulfided noble metal catalysts have high hydrogenation activity and lower acidity than the supported metal sulfides discussed previously.

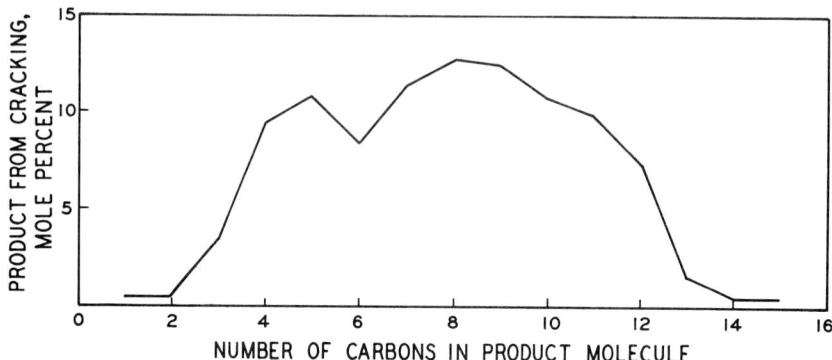

Figure 3. Products from hydrocracking n-hexadecane over platinum on silica-alumina at 371°C, 69 atm, and 53% conversion

Iso-to-normal ratios: iso-C_4/n-C_4 = 1.4; iso-C_5/n-C_5 = 2.3; iso-C_6/n-C_6 = 1.6.

Larson, MacIver, Tobin, and Flinn (35) studied the relationship between hydrogenation activity of supported platinum hydrocracking catalysts and catalyst acidity to determine optimum composition. They observed that with increasing platinum content a selective adsorption of platinum by acid sites causes a reduction in catalyst acidity.

Coonradt and Garwood (15, 16) report results from the hydrocracking of hexadecane over platinum on silica-alumina. They report extensive cyclization and isomerization. Cracking gives mainly two intermediate-sized paraffins with cracking at the center bond preferred. The distribution of products from cracking is shown in Figure 3. Cracking at the center bond is preferred, with little secondary cracking. The average carbon number of the product from cracking is higher than with the supported metal sulfides. Iso-to-normal ratios are close to equilibrium or relatively low compared with the supported metal sulfide catalysts. They suggest that one reason for the lower iso-to-normal ratios is that rapid hydrogenation of olefinic products prevents hydroisomerization. The studies of Platteeuw *et al.* (41) show that hydroisomerization of olefins does not occur over platinum on silica-alumina.

Coonradt and Garwood relate the amount of cyclization to the hydrogenation activity of the catalyst. The mechanism that they propose for hydrocracking with platinum on silica-alumina is an adaptation of that proposed for catalytic reforming by Mills, Heinemann, Milliken, and Oblad (37). Although cyclization does occur as a side reaction in catalytic cracking of high molecular weight paraffins, this reaction presumably occurs by a different mechanism (31).

Paraffins in Mixtures. Preferential adsorption of other types of hydrocarbons by catalytic sites may inhibit the reactions of paraffins in

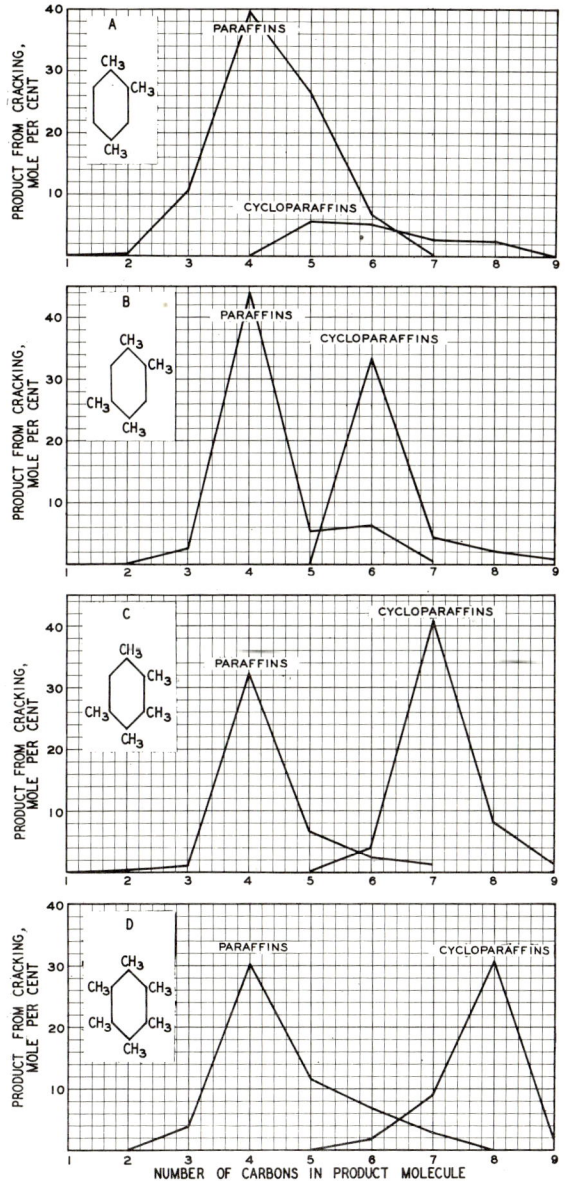

Figure 4. Hydrocracking of polymethylcyclohexanes

	Cyclohexane	Conversion to Cracking, %	Ring Yield, Mole %
A	1,2,4-Trimethyl	28.8	32
B	1,2,4,5-Tetramethyl	39.6	81
C	Pentamethyl	27.0	103
D	Hexamethyl	87.5	95

mixtures. For example, Doane (18) reports that added aromatics depress the rate of hydrocracking of hexadecane at 69 atm over a cobalt oxide–molybdenum oxide–silica-alumina catalyst. (The effect is not obtained if the pressure is doubled.) Beecher et al. (4) report that Decalin can suppress the hydrocracking of n-decane. Voorhies and Hatcher (54) report that cyclohexane is hydrocracked preferentially from a mixture of n-hexane and cyclohexane. Both of the latter studies were made with palladium mordenite catalysts; however, the selective cracking of the cycloparaffin appears to result from preferential adsorption rather than from the shape-selective properties of these zeolitic catalysts.

Reactions of Cycloparaffins

Acidic Catalyst—The Paring Reaction. The reactions of a number of cycloparaffins over a dual functional nickel sulfide on silica-alumina hydrocracking catalyst were reported by Egan, Langlois, and White (20). Figure 4 shows the distribution by carbon number of the products from cracking the following cyclohexanes: 1,2,4-trimethylcyclohexane (1,2,4-tri-MCH); 1,2,4,5-tetramethylcyclohexane (1,2,4,5-tetra-MCH); pentamethylcyclohexane (PMCH); and hexamethylcyclohexane (HMCH). The high selectivity of the cracking reaction of the last three cyclohexanes is striking. The principal acyclic product is isobutane in each case, and the principal cycloparaffin product contains four carbons less than the original cycloparaffin. Very little ring cleavage occurs. The yield of cyclic hydrocarbons from the cracking reaction is greater than 80 mole % of the cycloparaffin reactant (except for 1,2,4-tri-MCH). The small amount of methane in the product indicates that the production of lower molecular weight cycloparaffin is not a simple process of demethanation.

A reaction occurs that, in effect, pares methyl groups from cycloparaffin rings and eliminates them as branched paraffins, mainly isobutane, in such a way as to conserve rings. This has been named the paring reaction.

The selectivity of the cracking reaction to form one acyclic product and one cycloparaffin product decreases dramatically in going from a reactant with 10 or more carbon atoms to a reactant with only nine carbon atoms (1,2,4-tri-MCH). With the latter, the ring yield is low; there is almost as much pentane formed as butane, and there is no single predominant cycloparaffin product. Accompanying the change in selectivity is a corresponding change in rate of reaction. The rate of cracking of the C_{10} cycloparaffin (tetra-MCH) is over 100 times higher than that of the C_9 cycloparaffin (tri-MCH).

RATIO OF SINGLY BRANCHED TO UNBRANCHED PARAFFINS. A characteristic of the paring reaction of cycloparaffins is the unusually high ratio of singly branched paraffins to unbranched paraffins in the products. These ratios far exceed the thermodynamic equilibrium ratios as shown in Table II.

Table II. Observed and Equilibrium Ratios of Singly Branched to Unbranched Paraffins in the Products

Reactant	Temperature, °C	Paraffins	Ratio Observed	Equilibrium
1,2,4,5-Tetra-MCH	291	Butanes	48	1.1
		Pentanes	20	2.9
PMCH	234	Butanes	82	1.2
		Pentanes	34	3.4
HMCH	288	Butanes	25	1.1
		Pentanes	57	2.9

RATIO OF CYCLOPENTANES TO CYCLOHEXANES. The ratio of methylcyclopentane to cyclohexane exceeds the equilibrium ratio by threefold as seen in Table III. In contrast, the differences between the observed and equilibrium ratio with the C_7 and C_8 cycloparaffins are less pronounced.

Table III. Observed and Equilibrium Ratios of Cyclopentanes to Cyclohexanes in Products

Reactant	Temperature, °C	Carbon No. of Cycloparaffin	Ratio Observed	Equilibrium
1,2,4,5-Tetra-MCH	291	6	14.4	4.6
		7	0.8	1.2
PMCH	234	6	(>10)	3.1
		7	1.8	0.8
HMCH	288	6	(>10)	4.5
		7	0.8	1.2
		8	0.5[a]	0.8[a]
HMCH	232	6	(>10)	3.1
		7	0.8	0.8
		8	0.5[a]	0.6[a]

[a] Estimated

EFFECT OF STRUCTURE. Figure 5 shows the distribution of products by carbon number from five isomeric cycloparaffins containing 10 carbons; 1,2-diethyl-CH, n-butyl-CH, s-butyl-CH, isobutyl-CH, and tert-butyl-CH.

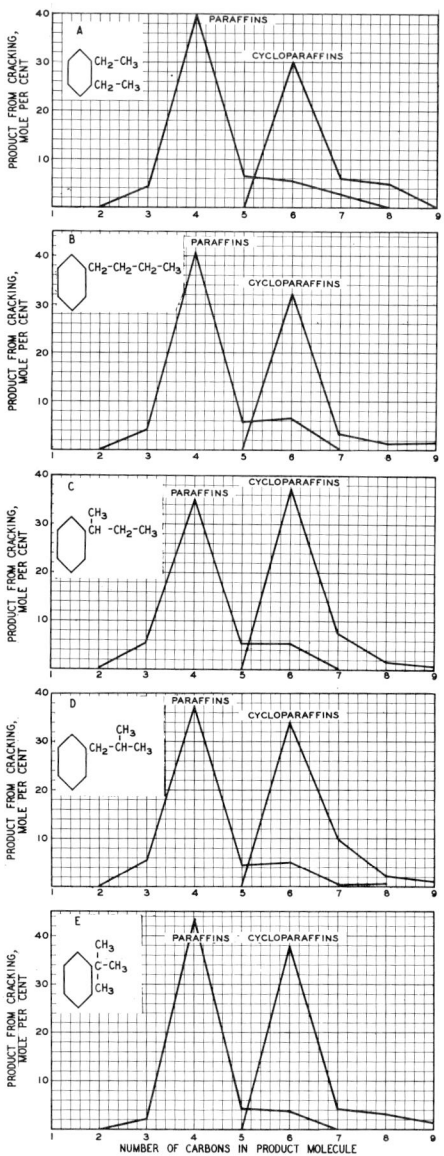

Figure 5. Hydrocracking of C_{10} cyclohexanes at 82 atm and 290°C

	Cyclohexane	Conversion to Cracked Products, %	Ring Yield, Mole %
A	1,2-Diethyl	38.0	81
B	n-Butyl	47.6	82
C	sec-Butyl	57.1	92
D	iso-Butyl	33.9	91
E	tert-Butyl	36.4	87

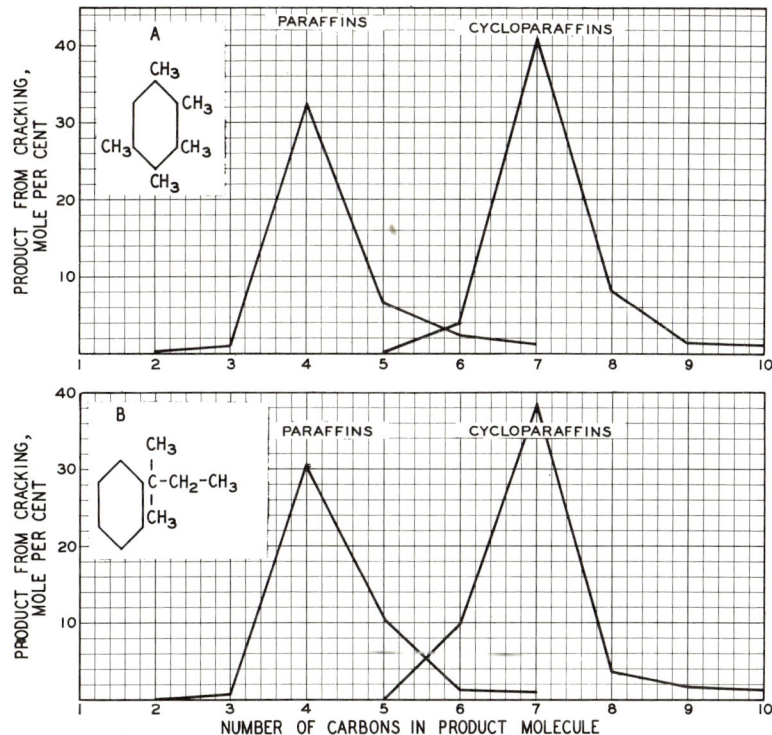

Figure 6. Hydrocracking of C_{11} cyclohexanes at 82 atm and 234°C
  A: PMCH
  B: tert-Amyl-CH

1,2,4,5-Tetramethyl-CH is shown in Figure 4. The similarity of the product distributions is apparent. Again, high ratios of isobutane to n-butane and of MCP to CH are observed. These results show that the composition of the product is essentially independent of the reactant structure. The ring yield in each case exceeds 80 mole %.

The similarity of the product distributions from two isomeric cycloparaffins containing 11 carbons, PMCH and *tert*-amyl-CH, is shown in Figure 6. The cracking of *tert*-amyl-CH is not a simple dealkylation of the C_5 side chain to produce isopentane. Instead isobutane and C_7 cycloparaffins are the principal products.

THE EFFECT OF TEMPERATURE. The paring reaction of the polymethylcyclohexanes becomes more selective at lower temperatures. For example, with tetra-MCH the yield of cycloparaffins increases from 56 to 81 mole % in lowering the reaction temperature from 348° to 291°C, and the yield of the predominant cycloparaffin, MCP, increases from 61 to 77%.

ISOMERIZATION PRODUCTS OF ALKYL CYCLOHEXANES. Extensive isomerization of the reactant precedes the cracking of cycloparaffins. Table IV presents the analyses of the C_{10} cycloparaffin isomers formed

Table IV. Composition of Mixture of Isomers, Weight Percent

Temperature, °C Cyclohexane used	291 1, 2, 4, 5-Tetra-M	286 Isobutyl
Amount unreacted, % of C_{10}-cycloparaffins in product	3.1	9.4
Extent of cracking, %	39.6	33.9
Analysis of mixture, excluding the reactant		
Cyclopentanes	56.5	46.7
Cyclohexanes		
tert-butyl	0.6	
isobutyl ⎱ sec-butyl ⎰	0.2	
1-methyl-3-isopropyl	1.8	1.9
1-methyl-4-isopropyl	0.8	1.3
1-methyl-2-isopropyl	0.5	1.1
1,2-diethyl		
1,3-diethyl		
1,4-diethyl		
1-methyl-3-n-propyl	9.6	12.7
1-methyl-4-n-propyl		
n-butyl		
1,3-dimethyl-5-ethyl	8.4	7.5
1-methyl-2-n-propyl	1.1	1.3
1,4-dimethyl-2-ethyl	3.8	4.1
1,3-dimethyl-4-ethyl	3.6	4.4
1,2-dimethyl-4-ethyl	6.8	7.6
1,3-dimethyl-2-ethyl	0.2	0.3
1,2-dimethyl-3-ethyl	0.9	1.3
1,2,4,5-tetramethyl		2.8
1,2,3,5-tetramethyl	4.0	5.4
1,2,3,4-tetramethyl	1.2	1.6
Cyclohexanes, total	43.5	53.2

from tetra-MCH and isobutyl-MCH. The similarity of these mixtures of isomers is striking. These data show that before 40% of the original cycloparaffin is cracked, a mixture of isomers is produced that approximates an equilibrium (or a steady-state) composition.

MECHANISM. A proposed mechanism for the paring reaction is given below. The first step is the formation of a carbonium ion intermediate, either by dehydrogenation followed by protonation of the double bond or by hydride ion transfer to another carbonium ion. The adsorbed

species then isomerize rapidly by a process including alternate ring contraction and expansion to a common mixture of carbonium ions on the catalyst surface. This is followed at a slower rate by selective cracking of certain favorable structures. The product distribution therefore is essentially independent of the structure of the original cycloparaffin (*see* chart on p. 53).

The high selectivity indicates that under the reaction conditions employed the types of carbonium ions that will crack are restricted. The product composition indicates that: (1) the predominant cracking occurs with tertiary ions which can produce another tertiary ion by β-scission (for example, I); and (2) that isomerization equilibrium of the products from cracking (paraffins and lighter cycloparaffins) is not established, presumably because of rapid displacement from the catalyst surface by heavier hydrocarbons.

The assumption that tertiary ion to tertiary ion cracking predominates is consistent with (1) the production of isobutane rather than n-butane and propane, (2) the production of MCP rather than CH, and (3) the failure of C_9 cycloparaffins to undergo the selective paring reaction. At least 10 carbon atoms are necessary to form the favorable intermediate. As reaction temperature is increased, the specificity for tertiary-tertiary cracking decreases as indicated by decreasing iso-to-normal ratios in the paraffin products.

The predominance of isobutane over higher molecular weight branched paraffins in the product is probably explained by simple statistical considerations. The number of favorable structures containing a four-carbon atom side chain is much greater than those containing five carbon atoms or more. As the molecular weight of the reactant is increased, this advantage decreases, and the production of heavier branched paraffins increases.

Methylcyclopentane is a primary product from a reaction involving tertiary-to-tertiary cracking while cyclohexane is not. Therefore, the high ratios of methylcyclopentane to cyclohexane shown in Table III are consistent with the proposed mechanism. Statistical considerations similar to those discussed in the preceding paragraph account for the lower ratios of cyclopentanes to cyclohexanes in the C_7 and C_8 products.

Stability of Cycloparaffin Ring. A notable feature of the hydrocracking of cycloparaffins is the stability of the cycloparaffin ring. Very little ring cleavage occurs. Additional evidence of the stability of the cycloparaffin ring comes from the hydrocracking of cyclododecane (59). This large ring hydrocarbon with no side chain conceivably could either undergo the paring reaction after contraction of the ring or it could undergo ring opening and then crack as an alkane to give essentially no cyclic products. Figure 7 shows that the product distribution obtained is essentially the

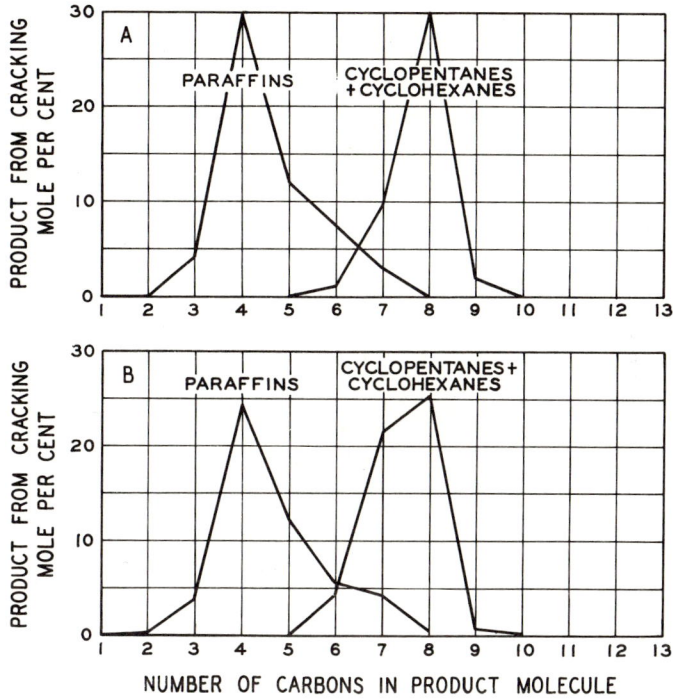

Figure 7. Product distribution from hydrocracking C_{12} cycloparaffins

		Cracking, %	Temp., °C
A	Hexamethylcyclohexane	87.5	234
B	Cyclododecane	91.0	296

same as that obtained from HMCH, indicating that the paring reaction is predominant. Even with this large cycloparaffin ring, there is essentially 100% ring conservation.

The reasons for this unusual stability of the cycloparaffin ring are not completely understood. Bond energies of the carbon–carbon bonds in the ring are approximately the same as those in the side chains. There are many structures such as II in which ring cleavage could occur by the favored β-scission of a tertiary ion to produce another tertiary ion. However, when ring cleavage of cation II occurs, the initial product is a cation containing an olefinic double bond, III.

This intermediate is still bound to the catalyst in the activated form. Subsequently, it can (1) desorb and hydrogenate to form a paraffin of the same carbon number as the reactant, (2) crack again to form lighter acyclic products, or (3) reverse the reaction and recyclize to form II or a similar cyclic species. If the last reaction is rapid compared with the first two, essentially no net ring cleavage will occur. The probability of recyclization is particularly high because the double bond is held in the immediate vicinity of the reactive cationic center by the carbon chain. In the case of the paring reaction, on the other hand, the olefinic fragment formed by the cracking reaction is free to diffuse away and be hydrogenated, thus effectively preventing the reverse reaction. It may be, therefore, that bonds in the cycloparaffin ring crack about as readily as those in the side chains. However, because of the favorable circumstances for reversing the reaction, the net loss of ring structures is small, and the side chain cleavage reaction which is essentially irreversible predominates.

Nonacidic Catalyst. The hydrogenolysis of a number of polymethylcyclopentanes with a platinum-on-alumina catalyst was studied by Gault and Germain (24, 25, 26). The principal reactions are demethanation to produce lower polymethylcyclopentanes and ring cleavage to produce paraffins. Ring cleavage occurs preferentially at the bond between two secondary carbon atoms, but all bonds are cleaved to a substantial extent. The relative ratios of bond cleavage depend on catalyst composition and temperature. Isomerization is a minor reaction.

The hydrogenolysis of a series of alkylcyclohexanes with a nickel-alumina catalyst was studied by Kochloefl and Bazant (33). They found the cyclohexane ring quite stable under the conditions studied. Methyl groups were removed successively from alkyl side chains without ring cleavage.

Reactions of Alkyl Aromatics

Acidic Catalysts. The hydrocracking of a number of alkyl aromatics with nickel sulfide on silica-alumina catalyst was reported by Sullivan, Egan, and Langlois (48, 49). The reactions occurring are isomerization, dealkylation, alkyl transfer, paring, and cyclization. Reaction products from hydrocracking aromatics, in general, exhibit wider variation and greater dependence on the specific structure of the reactant than is the case with cycloparaffin reactants.

ALKYLBENZENES WITH SIDE CHAINS OF THREE OR MORE CARBONS. The products from hydrocracking of alkylbenzenes containing side chains of three to five carbon atoms are relatively simple. Direct dealkylation is the primary cracking reaction. For example, *tert*-amylbenzene gives benzene and isopentane (20). n-Butylbenzene reacts to benzene and butane (22, 48). In the latter case, n-butane is the principal paraffin produced although substantial isomerization to isobutane also occurs. Alkyl group transfer or disproportionation to form benzene and dibutylbenzene is also important.

With larger alkyl side chains, the reaction product becomes more complex, and an important new reaction—cyclization—is observed (48). For example, the products from cracking n-decylbenzene are shown in Figure 8. Simple dealkylation to benzene and decane is still the most important reaction, but many other reactions also occur. All the paraffins from C_3 through C_{10} are produced in substantial quantities as well as

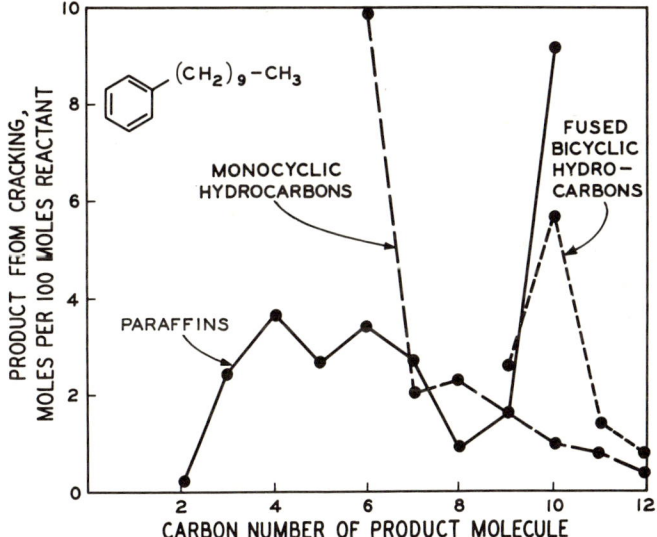

Figure 8. Hydrocracking of n-*decylbenzene at 288°C, 82 atm*

alkylbenzenes from 7 to 12 carbon number. Thus, in paraffin formation the bond between the aromatic ring and the α-carbon of the side chain is the most likely to be cleaved, but there is substantial cleavage of the other bonds in the side chain.

Also formed in considerable quantities are C_9–C_{12} polycyclic hydrocarbons, particularly Tetralins and indanes. The formation of these bicyclic species is surprising. Under the high hydrogen pressures employed in the hydrocracking reaction, dehydrocyclization is not favored thermodynamically. For example, in Table V, equilibrium calculations indicate that the cyclization of n-butylbenzene to form Tetralin and hydrogen is unfavorable.

Table V. Cyclization of Alkylbenzenes[a]

	Molal Ratio of Tetralin to Reactant at Equilibrium at 75 atm of Hydrogen and 82 atm Total Pressure		
Equation	227°C	327°C	427°C
n-Butylbenzene (g) = Tetralin (g) + Hydrogen (g)	0.001	0.005	0.019
n-Decylbenzene (g) = Tetralin (g) + 2-Methylpentane (g)	8400	5800	4200

[a] Calculated from data on n-butylbenzene (43), n-decylbenzene (43), 2-methylpentane (43), and Tetralin (19).

Consistent with these thermodynamic considerations, little Tetralin is found in the products from hydrocracking of n-butylbenzene. Data are not available for calculating the dehydrocyclization equilibrium for the higher alkylbenzenes (for example, n-decylbenzene to hexyltetralin plus hydrogen), but this cyclization reaction is probably unfavorable to about the same degree. If, however, cracking accompanies cyclization and the reaction proceeds by a mechanism in which no hydrogen is produced, it is thermodynamically possible (as shown in Table V) to form large amounts of Tetralin and indane-type hydrocarbons from monocyclic alkyl aromatics. These results suggest that the Tetralin and indane-type hydrocarbons are formed by a mechanism such as that shown below.

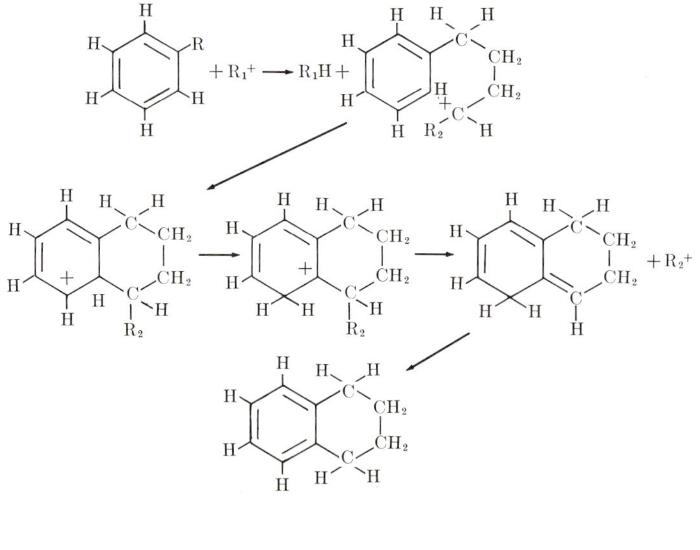

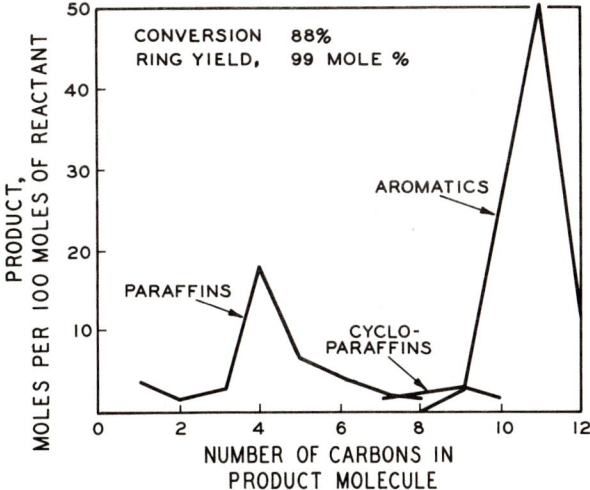

Figure 9. Hydrocracking of hexamethylbenzene at 349°C and 13.6 atm

HEXAMETHYLBENZENE—THE PARING REACTION. The products from hydrocracking of a polyalkylbenzene with small side chains, hexamethylbenzene, are shown in Figure 9 (49). Principal products are light isoparaffins and C_{10} and C_{11} methylbenzenes. Essentially no ring cleavage occurs, and hydrogenolysis to form methane is small. To account for the formation of the isoparaffin products, a process of isomerization leading to side chain growth followed by cracking of side chains of four or more carbon atoms is proposed. This mechanism (below) is similar to the paring reaction proposed for the hydrocracking of cycloparaffins.

According to this mechanism, xylene is a primary product. However, little xylene is found because of rapid methyl transfer from heavier aromatics such as hexamethylbenzene to form large amounts of C_{10} and C_{11} aromatics. Data from experiments with a reactant mixture of hexamethylbenzene and xylene confirm that alkyl transfer is extremely rapid and that the distribution of the polymethylbenzene species in the reaction mixture is essentially the equilibrium distribution.

Although the mechanism for the paring reaction of aromatics is formally similar to that for cycloparaffins, it is energetically less favorable. The isomerization process resulting in side chain growth involves contraction of the six-membered ring with a consequent loss of the resonance energy of the benzene ring. However, the plausibility of the cyclopentadienyl cations postulated as intermediates is strongly supported by the work of De Vries (17) and of Winstein and Battiste (60), who showed that hexamethylcyclopentadiene forms a stable cation which can be converted readily to pentamethylbenzene. These studies and the more recent work of Childs, Sakai, and Winstein (10, 11) indicate that intermediates such as IV and V may be represented more precisely by a nonclassical description. The cyclopentadienyl intermediates are stabilized apparently by the alkyl groups attached to the ring, and the side chain growth process occurs readily only with polyalkyl aromatics. Although the conversion of dimethylcyclohexane to ethylcyclohexane occurs readily, the corresponding conversion of xylene to ethylbenzene is very difficult. Under conditions at which the paring reaction occurs readily with hexamethylbenzene, no conversion of *o*-xylene to ethylbenzene is detected.

HYDROCRACKING REACTIONS OF CYCLOPARAFFINS AND ALKYLBENZENES ON ACIDIC CATALYSTS. The cracking of the cycloparaffins is preceded by rapid isomerization so that the products from cracking depend only on the molecular weight of the reactant and are essentially independent of its starting structure. Principal products are isobutane and a cycloparaffin containing four fewer carbon atoms. With alkyl aromatics, however, the product depends much more on the structure of the reactant. For example, *tert*-amylbenzene cracks to isopentane and benzene; whereas, *tert*-amylcyclohexane cracks to isobutane and C_7 cycloparaffins. This difference is the result of the rapid isomerization of the cycloparaffins before cracking compared with the aromatic which cracks before isomerization. A similar difference exists in the products from *n*-butylcyclohexane and *n*-butylbenzene. The principal paraffin from *n*-butylcyclohexane is isobutane, whereas, from *n*-butylbenzene it is *n*-butane. Thus, although skeletal isomerization does occur with the alkyl aromatics, it is appreciably slower relative to the cracking reaction than with the cycloparaffins.

Aromatic products from hydrocracking frequently undergo subsequent reactions, including alkyl transfer and migration of alkyl groups around the ring to form mixtures approximating the thermodynamic equilibrium. The cycloparaffin products from cracking, on the other hand, do not extensively react further, and compositions are generally far from the equilibrium distribution. The paraffin fragments formed from either aromatics or cycloparaffins do not undergo secondary reaction and are generally not in the equilibrium composition. With both classes of compounds, the ring structure is conserved, and very little ring cleavage occurs. Alkylaromatics with long side chains undergo substantial cyclization to form polycyclic products. This reaction is not observed with the cycloparaffins.

Nonacidic Catalysts. The hydrocracking of alkylaromatics over nonacidic catalysts is relatively simple and straightforward (5, 39, 42, 46). Little or no isomerization occurs. Successive removal of methyl groups from the side chains is the principal reaction. The larger side chains are attacked first. For example, the reaction of p-tert-butyltoluene over Rh on neutral alumina is reported to proceed sequentially as follows: p-tert-butyltoluene → p-proplytoluene → p-ethyltoluene → p xylene → toluene → benzene. In the hydrogenolysis of indane, the first reaction is the cleavage of the 1-2 (or 3-4) carbon–carbon bond of the five-

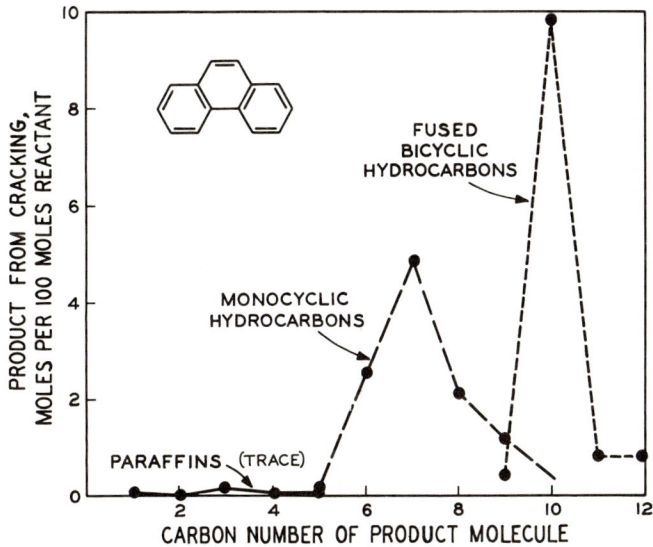

Figure 10. Hydrocracking of phenanthrene at 293°F and 82 atm

membered ring, followed by successive removal of the methyl groups from the resulting ethyltoluene. Thermal hydrogenolysis of alkylaromatics is similar to catalytic hydrogenolysis. For example, methylbenzenes are converted in high yield to benzene by thermal hydrogenolysis. Both catalytic and thermal hydrogenolysis are used commercially to produce benzene and naphthalene from alkylbenzenes and alkylnaphthalenes (3, 21, 36, 47, 57).

Reactions of Polycyclic Aromatics

The reactions of polycyclic aromatics over acidic hydrocracking catalysts are unexpected and surprisingly complex (48). This can be seen by considering the initial products (14% cracking conversion) obtained from the hydrocracking of phenanthrene shown in Figure 10. The principal product from hydrocracking is Tetralin, and the next most important product is methylcyclohexane. Partial hydrogenation is also important. Most of the C_{14} products are partially hydrogenated, and a large fraction contain only one aromatic ring. Surprisingly, only traces of paraffins were formed although one would expect one mole of butane for each mole of Tetralin formed.

The product distribution from phenanthrene indicates that 3 types of reactions are occurring:

(1) Saturation and cleavage of one of the terminal rings to form a paraffin and a bicyclic such as Tetralin. This reaction, which results in paraffin formation, is minor.

(2) Ring saturation and cleavage of the central ring. This accounts for the appreciable amounts of methylcyclohexane and ethylcyclohexane produced.

(3) An unusual cracking reaction that produces bicyclic hydrocarbons, principally Tetralin, without producing an equivalent amount of light paraffins. This is the predominant reaction. A suggested mechanism, shown below, leads to this type of product distribution.

 (a) Partial hydrogenation of the polycyclic aromatic.

 (b) Opening of one of the cycloparaffin rings to form a butyl side chain.

 (c) Alkyl transfer of the side chain to another reactant molecule.

 (d) Ring closure accompanied by hydrogenation to form a tetracyclic hydrocarbon.

 (e) Cracking of a central cycloparaffin ring to give a Tetralin and a cycloparaffin.

(a) 4 H₂ + [phenanthrene] ⟶ [octahydrophenanthrene]

(b) H₂ + [dihydrophenanthrene] ⟶ [tetralin]—C—C—C—C

(c) [tetralin]—C—C—C—C + [phenanthrene] ⟶ [tetralin] + [butylphenanthrene with C—C—C—C]

(d) 3 H₂ + [butylphenanthrene C—C—C—C] ⟶ [tricyclic with cyclopentane ring]—C

(e) 2 H₂ + [tricyclic with cyclopentane]—C ⟶ [tetralin] + [methylcyclopentane with C, C—C]

This mechanism accounts for the lack of formation of paraffins through the combination of a butyl group transfer and a cyclization reaction. It also accounts for the formation of Tetralin which is the major product at low conversions.

The relative extents of the three types of cracking reactions are shown on p. 64. To compute the distribution shown, all paraffins were assumed to be formed from Reaction 1. After deduction of an equivalent quantity of bicyclics, the remaining bicyclics were assumed to be formed by Reaction 3. Monocyclic compounds required to balance Equation 3 were deducted, and the remainder were assumed to be formed by Reaction 2.

	Moles of Phenanthrene Reacting by This Route
(1) Phenanthrene→ Bicyclic + Paraffin	0.3
(2) Phenanthrene → 2 Monocyclics	2.7
(3) 2 Phenanthrenes → 2 Bicyclics + Monocyclic	11.3

This computation shows that the complex process represented by Equation 3 is by far the most important initial mode of reaction. The result is a product that consists almost entirely of cyclic hydrocarbons. At higher conversion levels, more light paraffins are formed, but the processes leading to cyclic species are still predominant.

Hydrocracking of a number of other polycyclic aromatics was studied (48), and the same general type of products was observed. For the hydrocracking of Tetralin substantial quantities of tricyclic intermediates of the type postulated to be formed by butyl group transfer followed by cyclization were actually observed.

With paraffin, cycloparaffin, and alkyl aromatic reactants products obtained from hydrocracking with acidic hydrocracking catalysts are very similar to those obtained from catalytic cracking with acidic catalysts such as silica-alumina. For polycyclic aromatics, however, reaction products from catalytic cracking and hydrocracking are quite different. Catalytic cracking of phenanthrene with silica-alumina (1) produces only coke and small quantities of gases compared with the low molecular weight cyclic products produced in hydrocracking. In hydrocracking, the hydrogenation component serves to convert the refractory polycyclic aromatics into partially hydrogenated species which can then react readily.

Shape Selective Catalysts

A large amount of information has been published on catalytic cracking with molecular shape-selective catalysts, such as natural or synthetic crystalline aluminosilicates. Several excellent reviews have appeared on the chemistry of catalysis with zeolites (52, 53). Literature on hydrocracking of pure hydrocarbons and simple mixtures of hydrocarbons with zeolite-containing catalysts is limited. However, numerous patents on the use of zeolites in hydrocracking catalysts and published reports of commercial applications indicate a widespread use of this type of catalyst for hydrocracking. Both activity and selectivity advantages are reported.

Zeolitic catalysts are characterized by small crystalline cavities which can be entered only through pores of a uniform size. Only molecules of a certain size and shape can enter and leave these cavities. If the catalytic sites are within this pore structure, the reaction is restricted to those molecules that can enter.

For example, Weisz et al. (8, 9, 58) studied acid- and metal-catalyzed reactions of n-paraffins in the presence of branched paraffins and aromatics. They report that, by a proper balance of both shape-selective acid and metal functions, the n-paraffins can be hydrocracked from the mixture without simultaneously hydrocracking other paraffin isomers or by incurring the loss of aromatics through hydrogenation or hydrogenolysis.

Because of the small pores in zeolitic catalysts, reaction rates may be controlled by rates of diffusion of reactants and products. Beecher, Voorhies, and Eberly (4) studied hydrocracking of mixtures of n-decane and Decalin with mordenite catalysts impregnated with palladium. They found that acid leaching of the mordenite produces an aluminum-deficient structure of significantly higher catalytic activity. At least part of this improvement appears to be caused by the decrease in diffusional resistance. They observed that with this type of catalyst, the effective catalyst pore diameter appears to be smaller than calculated owing to the strong interaction or adsorption of hydrocarbon molecules on the pore walls.

Literature Cited

(1) Appleby, W. G., Gibson, J. W., Good, G. M., *Ind. Eng. Chem., Process Design Develop.* **1962**, 1, 102.
(2) Archibald, R. C., Greensfelder, B. S., Holzman, G., Rowe, D. H., *Ind. Eng. Chem.* **1960**, 52, 745.
(3) Asselin, G. F., Erickson, R. A., *Chem. Eng. Progr.* **1962**, 58 (4), 47.
(4) Beecher, R., Voorhies, A., Jr., Eberly, P., Jr., *Ind. Eng. Chem., Prod. Res. Develop.* **1968**, 7, 203.
(5) Beranek, L., Kraus, M., *Collection Czech. Chem. Commun.* **1966**, 31 (2), 566.
(6) Beuther, H., Larson, O. A., *Ind. Eng. Chem., Process Design Develop.* **1965**, 4, 177.
(7) Beuther, H., McKinley, J. B., Flinn, R. A., *Preprints, Div. Petrol. Chem., Am. Chem. Soc.* **1961**, 6 (3), A-75.
(8) Chen, N. Y., Maziuk, J., Schwartz, A. B., Weisz, P. B., *Oil Gas J.* **1968**, 66 (47), 154.
(9) Chen, N. Y., Weisz, P. B., *Chem. Eng. Progr. Symp. Ser.* **1967**, 63 (73), 86.
(10) Childs, R. F., Sakai, M., Winstein, S., *J. Am. Chem. Soc.* **1968**, 90, 7144.

(11) Childs, R. F., Winstein, S., *J. Am. Chem. Soc.* **1968**, 90, 7146.
(12) Ciapetta, F. G., Hunter, J. B., *Ind. Eng. Chem.* **1953**, 45, 147.
(13) *Ibid.*, p. 155.
(14) *Ibid.*, p. 162.
(15) Coonradt, H. L., Garwood, W. E., *Preprints, Division Petrol. Chem., Am. Chem. Soc.* **1967**, 12 (4), B-47.
(16) Coonradt, H. L., Garwood, W. E., *Ind. Eng. Chem., Process Design Develop.* **1964**, 3, 38.
(17) De Vries, L., *J. Am. Chem. Soc.* **1960**, 82, 5242.
(18) Doane, E. P., *Preprints, Div. Petrol. Chem., Am. Chem. Soc.* **1967**, 12 (4), B-139.
(19) Egan, C. J., *J. Chem. Eng. Data* **1963**, 8, 532.
(20) Egan, C. J., Langlois, G. E., White, R. J., *J. Am. Chem. Soc.* **1962**, 84, 1204.
(21) Fiegelman, S., Aristoff, E., *Hydrocarbon Process. Petrol. Refiner* **1965**, 44 (12), 147.
(22) Flinn, R. A., Larson, O. A., Beuther, H., *Ind. Eng. Chem.* **1960**, 52, 153.
(23) Frye, C. G., Barger, B. D., Brennan, H. M., Coky, J. R., Gutberlet, L. C., *Ind. Eng. Chem., Prod. Res. Develop.* **1963**, 2, 40.
(24) Gault, F. G., Germain, J. E., *Actes Deux. Congr. Intern. Catalyse, Paris 1960*, **1961**, 2, 2461.
(25) Gault, F. G., *Compt. Rend.* **1957**, 245, 1620.
(26) Gault, F. G., *Ann. Chim. (Paris)* **1960**, 5 (13), 645.
(27) Greensfelder, B. S., Voge, H. H., Good, G. M., *Ind. Eng. Chem.* **1949**, 41, 2573.
(28) *Ibid.*, **1945**, 37, 514.
(29) Haensel, V., Ipatieff, V. N., *Ind. Eng. Chem.* **1947**, 39, 853.
(30) Haensel, V., Pollitzer, E. L., Watkins, C. H., *Reprints, World Petrol. Congr., 6th*, **1963**, *Sec. III*, Paper 17.
(31) Hightower, J. W., Emmett, P. H., *Proc. Intern. Congr. Catalysis, 3rd, 1964*, **1965**, I, 688.
(32) Kennedy, R. M., "Catalysis," Vol. VI, p. 1, Reinhold, New York, 1958.
(33) Kochloefl, K., Bazant, V., *J. Catalysis* **1967**, 8, 250.
(34) Langlois, G. E., Sullivan, R. F., Egan, C. J., *J. Phys. Chem.* **1966**, 70, 3666.
(35) Larson, O. A., MacIver, D. S., Tobin, H. H., Flinn, R. A., *Ind. Eng. Chem., Prod. Res. Develop.* **1962**, 1, 300.
(36) Masamune, S., Fukuda, J., Kawatani, T., Malda, E., *Preprints, Div. Petrol. Chem., Am. Chem. Soc.* **1968**, 13 (4), A-7.
(37) Mills, G. A., Heinemann, H., Milliken, T. H., Oblad, A. G., *Ind. Eng. Chem.* **1953**, 45, 134.
(38) W. L. Nelson, *Oil Gas J.* **1967**, 65 (33), 130.
(39) Penchev, V., Davidova, N., Beranek, L., *Intern. Congr. Catalysis, 4th, Moscow*, **1968**, Preprint No. 16.
(40) Pier, M., *Z. Elektrochem.* **1949**, 53, 291.
(41) Platteeuw, J. C., de Ruiter, H., Van Zoonen, D., Kouwenhoven, H. W., *Ind. Eng. Chem., Prod. Res. Develop.* **1967**, 6, 76.
(42) Pollitzer, E. L., U. S. Patent **3,306,944** (Feb. 28, 1967).
(43) Rossini, F. D. *et al.*, "Selected Values of Physical and Thermodynamic Properties of Hydrocarbons and Related Compounds," Carnegie Press, Pittsburgh, 1953.
(44) Scott, J. W., Mason, H. F., Kozlowski, R. H., *Petrol. Refiner* **1960**, 39 (4), 155.
(45) Scott, J. W., Paterson, N. J., *World Petrol. Congr., 7th*, **1967**, Panel 17 (3), p. 33.
(46) Seeboth, H., Reiche, A., *Brennstoff Chem.* **1965**, 46 (11), 361.
(47) Stormont, D. H., *Oil Gas J.* **1961**, 59 (46), 160.

(48) Sullivan, R. F., Egan, C. J., Langlois, G. E., *J. Catalysis* **1964**, 3, 183.
(49) Sullivan, R. F., Egan, C. J., Langlois, G. E., Sieg, R. P., *J. Am. Chem. Soc.* **1961**, 83, 1156.
(50) Thomas, C. L., *J. Am. Chem. Soc.* **1944**, 66, 1586.
(51) Thomas, C. L., *Ind. Eng. Chem.* **1949**, 41, 2564.
(52) Turkevich, J., *Catalysis Rev.* **1967**, 1, 1.
(53) Venuto, P. B., Landis, P. S., "Advances in Catalysis and Related Subjects," Vol. XVIII, Academic, New York, 1968.
(54) Voge, H. H., Good, G. M., Greensfelder, B. S., *Ind. Eng. Chem.* **1946**, 38, 1033.
(55) Voorhies, A., Jr., Hatcher, W. J., Jr., *A.I.Ch.E. Ann. Meetg., 61st, Los Angeles, Calif.*, **1968**, Preprint 20A.
(56) Voorhies, A., Jr., Smith, W. M., "Advances in Petroleum Chemistry and Refining," Vol. VIII, p. 169, Interscience, New York, 1964.
(57) Weiss, A. H., Friedman, Lee, *Ind. Eng. Chem., Process Design Develop.* **1963**, 2, 163.
(58) Weisz, P. B., Frilette, V. J., Maatman, R. W., Mower, E. B., *J. Catalysis* **1962**, 1, 307.
(59) White, R. J., Egan, C. J., Langlois, G. E., *J. Phys. Chem.* **1964**, 68, 3085.
(60) Winstein, S., Battiste, M., *J. Am. Chem. Soc.* **1960**, 82, 5244.
(61) Wood, F. C., Eubank, O. C., Sosnowski, J., *Oil Gas J.* **1968**, 66 (25), 183.

RECEIVED January 12, 1970.

4

Secondary Reactions of Olefins in Pyrolysis of Petroleum Hydrocarbons

TOMOYA SAKAI, KAZUHIKO SOMA, YOICHI SASAKI,
HIROO TOMINAGA, and TAISEKI KUNUGI

Department of Synthetic Chemistry, Faculty of Engineering, University of Tokyo, Hongo, Tokyo, Japan

Thermal reactions of C_2H_4, C_3H_6, 1-C_4H_8, cis and trans-2-C_4H_8, 1,3-C_4H_6 and of these respective olefins with added C_4H_6 were studied at 500° ~ 800°C and atmospheric pressure. Primary and secondary products, cyclic compounds in particular, were analyzed in detail. Rate expressions of disappearance of olefins in moles/sec were obtained for C_2H_4, C_3H_6, 1-C_4H_8, cis- and trans-2-C_4H_8, 1,3-C_4H_6, and these olefins with added C_4H_6. Rate of formation of cyclic compounds in the reaction of C_2H_4 or C_3H_6 was directly proportional to the concentration of C_4H_6 formed. The plausible mechanism of formation of aromatics is the ring formation of olefins, or reactive species from olefins, with C_4H_6 followed by dealkylation and/or dehydrogenation.

The pyrolysis of paraffinic hydrocarbons is carried out commercially at temperatures of 750° ~ 800°C. With increasing cracking severity, yields of olefins such as ethylene, propylene, and butadiene increase, pass through their maxima, and then decrease, whereas, yields of hydrogen, methane, and aromatic hydrocarbons increase monotonically with increasing severity. This characteristic change in product distribution means the primary olefinic products are converted consecutively to hydrogen, methane, and aromatics in the advanced stages of the pyrolysis. Except for propylene, only a limited number of published studies exist dealing with the mechanism and kinetics of olefin pyrolysis. Certainly a detailed understanding of the reaction mechanism for formation of aromatics is lacking. This paper presents a detailed description of the change in product distribution obtained in olefin pyrolysis, with the expectation that this

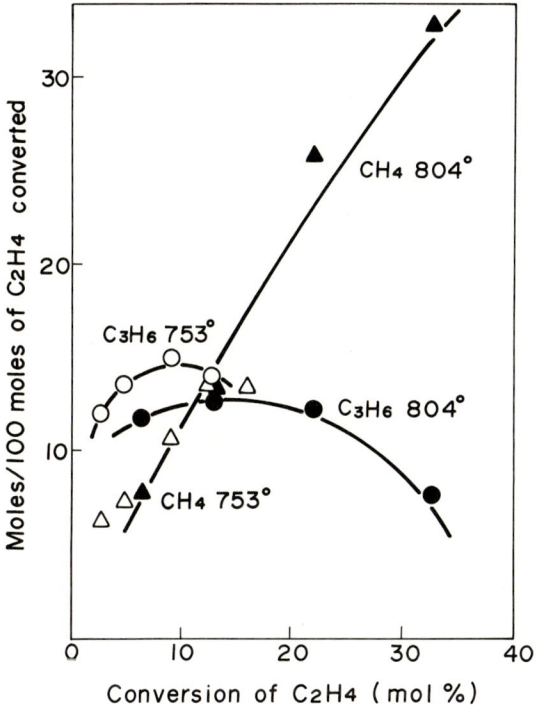

Figure 1. Selectivities of methane and propylene in thermal reaction of ethylene

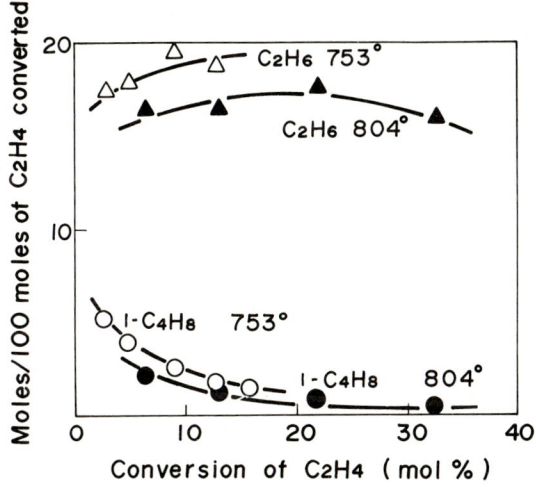

Figure 2. Selectivities of ethane and 1-butene in the thermal reaction of ethylene

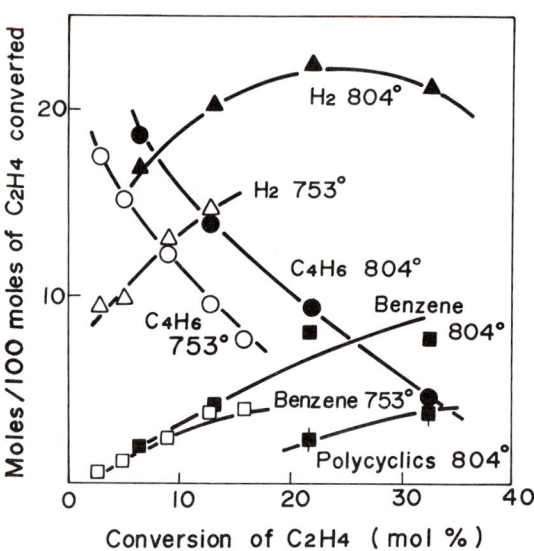

Figure 3. Selectivities of hydrogen, butadiene, benzene, and polycyclics in the thermal reaction of ethylene

study will give some insight into the mechanism of aromatic formation in industrial olefin manufacturing processes.

Experimental

Reagents. Purities of olefins used in this study, determined by gas chromatographic analysis, are shown below.

To determine the order of reaction, commercially available nitrogen of 99.95 mole % purity was used as diluent in some experiments. Trace amounts of oxygen in the nitrogen were removed over reduced copper gauze at *ca.* 400°C.

Olefin, Mole %		Impurity, Mole %			
Ethylene	99.92	methane	0.01,	ethane	0.07
Propylene	99.35	propane	0.65		
1-Butene	98.98	methane	0.033,	ethane	0.016,
		isobutane	0.189,	n-butane	0.105,
		cis-2-butene	0.023,	isobutylene	0.650
cis-2-Butene	99.33	propane	0.044,	propylene	0.034,
		1-butene	0.237,	$trans$-2-butene	0.35
Butadiene	99.41	propylene	0.04,	butenes	0.55

Reaction Apparatus. A conventional flow system was used for atmospheric pressure experiments. The reactor consisted of two quartz tubes with diameters 7.7 (outside of inner tube) and 15.2 (inside of outer

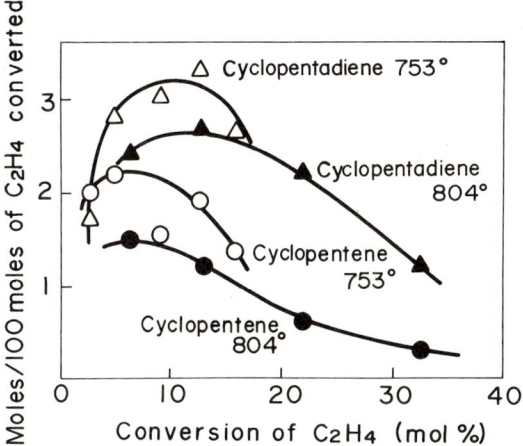

Figure 4. Selectivities of cyclopentene and cyclopentadiene in thermal reaction of ethylene

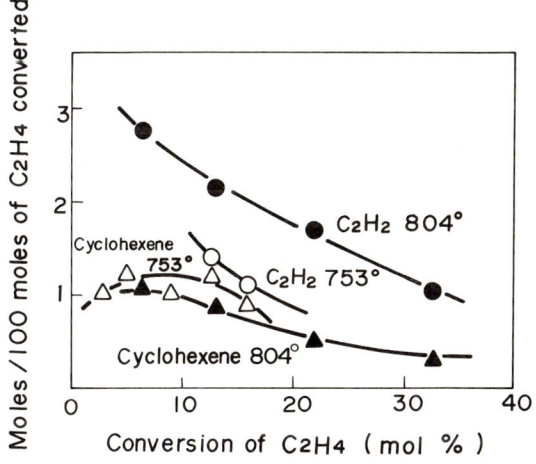

Figure 5. Selectivities of acetylene and cyclohexene in the thermal reaction of ethylene

tube) mm, respectively, and 300 mm long, which were placed in an electrically heated copper block 100 mm long. The reactants flowed down the annular space. The temperature was measured by an alumel–chromel thermocouple along the axis of the reactor. The average effective temperature and the equivalent reactor volume were calculated by the method of Hougen and Watson (8).

Product Analysis. Liquid product was collected in an ice-cooled trap. Exit gas was measured by a wet gas meter. All products were analyzed by gas chromatography. Some products were identified by infrared and/or mass spectra after fractionation by gas chromatography.

Results and Discussion

Thermal Reaction of Ethylene. Dahlgren and Douglas (5) reported that the primary products of the reaction were propylene, butenes, butadiene, and ethane at temperatures ranging from 480° to 582°C and at pressures from 9 to 137 mm Hg. However, analyses of reaction products were incomplete, and the primary products were not distinguished from the secondary ones.

In our study, selectivities of major and minor products were examined as plotted in Figure 1 to 5 vs. the conversion of ethylene. As for major products, selectivities of 1-butene and butadiene decrease sharply with ethylene conversion, whereas that of methane increases rapidly, and hydrogen increases only slightly. Higher temperature favors the formations of hydrogen, butadiene, and acetylene, whereas lower temperature favors formation of ethane, propylene, 1-butene, and cycloolefins. The amount of carbon deposit on the reactor wall was determined by combustion analysis to be negligibly small—*viz.*, below 0.04 wt % of converted ethylene. These figures show that hydrogen, ethane, acetylene, propylene, 1-butene, and butadiene are the primary products and that

Table I. Addition of Butadiene in the Thermal Reaction of Ethylene at 753°C

Butadiene in feed Ethylene, mole %	0.0	0.0	2.9	3.1	5.5
Residence time, sec	0.54	0.74	0.55	0.67	0.64
Product yield[a]					
H_2	0.211	0.805	0.311	0.846	0.775
CH_4	0.170	0.680	0.324	0.724	0.856
C_2H_6	0.545	1.34	0.676	1.36	1.24
C_3H_6	0.322	1.11	0.573	1.16	1.35
$n\text{-}C_4H_{10}$	0.032	0.082	0.037	0.074	0.069
$1\text{-}C_4H_8$	0.163	0.250	0.169	0.248	0.287
$1,3\text{-}C_4H_6$[b]	0.515	1.23	1.91	1.34	2.72
Cyclopentene	0.086	0.193	0.242	0.262	0.440
Cyclopentadiene	0.050	0.178	0.175	0.252	0.340
Cyclohexene	0.034	0.073	0.106	0.098	0.117
Cyclohexadiene	0.013	0.026	0.032	0.035	0.070
Benzene	0.041	0.117	0.128	0.180	0.307
Vinylcyclohexene and toluene	0.037	0.030	0.045	—	0.020

[a] Mole per 100 moles of feed.
[b] Butadiene in feed is included.

methane, cycloolefins, and aromatics are the secondary ones. Change in gas volume with conversion is small—i.e., less than 4%, which indicates that decomposition and polymerization of ethylene occur to nearly the same extent.

Marked decreases of the primarily formed butenes and butadiene with ethylene conversion suggest that these olefins play an important role forming secondary products. In fact, subsequent experiments showed that the addition of butadiene, 3–5 mole %, to ethylene accelerates the formation of cyclopentene, cyclohexene, cyclohexadiene, and benzene (Table I). It seems reasonable, therefore, to propose the following reaction scheme for the formation of cyclic compounds from olefins.

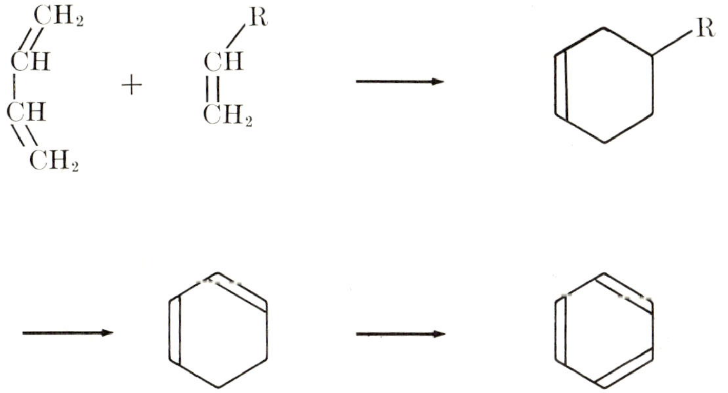

This scheme involves Diels-Alder reaction of olefin and butadiene as a main reaction, followed by successive dealkylation and/or dehydrogenation, and is in line with that of Wheeler and Wood (24). In this connection, the alternate proposal by Kinny and Crowley (10)—that a hexatriene intermediate is formed by polymerization and cyclics result from cyclization of this triene—is considered less likely.

In our experiment, in which the partial pressure of ethylene was varied at 854°C, rate of ethylene disappearance was determined to be 3/2 order with respect to ethylene. This result agrees with those of Burk et al. (4) and Dahlgren and Douglas (5). Moreover, 3/2 order kinetics is consistent with the reaction mechanism proposed by Kunugi et al. (12) based on the kinetic evaluation of the relevant elementary reactions. The rate constants can be obtained from Figure 6. The lines in this figure do not cross the abscissa at the origin, which suggests the existence of induction periods. From Arrhenius plots, the rate constant for the steady-state reaction is expressed as

$$k_{C_2H_4} = 10^{11.73} \exp\left(-\frac{49600}{RT}\right) \text{ ml}^{1/2}\text{mole}^{-1/2}\text{sec}^{-1}$$

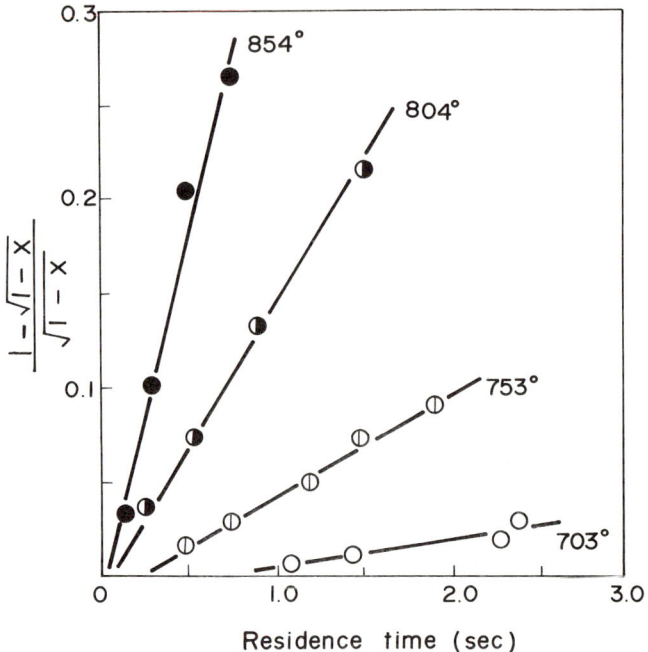

Figure 6. 3/2-Order kinetics for thermal reaction of ethylene

Thermal Reaction of Propylene. Thermal reaction of propylene has been studied extensively. Laidler and Wojciechowski (*16*) reported that main products were ethylene, methane, and hydrogen and that minor ones were ethane, propane, butenes, cyclopentadiene, cyclohexadiene, benzene, toluene, and diallyl at temperatures from 580° to 640°C pressures from 40 to 400 mm Hg in a static system. No allene was detected, which is in contrast to the results obtained at higher temperatures by Szwarc (*21*) and by Sakakibara (*19*). Reaction order was determined as 3/2, and the A-factor and activation energy were reported as $10^{13.34}$ ml$^{1/2}$ mole$^{-1/2}$ sec^{-1} and 56.7 kcal/mole, respectively. Kallend *et al.* (*9*) carried out a detailed analysis of the reaction product at 555° ~ 640°C and pressures 7 ~ 300 mm Hg. The main C_6 compounds present were 1,3- and 1,4-hexadiene. Methylcyclohexene and cyclohexadienes were not found.

In this paper the kinetics of the thermal reaction of propylene is described at temperatures ranging from 703° to 854°C, at atmospheric pressure and at residence times from 0.078 to 3.3 sec, with and without nitrogen dilution.

Product distributions are plotted in Figures 7 to 12. These data suggest that the primary products are ethylene, methane, hydrogen, butenes, butadiene, methylcyclopentene, hexadienes, acetylene, and ethane, and the secondary products are cyclopentene, cyclopentadiene, benzene, toluene, and polycyclic aromatics. All C_4 olefins were present, with 1-butene the major component. Isobutylene, *trans*- and *cis*-2-butene were also identified. Selectivity of polycyclic aromatic hydrocarbon formation was calculated in terms of naphthalene because the main part of these aromatics was found to be naphthalene. Trace amounts of propane, allene, methylacetylene, cyclohexane, cyclohexadiene, 4-methylcyclohexene, xylenes, and styrene were also identified. The amount of carbon deposit on the reactor wall was very small and was neglected.

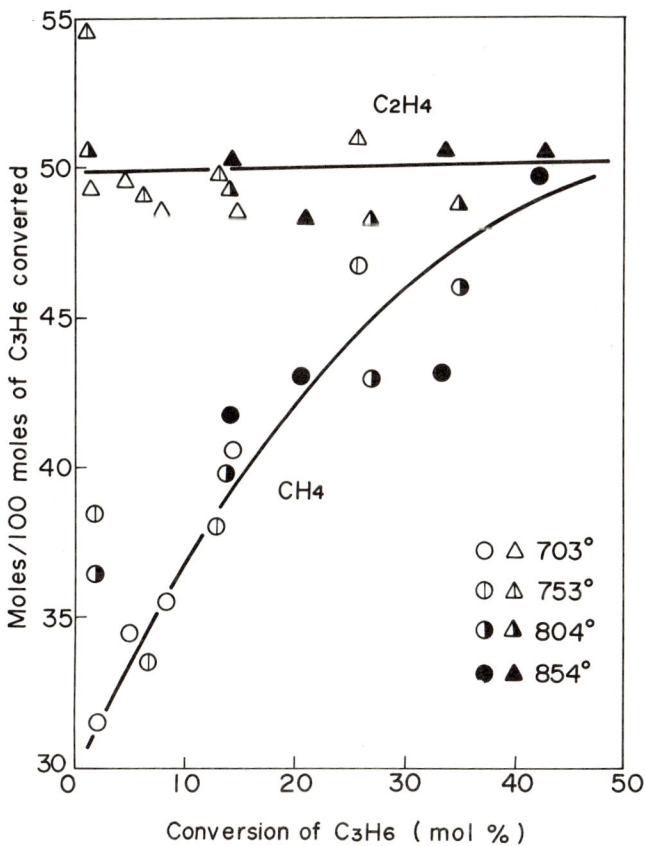

Figure 7. Selectivities of methane and ethylene in the thermal reaction of propylene

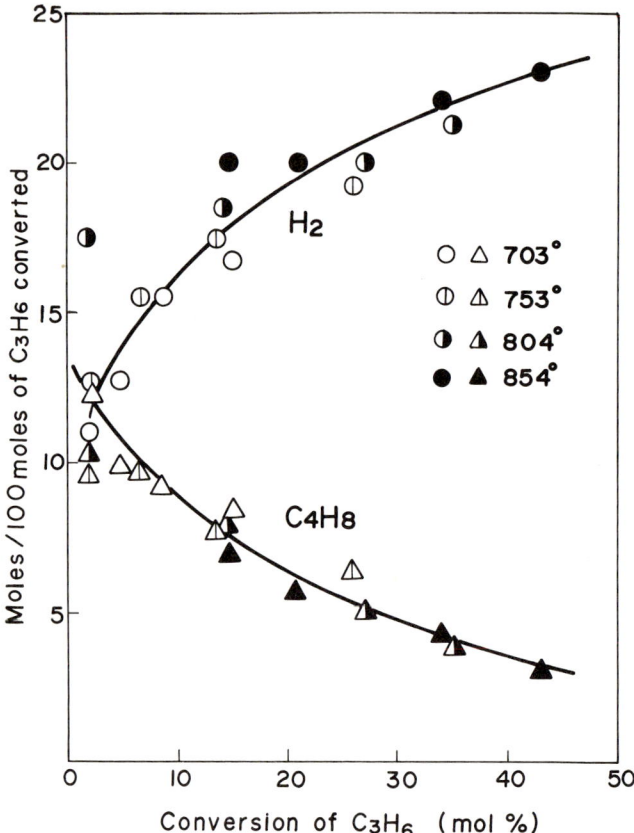

Figure 8. Selectivities of hydrogen and butenes in the thermal reaction of propylene

Except for ethane and acetylene, selectivity of the product formation is nearly independent of temperatures between 703° and 854°C. This coincides with the work of Kallend *et al.* (9) and of Hurd and Meinert (7). In addition, the above results are consistent with the conclusion deduced from the mechanism of thermal reaction of propylene which is proposed by Kunugi *et al.* (14) based on the kinetic evaluation of the relevant elementary reactions. Selectivity of the formation of hydrogen, methane, benzene, and polycyclics increases, whereas that of butenes and butadiene decreases with conversion. Product selectivity of cyclo-olefins and toluene decreases at higher conversions after passing through their maxima. These facts indicate that dehydrogenation, demethylation, and cyclization predominate at higher conversions to produce hydrogen, methane, benzene, and polycyclics. Selectivity of formation of ethylene

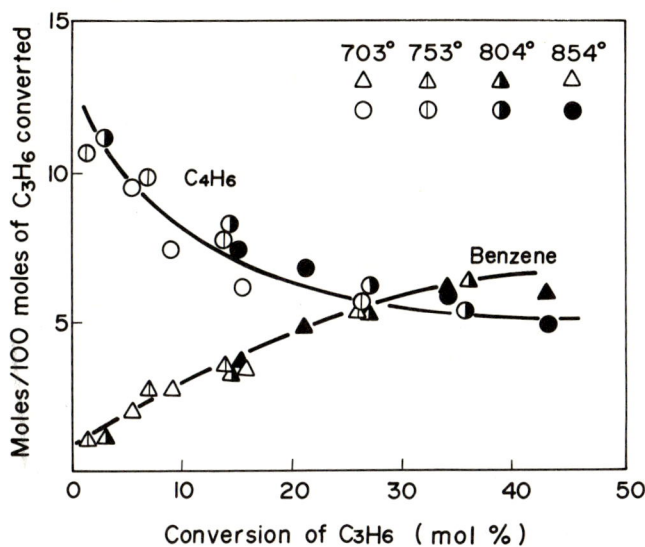

Figure 9. Selectivities of butadiene and benzene in the thermal reaction of propylene

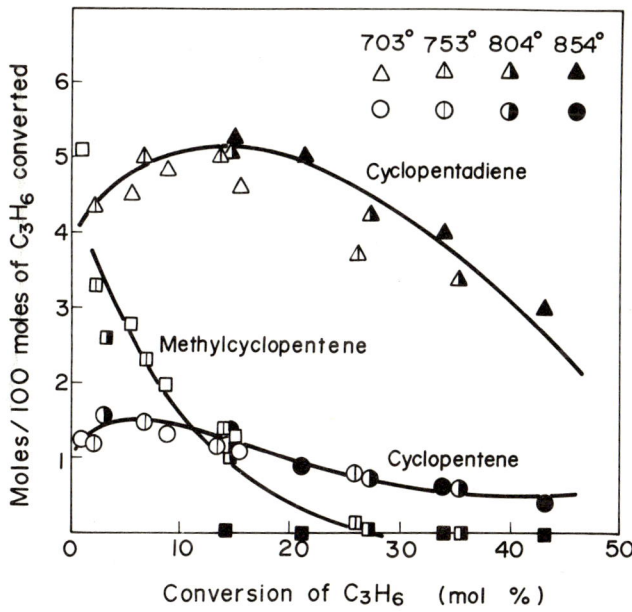

Figure 10. Selectivities of cyclopentene, cyclopentadiene, and methylcyclopentene in the thermal reaction of propylene

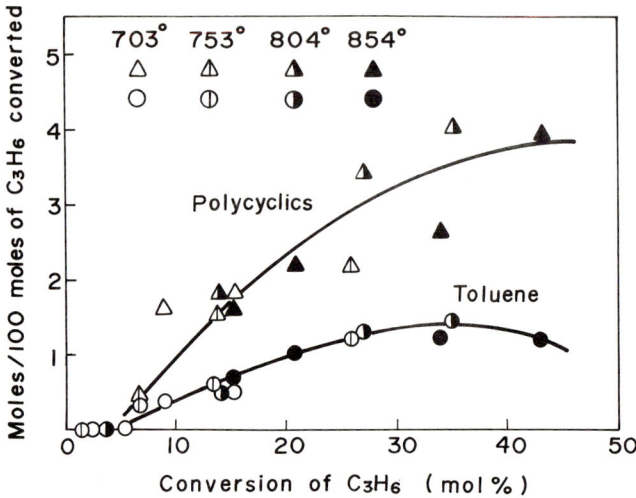

Figure 11. Selectivities of toluene and polycyclics in the thermal reaction of propylene

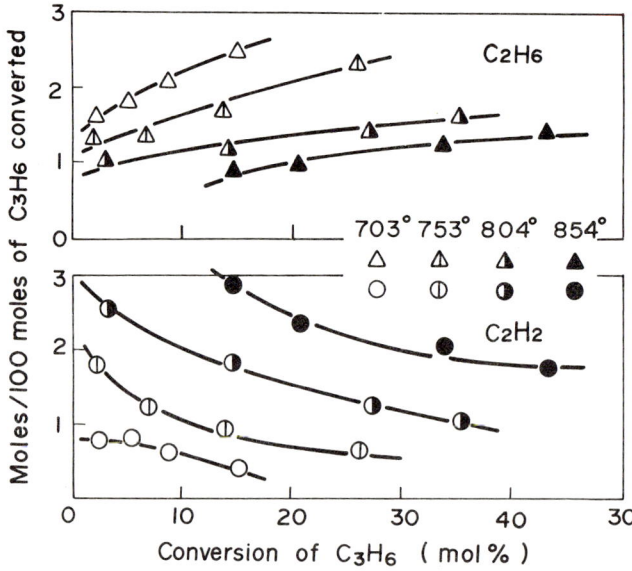

Figure 12. Selectivities of acetylene and ethane in the thermal reaction of propylene

shows little dependence on propylene conversion, which suggests that ethylene is relatively stable in the thermal reaction of propylene. Increase in selectivity of formation of methane at higher conversions may be caused by the decomposition of 1-butene. The main products in pyrolysis of 1-butene were methane and propylene at 640° ~ 760°C (*15*). Distribution of butene isomers was nearly the same as that found by Kallend *et al.* (*9*).

Selectivity of formation of methylcyclopentene decreases rapidly with conversion of propylene. In the thermal reaction of ethylene this compound was not identified. Formation of five-membered ring compounds—*i.e.*, methylcyclopentene, cyclopentene, and cyclopentadiene—may be attributed to allyl type radicals (*3, 14*).

$$\underset{\substack{CH_2 \\ \| \\ CH \\ \diagdown \\ CH_2}}{} + \underset{\substack{CHR \\ \| \\ CH_2}}{} \qquad (R: \quad CH_3, H)$$

$$\longrightarrow \cdot \underset{\substack{CH_2-CH-R \\ | \\ CH \\ \diagdown \\ CH_2-CH_2}}{}$$

$$\longrightarrow \underset{\substack{CH-CH-R \\ \| \\ CH \\ \diagdown \\ CH_2-CH_2}}{} \quad \text{or} \quad \underset{\substack{CH_2-CH-R \\ | \\ CH \\ \diagdown\diagdown \\ CH-CH_2}}{} + H\cdot$$

These products are consumed consecutively, probably to form benzene and polycyclic compounds. Toluene may also react consecutively to benzene. The ratio of toluene or xylenes to benzene was about twice that obtained in the thermal reaction of ethylene, respectively, at temperatures from 703° to 854°C and at conversions up to 40 mole %. The ratio of styrene to benzene was about one-third as large as that obtained in the thermal reaction of ethylene. Addition of butadiene in the thermal reaction of propylene increased the selectivity of cyclic compound formation, although the increase was smaller than in the case of ethylene. These facts support the mechanism for the formation of monocyclic aromatic compounds proposed by Wheeler and Wood (*24*); this is discussed in detail later.

Laidler and Wojciechowski (*16*) and Kallend *et al.* (*9*) reported the order of propylene disappearance as 3/2 and 7/5 of propylene, respectively. In this study, the 3/2-order kinetics was found (Figure 13). Existence of an induction period is observed, as noted by Kallend *et al.* (9) and Amano and Uchiyama (1). These periods are shorter than those observed in ethylene pyrolysis at the same temperatures. From Arrhenius plots, the rate constant for the steady state was calculated as

$$k_{C_3H_6} = 10^{15.06} \exp\left(-\frac{63200}{RT}\right) \text{ ml}^{1/2}\text{mole}^{-1/2}\text{sec}^{-1}$$

Thermal Reactions of Butenes. Among the C_4 olefins, 1-butene has been studied most extensively. Bryce and Kebarle (2) pyrolyzed 1-butene at 490° ~ 560°C in a static system, and the main gaseous products were methane, propylene, ethylene, and ethane. The main liquid products were cyclohexadiene, benzene, cyclopentene, cyclopentadiene, and toluene. The rates of formation of methane, propylene, ethylene, and ethane showed first-order dependence on the initial butene concentration. The activation energy for 1-butene disappearance was *ca.* 66 kcal/mole.

In this study, butene conversion was maintained at less than 15 mole % to investigate the primary reactions at temperatures from 640° to

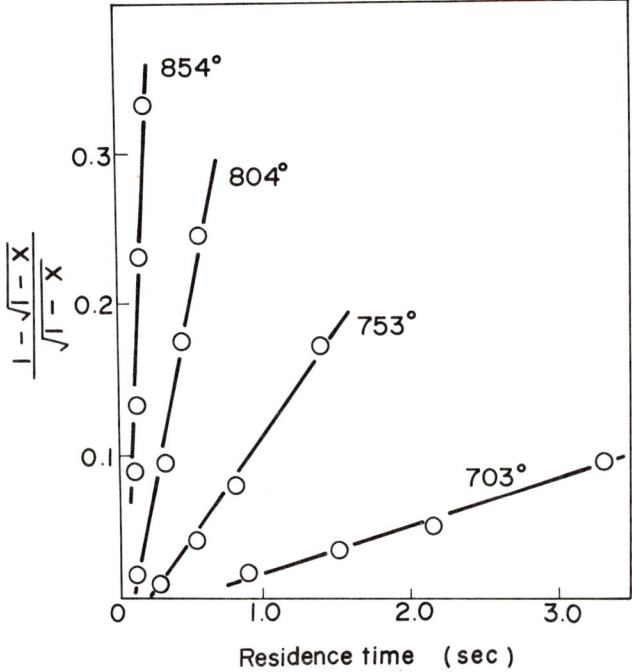

Figure 13. 3/2-Order kinetics for thermal reaction of propylene

680°C. Nitrogen was used as diluent at a nitrogen/1-butene ratio of about 3.

1-BUTENE. The main products were methane, propylene, butadiene, and ethylene with smaller amounts of ethane and *cis*- and *trans*-2-butene. Only trace amounts of hydrocarbons higher than C_5 were noticed at temperatures from 640° to 680°C and at conversion levels from 1.2 to 14.2 mole %. A considerable amount of liquid hydrocarbons was formed at 700°C and at higher conversions above 30 mole %.

Product distributions are listed in Table II. Selectivity of product formation remains constant at conversion levels from 1.2 to 14.2 mole % and at temperatures from 640° to 680°C. Nearly equal amounts of propylene and methane are formed at low 1-butene conversions. At higher conversions from 30 to 40 mole % and at 700°C, the selectivities of methane and ethylene increase, whereas those of ethane, propylene, and 2-butene decrease.

The over-all rate of 1-butene disappearance obeyed the first-order equation. Arrhenius plots give the rate constant as

$$k_{1-C_4H_8} = 10^{13.47} \exp\left(-\frac{61300}{RT}\right) \text{ sec}^{-1}$$

A free radical mechanism of the reaction was proposed, based upon the kinetic evaluation of the relevant elementary reactions (*13*), which was essentially the same as that proposed by Bryce and Kebarle (*2*). This mechanism accounts for the formation of the main products—*i.e.*, methane, propylene, butadiene, ethylene, and the minor products such as ethane and 2-butene isomers.

cis-2-BUTENE. The main product was *trans*-2-butene and the minor products were methane, propylene, and butadiene. Trace amounts of ethylene, 1-butene, and hydrogen were observed. Hydrocarbons higher than C_5 were not identified at temperatures from 640° to 700°C and at conversions around 30 mole %. Selectivity of product formation is listed in Table II.

trans-2-BUTENE. The main product was *cis*-2-butene, and the minor products were methane, propylene, and butadiene. Trace amounts of ethylene, 1-butene, and hydrogen were observed. Hydrocarbons higher than C_5 were not identified at temperatures of 640° ~ 700°C and at conversion levels up to 40 mole %. These results are essentially the same as those obtained with *cis*-2-butene. Selectivity of product formation is listed in Table II.

The over-all rate constants of *cis*- and *trans*-2-butene disappearance are calculated by the first-order rate equation at conversions below 30

Table II. Product Distributions in 1-Butene

Conversion, mole %	5	15	30	40
Selectivity[a]				
CH_4	46	46	51	53
C_2H_6	5	5	4	4
C_2H_4	21	21	27	29
C_3H_6	46	46	44	42
1-C_4H_8	—	—	—	—
cis-2-C_4H_8	5	5	3	3
trans-2-C_4H_8	5	5	3	3
C_4H_6	30	30	30	30

[a] Mole of products per 100 moles of converted butenes.

mole %. Arrhenius plots give the rate constants as

$$k_{cis-2-C_4H_8} = 10^{13.64} \exp\left(-\frac{56500}{RT}\right) \text{ sec}^{-1}$$

and

$$k_{trans-2-C_4H_8} = 10^{13.82} \exp\left(-\frac{57200}{RT}\right) \text{ sec}^{-1}$$

In the thermal reaction of 1-butene, considerable amounts of liquid hydrocarbons were produced at 700°C and at conversion levels around 30 mole % (13), whereas 2-butene isomers did not give significant yields of liquid hydrocarbons at comparable pyrolysis conditions. Thermal reaction of 1-butene is mainly the scission of the C–C bond that proceeds by a radical mechanism, while reaction of cis- and trans-2-butene involves isomerization that proceeds by a molecular mechanism (6, 17).

Thermal Reaction of Butadiene. Vaughan (23) reported that the thermal reaction of butadiene at 326° ~ 436°C and 1.5–720 mm Hg gave 4-vinylcyclohexene. Kistiakowsky and Ransom (11) studied this reaction at temperatures from 171° to 387°C and at pressures from 327 to 5000 mm Hg, obtaining the second-order rate constant of $10^{9.96} \exp(-23690/RT)$ ml mole^{-1} sec^{-1}. The activation energy increased with increasing temperature. This was attributed to the subsequent reaction of 4-vinylcyclohexene with butadiene to form 4,4'-octahydrodiphenyl. At temperatures from 420° to 650° and at pressures from 100 to 760 mm Hg, Rowley and Steiner (18) reported the rate constant as $10^{11.14} \exp(-26800/RT)$ ml mole^{-1} sec^{-1}. They also reported the rate constant for the reaction of ethylene and butadiene as $10^{10.48} \exp(-27500/RT)$ ml mole^{-1} sec^{-1}, at temperatures from 490° to 650°C and 1 atm. They preferred a cyclic configuration over a linear configuration for the activated complex, based on a statistical examination of the above kinetic parameters.

the Thermal Reaction of Butenes

	cis-2-Butene					trans-2-Butene			
10	20	30	40	50	10	20	30	40	
4	5	7	10	15	6	8	11	15	
—	—	—	—	—	—	—	—	—	
2	2.5	3	3.5	4	0.7	0.7	2.0	2.8	
4	5	7	10	15	6	8	11	15	
1	1	1	2	2	1	1	2	3	
—	—	—	—	—	85	85	77	67	
76	84	83	75	66	—	—	—	—	
5	7	9	13	15	6	8	11	15	

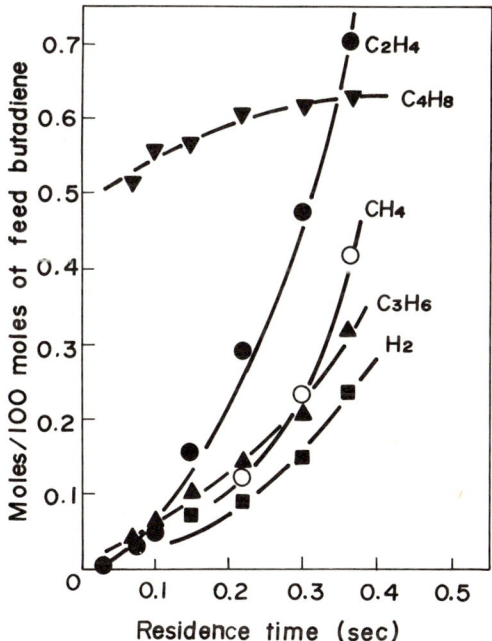

Figure 14. Product yields in the thermal reaction of butadiene at 650°C. Feed butadiene contains 0.55% mole % butenes as impurity.

In our study, butadiene was subjected to reaction in the temperature range 550° ~ 750°C, which is higher than that used in the above-mentioned papers. Residence times varied from 0.028 to 0.41 sec in the presence of about 66 mole % nitrogen as diluent. Most experiments were conducted at conversions below 2 mole % to investigate the primary reactions.

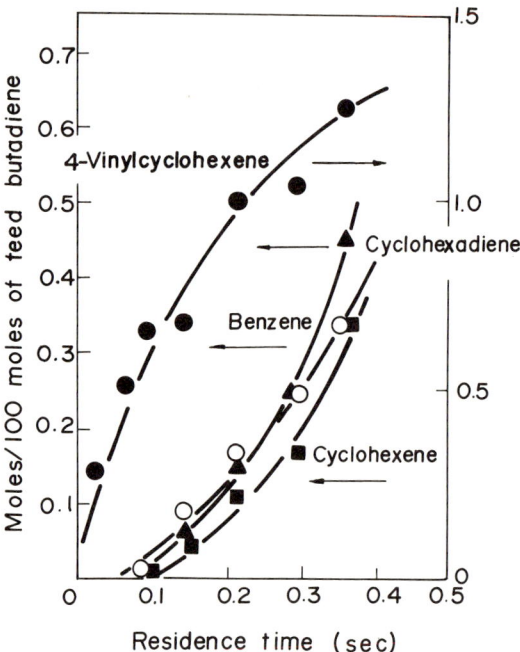

Figure 15. Product yields in thermal reaction of butadiene at 650°C

At 550° and 600°C, 4-vinylcyclohexene is a sole product, whereas at higher temperatures 650° ~ 750°C, the secondary products appear as shown in Figures 14 and 15. At residence times shorter than 0.10 sec at 650°C, the secondary reactions are negligible. At residence times longer than 0.10 sec, however, decomposition of the reactant and/or product is appreciable, producing ethylene, methane, propylene, and hydrogen. The amounts of methane and propylene are nearly equal as shown in Figure 14. This permits us to conclude that methane and propylene are produced by decomposition of butenes. Formation of cyclohexene, cyclohexadiene, and benzene becomes important in high conversion regions. Small amounts of styrene and cyclooctadiene were identified. Three unknown peaks were recorded on the gas chromatogram, considered to be cyclic C_5, straight chain C_6 and cyclic C_7 or C_8 hydrocarbons, respectively.

Although the reaction mechanism at high temperatures is not clear, the formation of the secondary products, such as cyclohexene, cyclohexadiene, and benzene, may be attributed to the devinylation and dehydrogenation of 4-vinylcyclohexene. The possibility of cyclohexene formation by the reaction of butadiene and product ethylene can be neglected because of the prohibitively low concentration of ethylene (discussed next).

The yields of ethylene and of the sum of cyclohexene, cyclohexadiene, and benzene are listed in Table III. At 650°C, the yield of ethylene is smaller than the sum of the yields of cyclohexene, cyclohexadiene, and benzene, but at 750°C the reverse is true. Both are approximately equal

Table III. Yields of Ethylene and C_6 Cyclic Compounds from Butadiene

Temp., °C	Residence Time, sec	Yield,[a] mole %		(A)/(B)
		A	B	
650	0.150	0.15	0.20	0.75
650	0.220	0.29	0.36	0.81
650	0.300	0.47	0.66	0.71
650	0.364	0.70	1.10	0.64
700	0.032	0.084	0.078	1.1
700	0.041	0.104	0.110	0.96
700	0.042	0.088	0.097	0.91
700	0.048	0.136	0.139	0.98
700	0.056	0.167	0.195	0.86
750	0.028	0.45	0.337	1.3
750	0.0289	0.50	0.369	1.4
750	0.0292	0.48	0.416	1.2
750	0.034	0.60	0.403	1.5

[a] A: ethylene. B: cyclohexene + cyclohexadiene + benzene.

at 700°C. This suggests that most of the ethylene is formed from 4-vinylcyclohexene at lower temperature, where the consumption of vinyl radicals to products other than ethylene—e.g., to 1-vinyl-3-butenyl radical with butadiene—may be expected, and that at higher temperatures part of the ethylene is formed by butadiene hydrogenolysis.

In Figure 16, conversions of butadiene at temperatures from 550° to 750°C are plotted, and a second-order rate equation is shown to hold. Here, the conversion of butadiene, x, is defined as twice the sum of the yields of 4-vinylcyclohexene, cyclohexene, cyclohexadiene, and benzene.

Arrhenius plots of the above data as well as those of Kistiakowsky and Ransom (11) at 171° ~ 387°C and of Rowley and Steiner (18) at 420° ~ 650°C resulted in a single straight line. Increase in activation energy at high temperature above 650°C, observed by Kistiakowsky and Ransom (11) and Rowley and Steiner (18), was not found in our experiments. The reason would be that the reverse reaction proceeded appreciably in our experiments at higher butadiene conversions. Wing Tsang (25) studied the decomposition of 4-vinylcyclohexene at 900° ~ 1050°K, by use of the shock wave, obtaining an A-factor as $10^{15.20}$ sec^{-1} and activation energy as 62 kcal/mole. Using his rate equation, it is calculated that

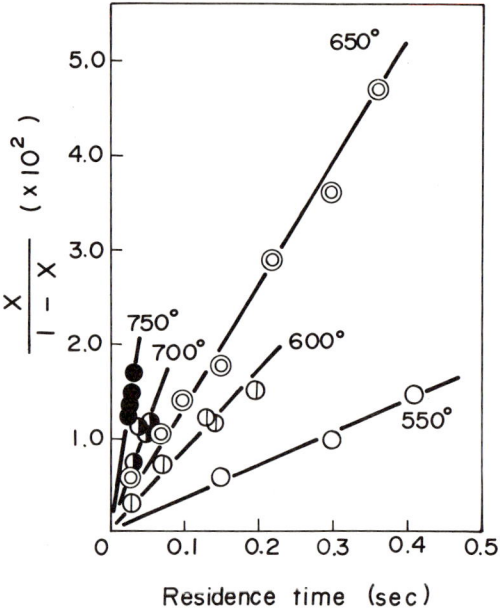

Figure 16. Second-order kinetics for thermal reaction of butadiene

the rate of decomposition of 4-vinylcyclohexene formed is about 1/100, 1/10, 1/3 that of butadiene dimerization at temperatures of 600°, 650°, and 700°C, respectively. Calibration of our experimental data obtained at 650° ∼ 700°C gives a rate constant almost identical to that of Rowley and Steiner (*18*) at 650° ∼ 700°C.

Thermal Reactions of Olefins with Butadiene. The rate constant of the reaction of ethylene with butadiene was reported by Rowley and Steiner (*18*), whereas that of propylene or butenes with butadiene has not been reported. Tarasenkova (*22*) reported that the thermal reaction of propylene with butadiene at 600°C gave toluene, the yield of which was twice as large as the yield of benzene plus xylenes. Moreover, the thermal reactions of 1-butene with butadiene and 2-butene with butadiene at 500° ∼ 550°C gave as main products ethylbenzene and *o*-xylene, respectively. The ratio of ethylbenzene to total xylenes was close to the ratio of 1-butene to 2-butene in the feed.

Thermal reactions of olefins with butadiene were examined in this study at temperatures from 510° to 670°C and with short residence times. Thermal reaction of the mixture ethylene–propylene–butadiene gave cyclohexene (CH), 4-methylcyclohexene (MCH), and 4-vinylcyclohexene (VCH) as primary products. The rates of formation of these compounds are as follows.

$$\frac{d[\text{CH}]}{dt} = 2k_{\text{CH}} [\text{C}_2\text{H}_4] [\text{C}_4\text{H}_6] \tag{1}$$

$$\frac{d[\text{MCH}]}{dt} = 2k_{\text{MCH}} [\text{C}_3\text{H}_6] [\text{C}_4\text{H}_6] \tag{2}$$

$$\frac{d[\text{VCH}]}{dt} = k_{\text{VCH}} [\text{C}_4\text{H}_6]^2 \tag{3}$$

Hence, the ratios of rate constants of formation of cyclohexene and 4-methylcyclohexene to that of 4-vinylcyclohexene are calculated by the following equations.

$$\frac{k_{\text{CH}}}{k_{\text{VCH}}} = \frac{[\text{C}_4\text{H}_6]}{2[\text{C}_2\text{H}_4]} \frac{d[\text{CH}]}{d[\text{VCH}]} \doteqdot \frac{[\text{C}_4\text{H}_6]}{2[\text{C}_2\text{H}_4]} \frac{\Delta[\text{CH}]}{\Delta[\text{VCH}]}$$

$$\frac{k_{\text{MCH}}}{k_{\text{VCH}}} = \frac{[\text{C}_4\text{H}_6]}{2[\text{C}_3\text{H}_6]} \frac{d[\text{CH}]}{d[\text{VCH}]} \doteqdot \frac{[\text{C}_4\text{H}_6]}{2[\text{C}_3\text{H}_6]} \frac{\Delta[\text{MCH}]}{\Delta[\text{VCH}]}$$

Thermal reaction of 1-butene, cis- and trans-2-butene or isobutylene with butadiene yielded 4-ethylcyclohexene, cis- and trans-4,5-dimethylcyclohexene, or 4,4-dimethylcyclohexene, respectively, accompanied by a larger amount of 4-vinylcyclohexene. Rate constants of these respective reactions were calculated by the method above.

Rate constants and their kinetic parameters obtained experimentally are summarized in Table IV. Results at temperatures above 600°C are not listed because small amounts of by-products were observed. Exceedingly large A-factors for butenes–butadiene reactions are not explained. Further examination of these values in connection with the large activation energies is needed.

The relative rates of formation of 4-vinylcyclohexene, cyclohexene, and 4-methylcyclohexene are about 10:4:1 at temperatures from 510° to 600°C, whereas those of 4-ethylcyclohexene, cis- and trans-4,5-dimethylcyclohexene, and 4,4-dimethylcyclohexene are less than 0.1 to that of 4-methylcyclohexene.

Formation of Cyclic Compounds. When we noted that the addition of small amounts of butadiene increased the yield of cyclics formed in the thermal reaction of ethylene and propylene, an effort was made to relate directly the formation of cyclics in thermal reaction of ethylene and propylene, respectively, to the Diels-Alder reaction between feed olefins and product butadiene. Reactions between product olefins and product butadiene were neglected owing to their small concentrations. Cyclics were deferred as the sum of C_6 rings with and without alkyl or vinyl groups.

Table IV. Second-Order Rate Constants of

Rate Constant, k × 10^{-2}
ml/(mole sec)

	510°C	520°C	530°C	550°C	570°C	590°C
C_2H_4	4.47	5.62	6.92	10.97	16.60	24.55
C_3H_6	1.39	1.66	2.09	3.24	5.02	7.59
C_4H_6	30.9	38.0	45.7	66.1	95.5	135.0
1-C_4H_8	0.04	—	0.17	0.56	1.02	2.14
cis-2-C_4H_8	0.03	—	0.11	0.42	—	1.18
trans-2-C_4H_8	0.42	0.65	0.83	1.15	1.74	2.95
iso-C_4H_8	0.18	—	0.36	0.83	1.78	3.47

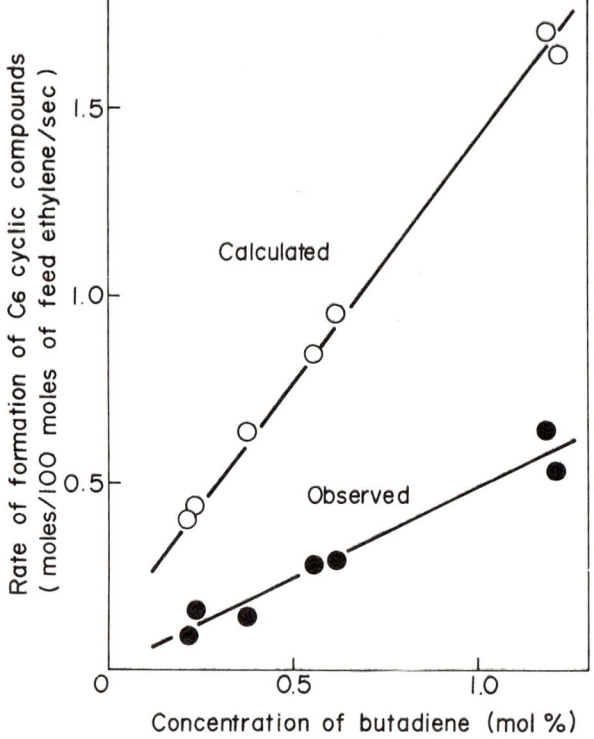

Figure 17. Rate of formation of C_6 cyclic compounds in thermal reaction of ethylene at 753°C

The calculated rates of formation for cyclics from ethylene at 750°C are shown in Figure 17 using the measured concentration of ethylene and butadiene from experiments in Equation 1. Also shown on this plot are the measured rates of formation for cyclic compounds obtained by graphical differentiation of the yield-to-residence time plots. Figure 17 shows

Thermal Reactions of Olefins with Butadiene

A-factor, $log_{10}A$	Activation Energy, kcal/mole
10.68	28.8
10.40	29.7
10.41	24.8
15.67	52.7
15.97	54.7
10.80	30.6
14.06	45.8

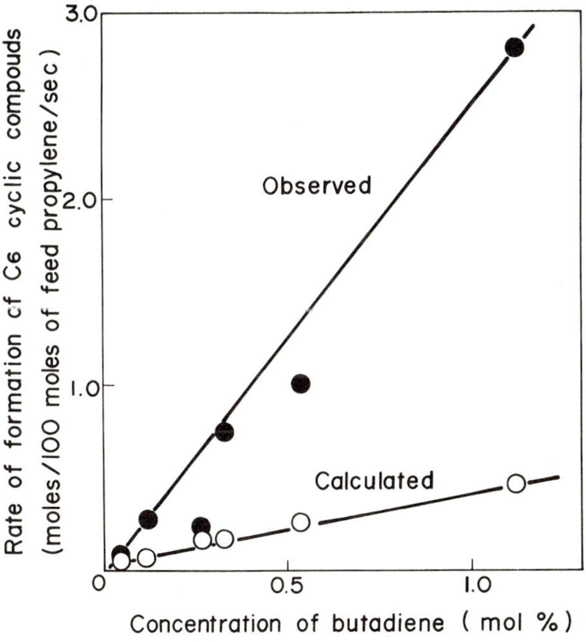

Figure 18. Rate of formation of C_6 cyclic compounds in thermal reaction of propylene at 753°C

the observed rate of cyclic formation increases linearly with butadiene concentration and lies below that calculated by Equation 1. Since the decomposition of cyclohexene was not included (20), the results support the premise that cyclics result from reaction of ethylene and butadiene.

A similar treatment of the propylene pyrolysis data at 750°C is shown in Figure 18. In this case, the calculated rates obtained by extrapolating Equation 2 to higher temperatures are much smaller than those determined by graphical differentiation of yield-to-residence time curves. The rate of formation of cyclics is linear with butadiene concentration. This suggests that the direct participation of butadiene is involved and species other than propylene also react with butadiene to form cyclics. At temperatures of 600°C and below, where the rate constant for formation of MCH was determined, species such as allyl radicals would not be present and their participation in the cyclization reactions would not be included in k_{MCH}.

The allyl radical may react with butadiene to produce cyclic olefins, which then convert to benzene or toluene:

$$\text{CH}_2=\text{CH}-\text{CH}_2\cdot \;+\; \text{CH}_2=\text{CH}-\text{CH}=\text{CH}_2 \longrightarrow \cdot\bigcirc \longrightarrow \text{Cyclic olefins}$$

A similar mechanism has been proposed above for the formation of five-membered ring compounds in the thermal reaction of propylene. In fact, for propylene the ratios of the yields of toluene plus xylenes to benzene were about twice as large as those in thermal reaction of ethylene.

To investigate the route to polycyclic compounds, the experiments were carried out on thermal reactions of benzene to diphenyl, of benzene with butadiene and of styrene with ethylene to Tetralin isomers, and of cyclohexene with butadiene to octalin. However, the rates of these reactions were too small to account for the formation of polycyclics in thermal reactions of both ethylene and propylene. The obtained result leads to the speculation that some reactive radicals should play an important role in polycyclic formation.

Acknowledgment

The authors thank Hiroaki Takashima and Noriko Kiyama for their assistance.

Literature Cited

(1) Amano, A., Uchiyama, M., *J. Phys. Chem.* **1963**, 67, 1242.
(2) Bryce, W. A., Kebarle, P., *Trans. Faraday Soc.* **1958**, 54, 1660.
(3) Bryce, W. A., Ruzicka, D. J., *Can. J. Chem.* **1960**, 38, 835.
(4) Burk, R. E., Baldwin, B. G., Whitacre, C. H., *Ind. Eng. Chem.* **1937**, 29, 326.
(5) Dahlgren, G., Douglas, J. E., *J. Am. Chem. Soc.* **1958**, 80, 5108.
(6) Danby, C. J., Spall, B. C., Stubbs, F. J., Hinshelwood, C., *Proc. Roy. Soc. (London)*, Ser. A **1955**, 228, 448.
(7) Hurd, C. D., Meinert, R. N., *J. Am. Chem. Soc.* **1930**, 52, 4978.
(8) Hougen, D. A., Watson, K. M., "Chemical Process Principles," p. 884, Wiley, New York, 1943.

(9) Kallend, A. S., Purnell, J. H., Shurlock, B. C., *Proc. Roy. Soc. (London),* Ser. A **1967**, 300, 120.
(10) Kinny, R. E., Crowley, D. J., *Ind. Eng. Chem.* **1954**, 46, 258.
(11) Kistiakowsky, G. B., Ransom, W. W., *J. Chem. Phys.* **1939**, 7, 725.
(12) Kunugi, T., Sakai, T., Soma, K., Sasaki, Y., *Ind. Eng. Chem., Fundamentals* **1969**, 8, 374.
(13) Kunugi, T., Sakai, T., Soma, K., Sasaki, Y., unpublished data.
(14) Kunugi, T., Soma, K., Sakai, T., unpublished data.
(15) Kunugi, T., Tominaga, H., Abiko, S., Uehara, K., Ohno, T., *Kogyo Kagaku Zasshi* **1967**, 70, 1477.
(16) Laidler, K. J., Wojciechowski, B. W., *Proc. Roy. Soc. (London),* Ser. A **1960**, 259, 257.
(17) Molera, M. J., Stubbs, F. J., *J. Chem. Soc.* **1952**, 381.
(18) Rowley, D., Steiner, H., *Discussions Faraday Soc.* **1951**, 10, 198.
(19) Sakakibara, Y., *Bull. Chem. Soc. Japan* **1964**, 37, 1262.
(20) Smith, S. R., Gordon, A. S., *J. Phys. Chem.* **1961**, 65, 1124.
(21) Szwarc, M., *J. Chem. Phys.* **1949**, 17, 284.
(22) Tarasenkova, E. M., *Tr. Leningrad Inzhener.-Ekon. Inst.* **1955**, (9), 123.
(23) Vaughan, W. E., *J. Am. Chem. Soc.* **1932**, 54, 3863.
(24) Wheeler, R. V., Wood, W. L., *J. Chem. Soc.* **1930**, 1819.
(25) Wing Tsang, *J. Chem. Phys.* **1965**, 42, 1805.

RECEIVED January 12, 1970.

5

Rate Modeling for the Butane–Butenes System

JOHN HAPPEL, MIGUEL A. HNATOW, and REIJI MEZAKI

Department of Chemical Engineering, New York University, Bronx, N. Y.

> *The stoichiometric number concept introduced by Horiuti is used to develop a generalized rate expression applicable to the vapor-phase catalytic dehydrogenation and hydrogenation in the butane–butenes system. The utilization of this concept provides a precise formulation of a potential term in the rate expression and thus simplifies particularly the rate modeling of a reaction system involving both forward and backward rates. Experiments were conducted using a differential flow type reactor under varying total pressures from 0.1 to 1.5 atm. Isothermal rate data were collected at 450°C. A commercial grade chromia–alumina catalyst was employed. These rate data, were best correlated by an empirical two-term rate model which was developed by adopting Langmuir type isotherms.*

This study deals with the development of a model for establishing a rate expression to describe the catalytic dehydrogenation of butane to produce an equilibrium mixture of mixed n-butenes. Classic approaches to this problem have been based on Langmuir kinetics as discussed in such pioneering books as the text of Hougen and Watson (16). Yang and Hougen (25) showed that in heterogeneous catalysis it was often possible to use a rate expression of the form

$$r = \frac{[\text{kinetic term}][\text{potential term}]}{[\text{adsorption term}]^n} \tag{1}$$

The kinetic models developed in all cases assumed the existence of a single rate-controlling step with all remaining mechanistic steps being in quasi-equilibrium. The potential term was first assumed to involve a simple difference of driving force terms so that the rate of dehydrogenation of butane would be given by

$$r = \frac{[\text{kinetic term}]\left[p_{\text{butane}} - \dfrac{p_{\text{butenes}}\, p_{\text{H}_2}}{K_{\text{eq}}}\right]}{[\text{adsorption term}]^n} \qquad (2)$$

However, Manes and co-workers noted (19, 20) that the potential term could assume a more complicated form in some cases. Horiuti (10, 11, 12, 13, 14) introduced the concept of the stoichiometric number of the rate-controlling step and showed how it could be used to develop a more rigorous form for the potential term. The stoichiometric number of each of the elementary steps which make up an over-all reaction is defined as the number of times it occurs for each occurrence of the over-all reaction.

Happel (6, 7) extended Horiuti's development further to show that it implies a generalized rate expression applicable to these reactions which do not require any assumptions as to the nature of the sites for adsorption or interaction among adsorbed species. The rate expression also was shown to be applicable to situations with one or more rate-controlling steps, provided they occur in a single path. Let us consider a solid catalyzed gaseous reaction system whose over-all reaction is shown by:

$$aA + bB + \ldots\ldots\ldots \rightleftarrows mM + nN + \ldots\ldots\ldots \qquad (3)$$

Happel's development leads to

$$r = r_f\left[1 - \left(\frac{a^m_M\, a^n_N\, ----\,.}{a^a_A\, a^b_B\, -----\,.\, K_{\text{eq}}}\right)^{1/v}\right] \qquad (4)$$

Where a denotes the activity of reaction component, and K_{eq} is the thermodynamic equilibrium constant of over-all reaction, v is the stoichiometric number of rate-determining step. If two or more rate-controlling steps exist in a single path, each having the same stoichiometric number, v is simply the common stoichiometric number of the rate-controlling steps. The term r_f is the forward rate and contains rate parameters and the partial pressures of various reaction components. In many cases r_f may be derived empirically and is not necessarily a single term expression as for the Langmuir-Hinshelwood rate models. Regardless of the nature of the rate-controlling step or steps, the stoichiometric number, if it exists, determines clearly the potential term of rate models. For complex reactions it is possible for the rate expression not to be characterized by a potential term. The use of the stoichiometric number concept is of particular value for analyzing reaction data which are obtained for a wide spectrum of experimental conditions. If it is necessary to establish a

Table I. Dehydrogenation

$$r = k\, p^m{}_{n\text{-butane}}$$

Source	Rate Model[a]	Temperature, °C
Dodd and Watson (3)	LH	600
Lyubarskii et al. (17)	PF	460–500
Happel et al. (5)	LH, PF[b]	450–560
Noda et al. (21)	LH, PF[b]	500
Carra et al. (2)	LH	510–550

[a] LH: Langmuir-Hinshelwood model, PF: power function model.
[b] In the present investigation the value of m was computed by utilizing the rate data reported.

mechanistic model which represents adequately both forward and backward rates—e.g., the use of Equation 4 appears to be of value.

A complete experimental program to obtain a valid rate expression should aim first at establishing whether it is possible to extract an appropriate potential factor from the over-all rate expression. Such a study by Happel and Atkins (8) showed that for butane dehydrogenation this is indeed the case.

The present study is aimed at the determination of the appropriate forward rate r_f to be used for the complete spectrum of partial pressures of butane, butenes, and hydrogen.

Previous Work

The catalytic dehydrogenation of n-butane and the catalytic hydrogenation of butenes have been studied widely owing to the fact that C_4 hydrocarbons are significant in the petrochemical industry. Mechanistic rate models as well as power function models have been postulated to represent the experimental data. Unfortunately some inconsistencies exist as to the best kinetic models obtained by various investigators. Part of these inconsistencies may be attributed to the difficulty in obtaining accurate rate data at a constant activity level of catalysts. The inconsistencies may also stem from the isomerization reactions of butenes which are found to occur simultaneously under experimental conditions of dehydrogenation and hydrogenation.

Table I summarizes the results of previous investigations on catalytic dehydrogenation of n-butane. In this table 2 models were used to correlate dehydrogenation rate data by various investigators: one is a power function model, and the other is a Langmuir-Hinshelwood model. The power function model can be obtained by applying the mass action law to describe rate data. Thus, the model presents the dependence of partial of *n*-Butane

Pressure, atm	Catalyst	m
0.2–3.0	chromia–alumina	—
0.1–1.0	chromia–alumina	1.0
0.2–3.3	chromia–alumina	0.5
0.06–1.80	chromia–alumina	0.75
0.1–0.5	chromia–alumina ($+Na_2O$ or $+Li_2O$)	—

pressures of reactants and products on reaction rates. The Langmuir-Hinshelwood model, as discussed previously, is generally represented by Equation 2 and is nonlinear with respect to rate parameters because adsorption terms contain the product terms of adsorption equilibrium constants and partial pressures of reactants and products. In cases where all the product terms are small compared with unity the Langmuir-Hinshelwood model can be reduced to a power function model. The power function model is largely empirical. The Langmuir-Hinshelwood model, on the other hand, possesses some mechanistic features as to reaction mechanism owing to the fact that the model is developed on the basis of the Langmuir isotherm.

In their pioneering work Dodd and Watson (3) correlated the dehydrogenation data by Langmuir-Hinshelwood rate models and found that a dual-site surface rate-controlling model is most plausible. Noda and co-workers (21) and more recently Carra and his colleagues (2) obtained essentially the same results as Dodd and Watson (3) for the chromia–alumina catalyzed dehydrogenation of *n*-butane. As Table I shows, somewhat different values for the orders of power function models are quoted in the references using this method of correlation.

Table II. Hydrogenation of Butenes

$$r = k \, p^n{}_{butenes} \, p^p{}_{H_2}$$

Source	Rate Models[a]	Temperature, °C	Catalyst	n	p
Twigg (24)	PF	60–135	Ni	0.5	0.5
Emmett and Gray (4)	PF	−29–20	iron	0.0	0.6
Taylor and Diebeler (23)	PF	50–135	Ni	0.5	0.5
Lyubarskii et al. (18)	PF	220–440	chromia–alumina	0.5	1.0
Happel et al. (5)	LH, PF[b]	430–570	chromia–alumina	0.5	0.5

[a] LH: Langmuir-Hinshelwood model, PF: Power function model.
[b] In the present investigation the values of n and p were computed by utilizing the rate data reported.

Table II shows the results of butenes hydrogenation reactions. An interesting feature of the power function model for the hydrogenation reaction is the pressure dependencies obtained. As shown in Table II some experimental results indicated consistently that the hydrogenation rate is proportional to the 0.5 power of hydrogen and butenes partial pressure. This may imply, as discussed by Bond (1), that hydrogen is dissociated upon adsorption and that butenes are adsorbed weakly on a portion of active sites owing to the steric hindrance of adsorbing butenes molecules.

Combining both the dehydrogenation and hydrogenation rate expressions Lyubarskii (18) obtained the following power function model to describe the net rate of dehydrogenation of n-butane.

$$r = k_f \frac{p_{butane}}{p^{0.5}_{butenes}} - k_b \, p^{0.5}_{butenes} \, p_{H_2} \tag{5}$$

This approach is of interest. Equation 5 not only describes the net rates for terminal conditions of dehydrogenation and hydrogenation reactions but also those rates under conditions near chemical equilibria. However, this expression is clearly not adequate for describing the net dehydrogenation rate when the feed contains no butenes because it would predict an infinitely fast dehydrogenation rate. Similar but more sophisticated modeling techniques have been used to investigate a model with which both dehydrogenation and hydrogenation data can be presented adequately. Happel and co-workers (5) studied the reactions over a chromia–alumina catalyst, measuring over-all reaction rates. For the dehydrogenation experiments butane pressures were approximately in the range 0.2–3.3 atm, and temperatures were in the range 450°–560°C. For the hydrogenation butenes partial pressures were varied from 0.07 to 0.81 atm, and hydrogen partial pressures were in the range 0.13–0.49 atm. Reaction temperatures included a range of 430° to 570°C. Experiments were conducted using a flow type differential reactor. Happel and co-workers (5) utilized a model which implies a single rate-controlling step. A Langmuir-Hinshelwood type correlation of the following form was employed:

$$r = r_f \left[1 - \left(\frac{p_{butenes} \, p_{H_2}}{p_{butane} \, K_{eq}} \right)^{1/2} \right] \tag{6}$$

where

$$r_f = \frac{k_f \, p^{1/2}_{butane}}{(1 + K_{butane} \, p^{1/2}_{butane} + K_{butenes} \, p^{1/2}_{butenes})^2} \tag{7}$$

K_{eq} is the thermodynamic equilibrium constant for butane dehydrogenation. K_{butane} and $K_{butenes}$ are arbitrary constants but may be considered as adsorption equilibrium constants of those species.

Equation 6 shows that the apparent stoichiometric number of the combined dehydrogenation and hydrogenation reaction is 2. In agreement with the observed stoichiometric number, as discussed by Happel (5), an elementary rate-controlling step should then occur twice for each time the following over-all reactions occur once:

$$C_4H_{10} \rightleftarrows C_4H_8 + H_2 \qquad (8)$$

Speculation regarding mechanism may lead to the conclusion that skeletal dissociation of n-butane and butenes occurs during adsorption. However, other studies by Okamoto and co-workers (22) indicated that such a dissociation is improbable. As noted above, Happel and Atkins (8) determined the stoichiometric number of this reaction system directly by isotopic tracing, using ^{14}C tagged butane and 1-butene as tracers. The tagged n-butane facilitated measurements of the forward rate of dehydrogenation reaction, and the tagged 1-butene was used to measure the reverse rate of the reaction. Experimental data were collected with a chromia–alumina catalyst identical to that of Happel's work (5), changing the reaction temperatures from 390° to 560°C and total pressures from 0.2 to 1.0 atm. Over this wide range of experimental conditions an apparent stoichiometric number was determined to be 1.14 ± 0.25. This implies that a single step whose stoichiometric number of unity is controlling or that two or more rate-controlling steps may be controlling, each with a stoichiometric number of unity. If ν is unity in Equation 4, the dehydrogenation and hydrogenation equation must be presented as

$$r = r_f \left(1 - \frac{p_{butenes}\, p_{H_2}}{p_{butane}\, K_{eq}}\right) \qquad (9)$$

Although the function r_f remains undetermined, it is obvious that the stoichiometric number of 2 from earlier work (5) was incorrectly estimated using a simple Langmuir model.

An explanation of the erroneous estimate of stoichiometric number may be that Equations 6 and 7 contain a number of flexible parameters such as ν, K_{butane}, $K_{butenes}$ and the exponents of partial pressures of both butane and butenes. In addition the equation is nonlinear with respect to these parameters, and the parameters are highly correlated. In those situations it is extremely difficult to obtain precise estimates of parameters, even with well-developed nonlinear regression computer programs. Uncertainty associated with the estimate of stoichiometric number may arise from the fact that the assumptions made when Equations 6 and 7 were developed are unwarranted for the system considered here. Models postulating either surface uniformity or the absence of interaction between adsorbed species may need to be modified. Further discussions concerning the value of stoichiometric number are presented elsewhere (6).

Experimental

The main consideration in the design and construction of the equipment has been that the measurements obtained reflect the effect of the process variables on the chemical phenomena occurring and that perturbations arising from physical processes such as diffusion, heat transfer, and wall effects be minimized.

Materials. Matheson technical grade *n*-butane or technical grade *n*-butenes was used in all experiments. The gases were purified further using adsorbents, filters, dryers, and fresh catalyst.

The catalyst, 20% chromia (Cr_2O_3) on alumina, is manufactured by the Houdry Process Corp. and is designated as Type A50. Typical properties are:

Surface area	55 m^2/gram
Total pore volume	0.31 cc/gram
Pellet density	1.42 grams/cc
Size	14–20 mesh
Calculated average pore diameter	225 A

Apparatus. The measurements were carried out in a flow reactor, illustrated in Figure 1, consisting of a stainless steel tube that provides for a catalyst chamber 1 inch in diameter by 3½ inches long. Six thermocouples inserted into a ¼-inch thermocouple well running down the center of the catalyst bed and three others located on the circumference of the bed measure the complete temperature profile of the reaction. The over-all temperature gradient during a run never exceeds 10°C. Ports located immediately in front of and behind the catalyst chamber draw samples of the gases entering and leaving the catalyst chamber. These ports also measure the pressure within the chamber and the pressure drop through the catalyst bed.

Peripheral equipment further purifies, meters, and preheats the hydrocarbon feed, the nitrogen used for flushing, the air for catalyst regeneration, and the hydrogen for catalyst reduction. Two preheaters are used to preheat the feed to proper reaction temperature. The heaters are programed to minimize hydrocarbon decomposition before it enters the reaction zone of the reactor. An electrically heated aluminum–bronze block serves as a heat sink which provides adequate temperature control to the catalyst bed. A second flow reactor, heated by an electric furnace and located in series with the main reactor, is an isomerization reactor that produces an equilibrium mixture of *n*-butenes.

Analysis. Analysis of 3 ml NTP samples of hydrocarbon gases was carried out by gas–liquid chromatography at 25°C using a model A

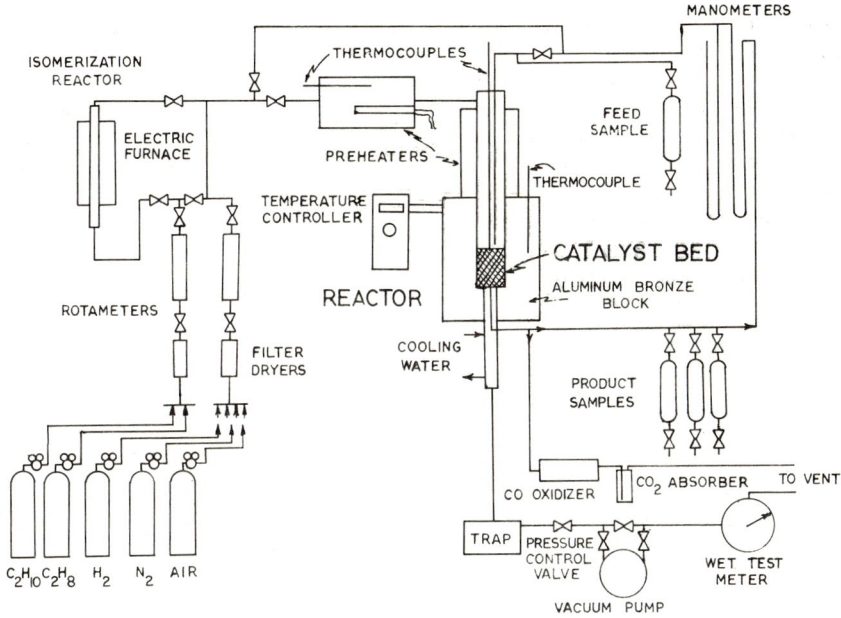

Figure 1. Vapor-phase catalytic dehydrogenation and hydrogenation apparatus

Varian Aerograph gas chromatograph. A 25-foot column filled with 20% by weight tributyl phosphate on 40–60 μ Chromosorb B was used with helium as carrier gas flowing at a rate of 50 ml/min. With this chromatographic column 12 components can be determined quantitatively, including methane, ethane, ethylene, propane, propylene, isobutane, *n*-butane, 1-butene, *trans*-2-butene, *cis*-2-butene, and 1-3-butadiene. Hydrogen was analyzed by passing a second sample over the tributyl phosphate chromatograph column, using nitrogen instead of helium as carrier gas. The procedure for analyzing the chromatograms was checked extensively using standard calibrated mixtures and mass spectroscopy.

Experiments. A standard method of pretreatment was developed to ensure constant catalyst activity. The procedure consisted of three steps. First, air was passed over the catalyst for 22 hours at a rate of 0.02 SCFM (standard cubic feet/minute) to regenerate the catalyst to a constant oxidized state. Hydrogen was then used to reduce the catalyst. The hydrogen pretreatment was studied, and an optimum period of 2 hours at a rate of 0.115 SCFM was determined. The experimental run then followed. This procedure was cyclic and repeated without interruptions. Standard runs were conducted at the beginning and at the end of a series of runs to ensure constant catalytic activity.

Three different kinds of rate data were collected for the butane–butene system: the dehydrogenation of *n*-butane with pure *n*-butane feed, the hydrogenation of butenes without *n*-butane, and the dehydrogenation or hydrogenation of butane–butenes in the presence of reaction products. We designate these data sets as the dehydrogenation, the hydrogenation, and the intermediate data, respectively. Experiments were conducted at a temperature of 450° ± 8°C and ranges in pressures and concentrations which would yield the most reliable data. In most cases it was possible to operate at less than 2% conversion with high selectivity, thus approaching differential reactor conditions. During each 10-minute run several sets of samples were taken at short intervals. These analyses showed no variation in catalyst activity with time and negligible formation of carbon and hydrocarbon dealkylate.

The existence and formation of butadiene severely reduced the activity of the catalyst. However, in our experiments, the ratio of butadiene to butenes was always less than 0.01. Furthermore, as shown previously, the activity of the catalyst remains constant during the experiments. For these reasons it is assumed that butadiene had no profound effect on the kinetics of the dehydrogenation and hydrogenation.

In the product samples of any experimental run we detected only n-butane, 1-butene, cis-2-butene, trans-2-butene, hydrogen, and a trace of butadiene. Thus, we assumed that no side reactions occurred during the reactions. Experimental data collected in this investigation are summarized elsewhere (9).

Mass Transfer Effects. Concerning exterior mass transfer to the catalyst, the method of Yang and Hougen (25) was applied to determine the partial pressure drops of the reactants across the gas film. The calculation indicated that for the highest reaction rates observed the partial pressure drops between the main gas stream and external surface of the catalyst did not exceed 0.1% of their ambient values. This shows that the effect of mass transfer across the gas film was negligible.

The effect of mass transfer in catalyst pores was determined by conducting experiments with different mesh sizes of catalyst particles. Three particle sizes (⅛-inch pellets, −14 +20 mesh, and −20 +25 mesh crushed catalyst) were used. The rates observed were in good agreement, indicating that the pore diffusion is not rate controlling under the experimental conditions used in this investigation.

Preliminary Correlations

In constructing a plausible reaction rate model for the dehydrogenation and hydrogenation reactions, first the pressure dependencies of these rates were investigated utilizing power function models. For the dehydrogenation the following function was fitted to our 26 dehydrogenation data points, and m was found to be 0.54 ± 0.07. The number following the ± sign represents bounds of approximate 95% confidence intervals.

$$r_f = k_f \, p^m_{butane} \tag{10}$$

A total of 31 hydrogenation data units were used to obtain the estimates of n and p of Equation 11:

$$r_b = k_b \, p^n_{butenes} \, p^p_{H_2} \tag{11}$$

The resulting estimates were, respectively, $n = 0.58 \pm 0.09$ and $p = 1.10 \pm 0.14$. The partial pressure dependencies of our hydrogenation rates were identical to those of Lyubarskii (18).

However, as noted above an equation like that of Lyubarskii (18), Equation 5, is not adequate for describing rates over the complete pressure range but serves only as a rough method of screening the data.

A sequential procedure was adopted to develop rate models which would overcome shortcomings of this type, employing modified Langmuir expressions empirically rather than power law models.

Development of Rate Models

An attempt was made to explore the possibilities for forming empirical rate models which have the following properties: (a) thermodynamic consistency, (b) agreement with the power function model, (c) agreement with the rate equation obtained by measuring the apparent stoichiometric number, (d) simultaneous and adequate representation of the dehydrogenation, hydrogenation, and intermediate data.

Let us examine the case of straightforward dehydrogenation of n-butane. It was assumed that the reaction produces an equilibrium mixture of butenes. The reaction is described by

$$n\text{-butane} \rightleftarrows \text{butenes} + \text{hydrogen} \tag{12}$$

The simplest possible relationship which satisfies the conditions (a) ~ (d) is

$$r_f = k_f\, p^{0.5}_{\text{butane}} \tag{13}$$

The forward rate of the dehydrogenation reaction r_f might be described by

$$r_f = \frac{k_f\, p_{\text{butane}}}{1 + K_{\text{butane}}\, p_{\text{butane}} + K_{\text{butenes}}\, p_{\text{butenes}} + K_{H_2}\, p_{H_2}} \tag{14}$$

K_{butane}, K_{butenes}, and K_{H_2} are, respectively, the adsorption equilibrium constants for butane, butenes, and hydrogen in a provisional adsorption type expression which accounts for multicomponent adsorption. For the terminal dehydrogenation conditions in which pure n-butane feed was converted to butenes and hydrogen to a very small degree, both butenes and hydrogen terms may be preliminarily neglected. This leads to

$$r_f = \frac{k_f\, p_{\text{butane}}}{1 + K_{\text{butane}}\, p_{\text{butane}}} \tag{15}$$

Equation 15 is a typical single-term Langmuir model. By combining Equations 9 and 14, at reasonably low total pressures we obtain the hydrogenation rate relationship

$$r_b = \frac{k_f \, p_{butenes} \, p_{H_2}}{(1 + K_{butenes} \, p_{butenes} + K_{H_2} \, p_{H_2}) \, K_{eq}} \tag{16}$$

For the dehydrogenation of n-butane our determined value of the thermodynamic equilibrium constant K_{eq} was 0.0133 atm at our reaction temperature of 450°C. Hence we have

$$k_b = \frac{k_f}{K_{eq}} = 75 \, k_f \tag{17}$$

Equation 16 indicates that at sufficiently low total pressures, the hydrogenation rate must be substantially faster than that of the forward reaction. On the contrary the observed hydrogenation rates measured at low total pressures were the same order of magnitude as the dehydrogenation rates. Thus, the rate model of Equations 14 and 16 seems inadequate. From this comparison of the magnitudes of the dehydrogenation rates with those of the hydrogenation rates, it may be concluded that the single-term models of Langmuir type fail to describe adequately the terminal dehydrogenation and hydrogenation rate data. Modification of Langmuir type model using power function and Freundlich equation terms was also unsuccessful.

The preceding studies of rate models lead to the consideration of two-term models to correlate the rate data. Several two-term rate models including Langmuir and Freundlich type equation were examined. From this study Equation 18 was found to correlate best the terminal rate data.

For the over-all rate, r

$$r = \left[\frac{k_{1f} \, p_{butane}}{(1 + K_{butane} \, p_{butane} + K_{H_2} \, p_{H_2} + K_{ae} \, p_{butane} \, p_{butenes})^2} \right.$$

$$\left. + \frac{k_{2f} \, p_{butane}}{(1 + K'_{butane} \, p_{butane} + K'_{ae} \, p_{butane} \, p_{butenes})^2} \right] \left[1 - \frac{p_{butenes} \, p_{H_2}}{p_{butane} \, K_{eq}} \right] \tag{18}$$

The 26 intermediate data which were collected by using the mixed feeds of n-butane–butenes–hydrogen system and the existing data were used to test the adequacy of rate Equation 18. Equation 18 was in a good agreement with most of the intermediate data. However the seven dehydrogenation data which were obtained in the presence of butenes showed noticeable systematic deviations from the predicted rates. Butenes seem to reduce the dehydrogenation rate more sharply than predicted from Equation 18. To take into account the effect of butenes and to randomize the deviations, we introduced an empirical exponential term into Equation 18 and obtained the following expression:

$$r = \left[\frac{k_{1f}\, p_{butane}\, e^{-K_{butenes}\, p_{butenes}}}{(1 + K_{butane}\, p_{butane} + K_{H_2}\, p_{H_2} + K_{ae}\, p_{butane}\, p_{butenes})^2} \right.$$

$$\left. + \frac{k_{2f}\, p_{butane}}{(1 + K'_{butane}\, p_{butane} + K'_{ae}\, p_{butane}\, p_{butenes})^2} \right] \left[1 - \frac{p_{butenes}\, p_{H_2}}{p_{butane}\, K_{eq}} \right] \quad (19)$$

Instead of using the exponential term, it is mathematically possible to add a term to the denominator, consisting of the product of an adsorption constant and the partial pressure of butenes. However, the constant was estimated to be about 200, and this product term, if it is added, results in the prediction of excessively low dehydrogenation rates where the partial pressures of butenes are very low. Equation 19 is, to some extent, empirical, but it seems to be adequate for expressing the dehydrogenation and hydrogenation data utilized in this investigation, fulfilling all four conditions listed in the beginning of this section.

Correlation of Data

The eight rate parameters of Equation 19 were estimated with a non-linear least-squares routine. Table III shows the best point estimates of parameters along with approximate 95% confidence intervals. The table also shows some statistical data. We can say that the most probable values of the parameters for Equation 19 will be found within the in-

Table III. Parameter Estimates for Equation 19

k_{1f}	0.0106	± 0.0018[a]
k_{2f}	0.00071	± 0.00015
K_{butane}	0.06	± 0.18
K'_{butane}	0.24	± 0.19
$K_{butenes}$	12.6	± 7.1
K_{H_2}	5.11	± 3.91
K_{ae}	5.31	± 6.24
K'_{ae}	1.75	± 0.63
Sum of squares of residual rates		3.952×10^{-5}
Mean square		5.269×10^{-7}
Average deviation, %		
dehydrogenation data		10.50
hydrogenation data		10.12
intermediate data		13.14
total		11.37

[a] Approximate 95% confidence intervals.

tervals shown in 95% of cases for new sets of data points. It should be noted from Table III that some of the parameter values can be negative owing to the relatively large confidence intervals. In a case where negative parameters are obtained, Equation 19 is completely empirical. In Figure 2 the calculated reaction rates are plotted against observed rates.

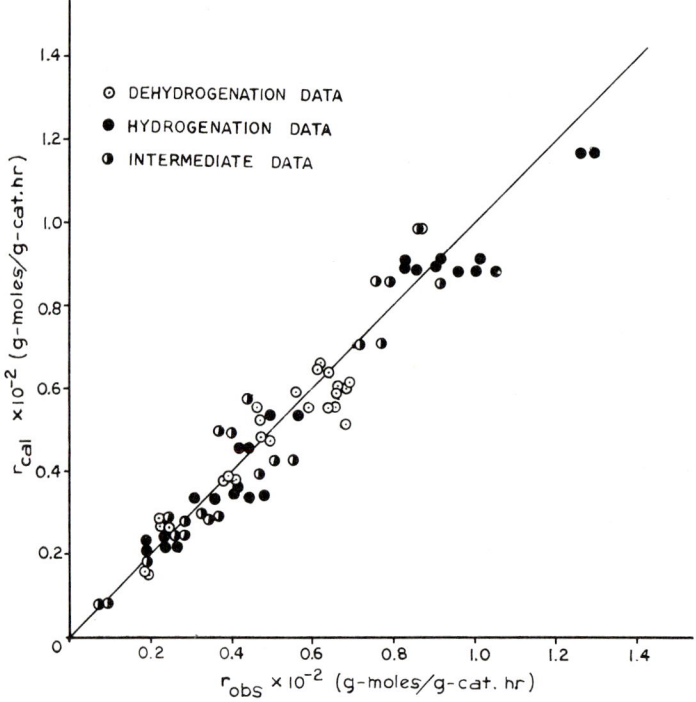

Figure 2. Observed and calculated rates for the catalytic dehydrogenation and hydrogenation of butane–butenes

One crucial problem in using the nonlinear least squares technique is that estimates of the totally unknown parameters should be supplied to begin the computation. Unfortunately, the ease of convergence, the converged parameter values, and the sum of squares of residuals at convergence depend heavily on these initial parameter estimates. Needless to say, difficulties in finding good initial estimates increase with the number of parameters involved and the complexity of rate models.

In obtaining appropriate initial estimates of parameters for Equation 19 the terminal differential rate data collected for both the dehydrogenation and the hydrogenation reactions were used advantageously. The use of these data simplifies the calculation because partial pressures of both butenes and hydrogen can be neglected for the dehydrogenation reaction. For the hydrogenation reaction, on the contrary, the partial pressures of n-butane are assumed to be insignificantly small. For the two terminal experimental conditions, thus, we obtain the following rate expressions: for the terminal dehydrogenation

$$r_\text{f} = \frac{k_\text{1f}\, p_\text{butane}}{(1 + K_\text{butane}\, p_\text{butane})^2} \tag{20}$$

and for the terminal hydrogenation

$$r_\text{b} = \frac{k_\text{1f}\, p_\text{butenes}\, p_{\text{H}_2}}{(1 + K_{\text{H}_2}\, p_{\text{H}_2})^2\, K_\text{eq}} + \frac{k_\text{2f}\, p_\text{butenes}\, p_{\text{H}_2}}{K_\text{eq}} \tag{21}$$

Equation 20 is nonlinear with respect to two rate parameters, but it is readily transformed into a linear form. Thus a linear regression technique can be used to obtain initial estimates of these two parameters. Equation 21 cannot be linearized. However it possesses only two unknown parameters instead of eight parameters of Equation 19. The estimates obtained from Equations 20 and 21 supply reasonably accurate initial estimates of these four parameters. Hence, the subsequent nonlinear estimation of all eight parameters of Equation 19 can be simplified substantially.

The following objective function was minimized with respect to 8 parameters to obtain the best estimates of parameters.

$$\sum_{i=1}^{N} \left(\frac{r_i - \hat{r}_i}{r_i} \right)^2 \tag{22}$$

Where r_i and $\hat{r}_i$ are, respectively, the observed and calculated reaction rates. The objective function of this type can frequently minimize the difficulty which stems from the nonhomogeneity of error variance.

To examine the convergence of nonlinear estimation routine, the converged values of parameters (see Table III) were perturbed by 10% of the converged values. These perturbed values were used as initial estimates of the parameters for the nonlinear least squares. The least-squares routine terminated at the exactly same parameter values as those presented in Table III. This indicates that the estimation routine converges at least at a local minimum on the sum of squares surface.

Discussion

The adequacy of the rate model of Equation 19 was examined by two methods: examination of residual rates and analysis of variance. The residuals of rates obtained from these equations are distributed randomly and showed no systematic correlation with any of the dependent and independent variables. As can be seen from Figure 2, moreover, three different data sets (the dehydrogenation, the hydrogenation, and the intermediate) are also scattered randomly along the diagonal line. According to the analysis of variance the mean square of lack of fit is 5.54×10^{-7} with 71 degrees of freedom, and the mean square of experimental error is 2.36×10^{-7} with four degrees of freedom. The ratio of these variances is only 2.34. The upper 5% point for the F distribution with 71 and four degrees of freedom is 5.68, which indicates that the model is adequate.

In the light of the rate models obtained it may be of interest to speculate as to possible reaction mechanisms in the dehydrogenation and hydrogenation of butane–butenes system. The system may be described tentatively by lumping the *cis*- and *trans*-2-butenes as,

$$n\text{-butane} \rightleftarrows 1\text{-butene}$$
$$\updownarrow \qquad\qquad\qquad (23)$$
$$n\text{-butane} \rightleftarrows 2\text{-butenes}$$

We have observed that the rate of the isomerization reaction between 1-butene and 2-butenes is rapid compared with those of the dehydrogenation and hydrogenation under the experimental conditions used in the present investigation and that *cis*- and *trans*-2-butene are in thermodynamic equilibrium. Let us also assume that the dehydrogenation and hydrogenation for *n*-butane–1-butene (step 1) and those for *n*-butane–2-butenes (step 2) proceed with different mechanisms on different catalyst sites. Then, the over-all rate might be written as

$$r = r_{1f}\left[1 - \frac{p_{1\text{-butene}}\, p_{H_2}}{p_{\text{butane}}\, K_{eq\,1}}\right] + r_{2f}\left[1 - \frac{p_{2\text{-butenes}}\, p_{H_2}}{p_{\text{butane}}\, K_{eq\,2}}\right] \quad (24)$$

where r_{1f} and r_{2f} are, respectively, the dehydrogenation rates of step 1 and 2. $K_{eq\,1}$ and $K_{eq\,2}$ are thermodynamic equilibrium constants, and they are defined by

$$K_{eq\,1} = \left.\frac{p_{1\text{-butene}}\, p_{H_2}}{p_{\text{butane}}}\right|_{eq} \quad (25)$$

and

$$K_{eq\,2} = \left.\frac{p_{2\text{-butenes}}\, p_{H_2}}{p_{\text{butane}}}\right|_{eq} \quad (26)$$

The subscript, eq denotes equilibrium partial pressures. Combining Equation 25 and 26 we obtain

$$\left.\frac{p_{butane}}{p_{H2}}\right|_{eq} = \frac{p_{1\text{-butene}}|_{eq}}{K_{eq\,1}} = \frac{p_{2\text{-butenes}}|_{eq}}{K_{eq\,2}} \tag{27}$$

Using Equation 28 and assuming that 1-butene and 2-butenes are in thermodynamic equilibrium

$$K_{eq} = \left.\frac{p_{n\text{-butenes}}\, p_{H2}}{p_{butane}}\right|_{eq} \tag{28}$$

Equation 24 can be reduced to

$$r = (r_{1f} + r_{2f})\left[1 - \frac{p_{n\text{-butenes}}\, p_{H2}}{p_{butane}\, K_{eq}}\right] \tag{29}$$

Note that Equation 29 is quite similar to Equation 18.

The two rate constants of Equation 19, k_{1f} and k_{2f}, are, respectively, proportional to the net dehydrogenation rate of step 1 and that of step 2. Thus, if experimental conditions are chosen so that the isomerization reaction rate between 1-butene and 2-butenes is slow compared with the two dehydrogenation rates of n-butane, then the concentration distributions of 1-butene and 2-butenes in the reaction products provide information concerning the relative value of k_{1f} and k_{2f}. To fulfill the above experimental conditions, we performed several experimental runs with a very low partial pressure of n-butane at differential conversion levels. Analyses of these experimental data indicated that the value of k_{1f} may be approximately 10 times larger than that of k_{2f}, provided that the reaction scheme shown by Equation 23 is correct. From Table III the ratio of k_{1f}/k_{2f} can be calculated to be 15. The separation of these reactions will be studied further.

The kinetic model represented by Equation 19 which was developed by adopting modified Langmuir type isotherms seems capable of reconciling all thermodynamic and kinetic restrictions imposed on the terminal hydrogenation and the dehydrogenation reaction rates. It was also demonstrated that the single-term Langmuir equations do not suffice to take into account the observed rate data for both hydrogenation and dehydrogenation. The rate expression presented in this investigation, however, must be regarded as empirical. It should also be noted that not much data have been obtained with substantial amounts of butenes or hydrogen present during dehydrogenation or with butane present during hydrogenation. Additional accurate experimental data are being obtained at these intermediate conditions with all three species present. It may be necessary to use additional parameters to obtain satisfactory representation of r_f under these conditions. The present rate model may then be modified while still meeting the conditions for thermodynamic consistency and the other kinetic restrictions discussed above.

Acknowledgment

The contribution of the National Science Foundation is acknowledged for partial support of the work reported here.

Nomenclature

$a, b, \ldots m, n$	=	stoichiometric coefficients of reaction components $A, B, \ldots M, N$
a_I	=	activity of reaction component I
k	=	rate constant for power function models
k_{1f}	=	forward reaction rate constant (gram-moles/gram-catalyst hr, atm)
k_{2f}	=	forward reaction rate constant (gram-moles/gram-catalyst hr, atm)
k_b	=	backward reaction rate constant
k_f	=	forward reaction rate constant
K_{ae}, K_{ae}'	=	constant (atm^{-2})
K_{eq}, K_{eq1}, K_{eq2}	=	thermodynamic equilibrium constant (atm)
K_I	=	constant for reaction component I or adsorption equilibrium constant for reaction component I (atm^{-1})
m	=	constant
n	=	constant
p	=	constant
N	=	number of observations
p_I	=	partial pressure of reaction component I (atm)
r	=	reaction rate (gram-moles/gram-catalyst hr)
$\hat{r}$	=	calculated reaction rate (gram-moles/gram-catalyst hr)
r_{1f}, r_{2f}	=	forward reaction rate (gram-moles/gram-catalyst hr)
r_b	=	backward reaction rate (gram-moles/gram-catalyst hr)
r_f	=	forward reaction rate (gram-moles/gram-catalyst hr)
v	=	stoichiometric number

Literature Cited

(1) Bond, G. C., "Catalysis by Metals," Academic Press, New York, 1962.
(2) Carra, S., Forni, L., Vintani, C., *J. Catalysis* **1967**, 9, 154.
(3) Dodd, R. H., Watson, K. M., *Trans. Am. Inst. Chem. Engrs.* **1946**, 42, 263.
(4) Emmett, P. H., Gray, J. B., *J. Am. Chem. Soc.* **1944**, 66, 1338.
(5) Happel, J., Blanck, H., Hamill, T. D., *Ind. Eng. Chem., Fundamentals* **1966**, 5, 289.
(6) Happel, J., *Chem. Eng. Symp. Ser.* **1967**, 63 (72), 31.
(7) Happel, J., *J. Res. Inst. Catalysis (Hokkaido Univ.)* **1968**, 16, 305.

(8) Happel, J., Atkins, R. S., *Ind. Eng. Chem., Fundamentals* **1970**, 9, 11.
(9) Hnatow, Miguel A., Ph.D. Thesis, New York University (1970).
(10) Horiuti, J., *J. Res. Inst. Catalysis (Hokkaido Univ.)* **1948**, 1, 8.
(11) Horiuti, J., "Chemical Kinetics," Iwanami Book Co., Tokyo, 1950.
(12) Horiuti, J., *Chem. Ind.* **1956**, 9, 355.
(13) Horiuti, J., *J. Res. Inst. Catalysis (Hokkaido Univ.)* **1957**, 5, 1.
(14) Horiuti, J., *J. Catalysis* **1962**, 1, 199.
(15) Hougen, O. A., Watson, K. M., *Ind. Eng. Chem.* **1943**, 35, 529.
(16) Hougen, O. A., Watson, K. M., "Chemical Engineering Principles," Part III, Wiley, New York, 1949.
(17) Lyubarskii, G. D., Merilaynen, S. K., Pshezhetskii, S. Ya., *Zh. Fiz. Khim.* **1954**, 28, 1272.
(18) Lyubarskii, G. D., Ermakova, S. K., Pshezhetskii, S. Ya., *Zh. Fiz. Khim.* **1957**, 31, 1492.
(19) Manes, M., Hofer, L. J., Weller, S., *J. Chem. Phys.* **1950**, 18, 1355.
(20) Manes, M., Hofer, L. J., Weller, S., *J. Chem. Phys.* **1954**, 22, 1612.
(21) Noda, S., Hudgins, R. R., Silveston, P. L., *Can. J. Chem. Eng.* **1967**, 45, 294.
(22) Okamoto, Y., Happel, J., Koyama, H., *Bull. Chem. Soc. Japan* **1967**, 40, 2333.
(23) Taylor, T. I., Dibeler, V. H., *J. Phys. Colloid Chem.* **1951**, 55, 1036.
(24) Twigg, G. H., *Proc. Roy. Soc. (London)* **1941**, A178, 106.
(25) Yang, K. H., Hougen, O. A., *Chem. Eng. Progr.* **1950**, 46, 146.

RECEIVED January 12, 1970.

6

Some Problems of the Cracking Mechanism of Hydrocarbons

R. S. MAGARIL

Tyumen Industrial Institute, Tyumen, U.S.S.R.

> *The model of the cracking mechanism of olefins proposed by Voevodsky should lead to a strong dependence of olefin cracking products on temperature and pressure. This was not observed. The author proposes a chain olefin cracking mechanism when chains are being formed via a series of displacement and addition reactions of the radicals on π-bonds. Some influence of the paraffinic and cycloparaffinic products on the kinetics and mechanism is considered. Decomposition of olefins (primary cracking products) defines the kinetic of summary reaction.*

The cracking of olefinic hydrocarbons has not been as well studied as the cracking of paraffins. V. V. Voevodsky theorized (25) that the allyl radicals, in addition to decomposition and addition to double bonds, can enter into disproportion with the starting olefin. The reaction results in the formation of diene and alkyl radicals of the type shown in Reaction 1 (Voevodsky Reaction)

$$CH_2=CH-\dot{C}H-CH_3 + CH_2=CH-CH_2-CH_3 \rightarrow$$
$$CH_2=CH-CH=CH_2 + CH_3-\dot{C}H-CH_2-CH_3$$
$$(\dot{C}H_2-CH_2-CH_2-CH_3)$$

As a result, some chain reactions occur simultaneously, the decomposition of allyl type radicals competing with reactions like Reaction 1. Hence, the cracking of 1-pentene is described by Voevodsky in the following way:

(1) $C_5H_9\cdot \rightarrow CH_3\cdot + C_4H_6$ ⎫
(2) $CH_3\cdot + C_5H_{10} \rightarrow CH_4 + C_5H_9\cdot$ ⎬ Chain I

(3) $C_5H_9 \cdot + C_5H_{10} \rightarrow$

$\quad C_5H_8 + \dot{C}H_2-CH_2-CH_2-CH_2-CH_3 \; (R_1 \cdot)$ $\Bigg\}$ Chain II

(4) $\dot{C}H_2-CH_2-CH_2-CH_2-CH_3 \rightarrow 2C_2H_4 + CH_3 \cdot$

(5) $CH_3 \cdot + C_5H_{10} \rightarrow CH_4 + C_5H_9 \cdot$

(6) $\dot{C}H_2-CH_2-CH_2-CH_2-CH_3 \rightarrow$

$\quad CH_3-CH_2-\dot{C}H-CH_2-CH_3 \; (R_2 \cdot)$ $\Bigg\}$ Chain IIa

(7) $CH_3-CH_2-\dot{C}H-CH_2-CH_3 \rightarrow C_4H_8 + CH_3 \cdot$

(8) $C_5H_9 \cdot + C_5H_{10} \rightarrow C_5H_8 + CH_3-\dot{C}H-CH_2-CH_2-CH_3$

(9) $CH_3-\dot{C}H-CH_2-CH_2-CH_3 \rightarrow C_3H_6 + C_2H_5 \cdot$ $\Bigg\}$ Chain III

(10) $C_2H_5 \cdot + C_5H_{10} \rightarrow C_2H_6 + C_5H_9 \cdot$

(11) $C_2H_5 \cdot + C_5H_{10} \rightarrow C_2H_4 + C_5H_{11} \cdot$

Using the data of Arbusov (1) on the composition of the products of 1-pentene cracking at 600°C (Table I), Voevodsky found that the ratio of the probability of $C_5H_9 \cdot$ entering into Reactions 1, 3, and 8 is 0.1, and 0.51; the probability of isomerization of $R_1 \cdot$ to $R_2 \cdot$ is 0.58, and the probability for the rate of Reaction 10 to be $K_{10}/K_{10} + K_{11}$ is 0.34. He obtained satisfactory agreement with the experimental data (1) of the composition of the cracking products for 1-butene, 1-hexene, and 4-methyl-1-pentene, using the concepts of probabilities of different possible reactions of radicals.

The same calculations were done for 2-pentene. However, a more detailed examination throws doubt on the validity of Voevodsky's mechanism.

In calculating the relative rates of the parallel chain processes I-III, Voevodsky did not take into account the fact that butadiene in the gas condenses easily into liquid products according to the Diels-Alder reaction. However, it is clear that the amount of methane, forming in chain process I, is greater (if butadiene condenses) or equal to the amount of butadiene formed.

Hence, the methane concentration in the gas from 1-pentene cracking must correspond to the scheme:

$$[CH_4] \geq [C_4H_6] + [C_4H_8] + \frac{1}{2}([C_2H_4] + [C_2H_6] - [C_3H_6])$$

In fact, at 600°C for 1-pentene (data used by Voevodsky as an initial point)

$$[CH_4] = 18.2\% < [C_4H_6] + [C_4H_8] + \frac{1}{2}([C_2H_4] + [C_2H_6] - [C_3H_6]) = 29.6\%$$

Table I. Composition of the

Hydrocarbon	T, °C	Partial Pressure of Hydrocarbons[a]
1-C_5H_{10}	600	0.055
	650	0.055
1-C_6H_{12}	650	0.041
	700	0.038
4-Methylpentene-1	550	0.093
	600	0.049
	650	0.038
2-C_5H_{10}	600	0.06
	650	0.043

[a] In calculating the hydrocarbon pressure the total pressure of the hydrocarbon + H_2O is assumed to equal atmospheric pressure. Actually, it was 60–100 mm lower.

the amount of methane formed is considerably less than predicted by theory.

Such discrepancy in the composition of the gases from the cracking of 1-pentene compared with that predicted by the reaction scheme gives rise to doubt about the validity of using the relative probabilities of different reaction directions, which have been determined from the composition of the gas obtained in 1-pentene cracking, for calculating the composition of products obtained in the cracking of other olefins.

Consider the possible ratio of the kinetic constants of Reactions 1, 3, and 8. Using the values of 77, 94, and 90 kcal/mole for the bond-breaking energies of $CH_2\!=\!CH\!-\!CH(C_2H_5)\!-\!H$, and $CH_3CH_2CH_2CH(CH_3)\!-\!H$, we obtained $Q_1 = -39$, $Q_3 = -11$, and $Q_8 = -7$ kcal/mole, respectively.

Defining the activation energies according to Polyani-Semenov rule (20) we find

$$V_1 : V_3 : V_8 = 10^{13} \, e - \frac{40000}{RT} : P_3 \times 10^{-10} \, e - \frac{20000}{RT} \times [C_5H_{10}] :$$

$$P_8 \times 10^{-10} \, e - \frac{17000}{RT} \times [C_5H_{10}]$$

and since $[C_5H_{10}] \approx 10^{17}$ cm^{-3}, at 600°C, $V_1:V_3:V_8 = 1: 0.1\, P_3: 0.56\, P_8$; from the ratio found by Voevodsky, $V_1:V_3:V_8 = 1: 3, 8: 5.1$ and hence $P_3 = 40$ and $P_8 = 9$.

Products from Olefin Cracking (1)

				Gas Composition, % total				
H_2	CH_4	C_2H_4	C_2H_6	C_3H_6	C_3H_8	$1,3\text{-}C_4H_6$	C_4H_8	$iso\text{-}C_4H_8$
2.6	18.2	30.9	7.0	20.1	0.5	13.7	7.7	–
3.1	17.6	32.9	6.9	21.2	0.6	11.9	5.6	–
1.8	11.2	40.3	2.3	30.2	–	4.3	9.9	–
2.4	11.9	40.0	1.4	28.0	–	5.8	10.5	–
6.2	10.2	14.7	3.2	60.6	–	0.6	3.5	1.0
6.9	9.3	17.0	1.2	59.3	0.3	0.9	4.1	1.0
7.9	8.0	16.3	0.9	58.4	0.2	1.9	5.3	1.1
6.6	47.9	14.5	5.0	5.4	–	15.1	5.5	–
6.7	40.4	17.2	4.1	10.0	–	15.7	6.0	–

It is improbable that the value of the preexponential coefficient would exceed the frequency factor by one or two orders of magnitude. Possibly the reason for the abnormally high steric coefficient in Reaction 1 is its deviation from the Polyani-Semenov rule, although the activation energy calculated by Moin (15) gives the same result within the limits of the precision of this method. In any case, the low limit of the activation energy of the Voevodsky reaction is its thermal effect. Accepting the fact that Reactions 3 and 8 proceed without an activation barrier, $P_3 \approx 0.2$ and $P_2 \approx 0.03$.

The resulting values of the steric coefficients are one to three orders higher than expected, but they are not unrealistic. However, under the accepted conditions at 650°C the ratio of the speeds for processes I-III must change until 1:1, 5:1.8. In fact the change calculated from experimental data is considerably less than predicted. Methane and butadiene in the gas products do not increase; on the contrary they decrease slightly (Table I).

Let us examine the case of 1-hexene. According to Voevodsky, its cracking scheme may be represented as follows:

$$H_2C = CH-\dot{C}H-CH_2-CH_2-CH_3 \rightarrow C_4H_6 + C_2H_5\cdot$$

$$C_2H_5\cdot + C_6H_{12} \begin{cases} \nearrow C_2H_6 + C_6H_{11}\cdot \\ \searrow C_2H_4 + C_6H_{13}\cdot \end{cases} \text{Chain I}$$

$H_2C = CH-\dot{C}H-CH_2-CH_2-CH_3 \rightarrow$
$\quad CH_2 = CH-CH_2-CH_2-\dot{C}H-CH_3$
$\quad CH_2 = CH-CH_2-CH_2-CH-\dot{C}H-CH_3 \rightarrow C_3H_6 + C_3H_5\cdot$
$\quad C_3H_5\cdot + C_6H_{12} \rightarrow C_3H_6 + C_6H_{11}\cdot$ } Chain II

$C_6H_{11}\cdot + C_6H_{12} \rightarrow C_6H_{10} +$
$\quad \dot{C}H_2-CH_2-CH_2-CH_2-CH_2-CH_3$
$\quad \dot{C}H_2-CH_2-CH_2-CH_2-CH_2-CH_3 \rightarrow 2C_2H_4 + C_2H_5\cdot$
$\quad\quad\quad\quad\quad\quad \nearrow C_2H_6 + C_6H_{11}\cdot$
$\quad C_2H_5\cdot + C_6H_{12}$
$\quad\quad\quad\quad\quad\quad \searrow C_2H_4 + C_6H_{13}\cdot$ } Chain III

$\dot{C}H_2-CH_2-CH_2-CH_2-CH_2-CH_3 \rightarrow$
$\quad CH_3-CH_2-\dot{C}H-CH_2-CH_2-CH_3$
$\quad CH_3-CH_2-\dot{C}H-CH_2-CH_2-CH_3 \rightarrow C_4H_8 + C_2H_5\cdot$
$\quad\quad\quad\quad\quad\quad \nearrow C_2H_6 + C_6H_{11}\cdot$
$\quad C_2H_5\cdot + C_6H_{12}$
$\quad\quad\quad\quad\quad\quad \searrow C_2H_4 + C_6H_{13}\cdot$ } Chain IV

$C_6H_{11}\cdot + C_6H_{12} \rightarrow C_6H_{10} +$
$\quad CH_3-\dot{C}H-CH_2-CH_2-CH_2-CH_3$
$\quad CH_3-\dot{C}H-CH_2-CH_2-CH_2-CH_3 \rightarrow C_3H_6 + C_2H_4 +$
$\quad\quad CH_3\cdot$ or $(C_3H_6 + C_3H_7\cdot \rightarrow C_3H_8$
$\quad CH_3\cdot + C_6H_{12} \rightarrow C_6H_{11}\cdot + CH_4$ } Chain V

In accordance with this scheme the ratio of the decomposition reaction rates of the radical $C_6H_{11}\cdot$ entering into the Voevodsky reaction can be expressed through the concentration of the product in the cracking gases in the following way:

$$\frac{\frac{1}{2}([C_3H_6] - [CH_4]) + [C_4H_6]}{[CH_4] + [C_4H_8] + \frac{1}{3}([C_2H_4] + [C_2H_6] - [CH_4] - [C_4H_8] - [C_4H_6])}$$

Using the composition of the cracking gases from 1-hexene at 650° and 700°C the ratio of rates is −0.515. From this value we find that the activation energy for the decomposition of $C_6H_{11}\cdot$ and for the Voevodsky reaction for this radical are equal. This, of course, is impossible because the ratio of the rates would be

$$\frac{10^{13}}{P \times 10^{-10} \times 10^{17}} = \frac{10^6}{P}$$

and the Voevodsky reaction would not be observed at all.

For 4-methyl-1-pentene the cracking scheme is analogous to the following scheme (25):

$$\left.\begin{array}{l} H_2C = CH-\dot{C}H-\underset{\underset{CH_3}{|}}{C}H-CH_3 \to C_5H_8 + CH_3\cdot \\ CH_3\cdot + C_6H_{12} \to CH_4 + C_6H_{11}\cdot \end{array}\right\} \text{Chain I}$$

$$\left.\begin{array}{l} CH_2 = CH-\dot{C}H-\underset{\underset{CH_3}{|}}{C}H-CH_3 \to CH_2 = CH-CH_2-\underset{\underset{CH_3}{|}}{C}H-\dot{C}H_2 \\ CH_2 = CH_2-CH_2-\underset{\underset{CH_3}{|}}{C}H-\dot{C}H_2 \to C_3H_6 + C_3H_5\cdot \\ C_3H_5\cdot + C_6H_{12} \to C_3H_6 + C_6H_{11}\cdot \end{array}\right\} \text{Chain II}$$

$$\left.\begin{array}{l} C_6H_{11}\cdot + C_6H_{12} \to C_6H_{10} + \dot{C}H_2-CH_2-CH_2-\underset{\underset{CH_3}{|}}{C}H-CH_3 \\ \dot{C}H_2-CH_2-CH_2-\underset{\underset{CH_3}{|}}{C}H-CH_3 \to C_2H_4 + C_3H_6 + CH_3\cdot \\ CH_3\cdot + C_6H_{12} \to CH_4 + C_6H_{11}\cdot \end{array}\right\} \text{Chain III}$$

$$\left.\begin{array}{l} C_6H_{11}\cdot + C_6H_{12} \to C_6H_{10} + CH_3-\dot{C}H-CH_2-\underset{\underset{CH_3}{|}}{C}H-CH_3 \\ CH_3-\dot{C}H-CH_2-\underset{\underset{CH_3}{|}}{C}H-CH_3 \to C_3H_6 + CH_3-\dot{C}H-CH_3 \\ CH_3-\dot{C}H-CH_3 + C_6H_{12} \to C_3H_8 + C_6H_{11}\cdot \\ \phantom{CH_3-\dot{C}H-CH_3 + C_6H_{12}} \searrow C_3H_6 + C_6H_{13}\cdot \end{array}\right\} \text{Chain IV}$$

$$CH_3-\dot{C}H-CH_2-\overset{\overset{\displaystyle CH_3}{|}}{CH}-CH_3 \rightarrow CH_3-CH_2-CH_2-\overset{\overset{\displaystyle CH_3}{|}}{\underset{\displaystyle \cdot}{C}}-CH_3$$

$$CH_3-CH_2-CH_2-\overset{\overset{\displaystyle CH_3}{|}}{\underset{\displaystyle \cdot}{C}}-CH_3 \rightarrow \text{iso-}C_4H_8 + C_2H_5\cdot$$

$$C_2H_5\cdot + C_6H_{12} \diagup^{\displaystyle C_2H_6 + C_6H_{11}\cdot}_{\displaystyle C_2H_4 + C_6H_{13}\cdot}$$

} Chain V

By raising the temperature from 550° to 650°C and simultaneously reducing the pressure by 2.5 atm, the role of the monomolecular decomposition reaction should become more important, thus increasing the methane and propylene content and decreasing the ethylene content; the gas composition changes only slightly if we take into account the error in defining the components.

For 2-pentene the following chain processes proceed simultaneously (25):

$$H_2C-HC = CH-CH_2-CH_3 \rightarrow CH_2 = CH-\dot{C}H-CH_2-CH_3$$
$$CH_2 = CH-\dot{C}H-CH_2-CH_3 \rightarrow C_4H_6 + CH_3\cdot$$
$$CH_3\cdot + C_5H_{10} \rightarrow CH_4 + C_5H_9\cdot$$

} Chain I

$$C_5H_9\cdot + C_5H_{10} \rightarrow C_5H_8 + CH_3-\dot{C}H-CH_2-CH_2-CH_3$$
$$CH_3-\dot{C}H-CH_2-CH_2-CH_3 \rightarrow C_3H_6 + C_2H_5\cdot$$

$$C_2H_5\cdot + C_5H_{10} \diagup^{\displaystyle C_2H_6 + C_5H_9\cdot}_{\displaystyle C_2H_4 + C_5H_{11}\cdot}$$

} Chain II

$$C_5H_9\cdot + C_5H_{10} \rightarrow C_5H_8 + CH_3-CH_2-\dot{C}H-CH_2-CH_3$$
$$CH_3-CH_2-\dot{C}H-CH_2-CH_3 \rightarrow C_4H_8 + CH_3\cdot$$
$$CH_3\cdot + C_5H_{10} \rightarrow CH_4 + C_5H_9\cdot$$

} Chain III

$$CH_3-CH_2-\dot{C}H-CH_2-CH_3 \rightarrow \dot{C}H_2-CH_2-CH_2-CH_2-CH_3$$
$$\dot{C}H_2-CH_2-CH_2-CH_2-CH_3 \rightarrow 2\,C_2H_4 + CH_3\cdot$$
$$CH_3\cdot + C_5H_{10} \rightarrow CH_4 + C_5H_9\cdot$$

} Chain IV

The ratio of the rate of chain process I to the rate of the rest of the chain processes resulting from the Voevodsky reaction can be represented by the concentration of the components in the gas as follows:

$$\frac{[CH_4] - [C_4H_8] - \frac{1}{2}([C_2H_4] + [C_2H_6] - [C_3H_6])}{[C_3H_6] + [C_4H_8] + \frac{1}{2}([C_2H_4] + [C_2H_6] - [C_3H_6])}$$

From the composition of the gas from the cracking of 2-pentene one can determine that the ratio of the rates at 600°C is 1.98, and at 650°C it is 1.33. By raising the temperature 50° and decreasing the pressure 1.5 times, the rate of the decomposition of C_5H_9· decreases compared with the rate of its entering into Reaction 1; this is, of course, improbable.

The accuracy of these calculations should not be exaggerated since small errors in determining gas composition can cause great errors in evaluating the ratio. However, it is clear that according to the cracking mechanism suggested by Voevodsky, temperature increase and pressure decrease should considerably increase the decomposition role of allyl type radicals entering into the bimolecular reaction (25) and thus alter the cracking gas composition. However, the gas composition changes very little with temperature and pressure change.

The compounds we studied show that the theoretical Reaction 1 is almost never realized. At the same time the true cracking mechanism of olefinic hydrocarbons should include the formation of alkyl radicals.

We think that it proceeds in the following way. The active radicals (H, CH_3·, C_2H_5·) enter into the substitution reaction of an olefin molecule at the weakest C—H bonds and cause the formation of allyl type radicals, which have an activation energy of 6–8 kcal/mole (25). In addition, the active radicals enter into the addition to the double bonds, resulting in alkyl radicals:

$$H + I\text{-}C_5H_{10} \nearrow\!\!\!\!\!\!\!\!\searrow \begin{array}{l} \dot{C}H_2CH_2CH_2CH_2CH_3 \\ CH_3\dot{C}HCH_2CH_2CH_3 \end{array}$$

$$CH_3\cdot + I\text{-}C_5H_{10} \nearrow\!\!\!\!\!\!\!\!\searrow \begin{array}{l} CH_3CH_2\dot{C}HCH_2CH_2CH_3 \\ \dot{C}H_2CH\ CH_2CH_2CH_3 \\ \phantom{\dot{C}H_2CH\ }|\\ \phantom{\dot{C}H_2CH\ }CH_3 \end{array}$$

The addition reaction has an activation energy of 2–4 kcal/mole (4, 13, 14, 18, 26).

Thus, substitution and addition compete when the active radicals interact with olefinic molecules. Temperature change has little influence on the ratio of the reaction rates since the difference in the activation energies of the reactions is small, about 3–4 kcal/mole. Pressure change does not influence the ratio of the reaction rates since both reactions are bimolecular. For more precise studies, it is necessary to consider the substitution of the radicals into bonds other than the weakest bond, C—H.

Thus, for 1-pentene, the reactions

$$CH_3\cdot + C_5H_{10} \rightarrow CH_4 + CH_2 = CH—\dot{C}H—CH_2 — CH_3 \text{ and}$$

$$CH_3\cdot + C_5H_{10} \rightarrow CH_4 + CH = CH—CH_2CH_2\cdot CH_2$$

have a difference in their heat effects of about 17 kcal/mole and a difference in their activation energies of about 4 kcal/mole; this is not significant at high temperatures. For example, the cracking mechanism of 1-pentene, is thought to proceed as follows:

$$CH_2 = CH_2—\dot{C}H—CH_2—CH_3 \rightarrow C_4H_6 + CH_3\cdot$$

$$CH_3\cdot + C_5H_{10} \begin{cases} CH_4 + CH_2=CH_2—\dot{C}H—CH_2—CH_3 \\ CH_4 + CH_2=CH_2—CH_2—CH_2—\dot{C}H_2 \\ CH_3—CH_2—\dot{C}H—CH_2—CH_2—CH_3 \\ \dot{C}H_2—CH—CH_2—CH_2—CH_3 \\ \phantom{\dot{C}H_2—}| \\ \phantom{\dot{C}H_2—}CH_3 \end{cases}$$

$$CH_2=CH_2—CH_2—CH_2—CH_2 \rightarrow C_2H_4 + C_3H_5\cdot$$

$$CH_3—CH_2—\dot{C}H—CH_2—CH_2—CH_3 \rightarrow C_4H_8 + C_2H_5\cdot$$

$$\boxed{\dot{C}H_2—CH_2—\dot{C}H_2—CH_2—CH_2—CH_3} \rightarrow 2C_2H_4 + C_2H_5\cdot$$

$$CH_3—CH_2—\dot{C}H_2—CH_2—CH—CH_3 \rightarrow C_3H_6 + C_2H_4 + CH_3\cdot$$

$$\dot{C}H_2—CH—CH_2—CH_2—CH_3 \rightarrow C_3H_6 + C_2\dot{H}_4 + CH_3\cdot$$
$$\phantom{\dot{C}H_2—}|$$
$$\phantom{\dot{C}H_2—}CH_3$$

$$\downarrow \qquad\qquad\qquad \nearrow \text{iso-}C_5H_{10}\text{-I} + CH_3\cdot$$

$$CH_3—CH—\dot{C}H—CH_2—CH_3$$
$$|\qquad\qquad\qquad \searrow$$
$$CH_3 \qquad\qquad\quad n\text{-}C_5H_{10}\text{-}2 + CH_3\cdot$$

$$C_2H_5\cdot + C_5H_{10} \to \begin{array}{l} C_2H_6 + C_5H_9\cdot \\ CH_3-CH_2-CH_2-\dot{C}H-CH_2-CH_2-CH_3 \\ \dot{C}H_2-CH-CH_2-CH_2-CH_3 \\ \phantom{\dot{C}H_2-CH-}|\\ \phantom{\dot{C}H_2-CH-}CH_2 \\ \phantom{\dot{C}H_2-CH-}|\\ \phantom{\dot{C}H_2-CH-}CH_3 \end{array}$$

$$CH_3-CH_2-CH_2-\dot{C}H-CH_2-CH_2-CH_3$$
$$\boxed{CH_3-\dot{C}H-CH_2-CH_2-CH_2-CH_2-CH_3} \to C_3H_6 + C_2H_4 + C_2H_5\cdot$$
$$\dot{C}H_2-CH_2-CH_2-\dot{C}H_2-CH_2-CH_2-CH_3 \to 3C_2H_4 + CH_3\cdot$$

$$\dot{C}H_2-CH-CH_2-CH_2-CH_3 \to C_4H_8 + C_2H_4 + CH_3\cdot$$
$$\phantom{\dot{C}H_2-}|$$
$$\phantom{\dot{C}H_2-}CH_2$$
$$\phantom{\dot{C}H_2-}|$$
$$\phantom{\dot{C}H_2-}CH_3$$

$$CH_3-CH-\dot{C}H-CH_2-CH_3 \to C_6H_{10} + CH_3\cdot$$
$$|$$
$$CH_2$$
$$|$$
$$CH_3$$

$$C_3H_5\cdot + C_5H_{10} \to C_3H_6 + C_5H_9\cdot$$

This mechanism explains the composition of the gas produced in the cracking of 1-pentene. According to this scheme alkyl radicals are formed, some of which correspond to the starting olefins and some with a greater number of carbon atoms. This makes it possible to explain some experimental results which cannot be explained by the Voevodsky mechanism. For example, isobutene cracking yields a large amount of ethylene (6, 9, 10), and tracer methods proved that some of the ethylene resulted directly from isobutene (9). On decomposition, the radical

$$\begin{array}{c} CH_3-CH-CH_3 \\ | \\ \dot{C}H_2 \end{array}$$

can give only propylene and $CH_3\cdot$. If the methyl radical is added to isobutene, we get the radical

$$\begin{array}{c} CH_3 \\ | \\ CH_3-CH_2-\dot{C}-CH_3\cdot \end{array}$$

This can be isomerized into

$$\dot{C}H_2-CH_2-\underset{\underset{\displaystyle CH_3}{|}}{CH}-CH_3$$

and then it decomposes into ethylene and propylene

$$(or-C_3H_7\cdot + C_4H_8 \rightarrow C_3H_8 + C_4H_7\cdot)$$

A considerable quantity of isobutene is formed during the cracking of 2,3-dimethyl-2-butene (2). On decomposition, the radical

$$CH_3-\underset{\underset{\displaystyle }{|}}{\overset{\overset{\displaystyle CH_3}{|}}{C}}\cdot\text{---}\overset{\overset{\displaystyle CH_3}{|}}{CH}-CH_3$$

can give only the iso-C_5H_{10} + $CH_3\cdot$, or it can isomerize to C_3H_6 + iso-$C_3H_7\cdot$. If a methyl radical is added to the 2,3-dimethyl-2-butene, we get the radical

$$CH_3-\underset{\underset{\displaystyle CH_3}{|}}{\overset{\overset{\displaystyle CH_3}{|}}{C}}\text{---}\cdot\overset{\overset{\displaystyle CH_3}{|}}{C}-CH_3$$

which can isomerize into the radical

$$\dot{C}H_2-\underset{\underset{\displaystyle CH_3}{|}}{\overset{\overset{\displaystyle CH_3}{|}}{C}}\text{---}\overset{\overset{\displaystyle CH_3}{|}}{CH}-CH_3$$

the latter decomposing into iso-C_4H_8 + $C_3H_7\cdot$.

Influence of Olefins on the Kinetics of the Cracking of Paraffin and Cycloparaffin Hydrocarbons

In considering the thermal decomposition of paraffin and cycloparaffin hydrocarbons it is significant to know the influence of the resulting olefins on the reaction kinetics. The inhibition of cracking by propene and isobutene is discussed widely in literature and is associated with the fact that the energy of bond breaking of C—H conjugated with double bonds is lower than the C—H bond in methylene groups (about 77 and 86 kcal/mole, respectively). As a result the active radicals (H, $CH_3\cdot$, $C_2H_5\cdot$) are rapidly becoming allyl in character in the reactions of

$$R\cdot + C_3H_6 \rightarrow RH + C_3H_5\cdot.$$

If, however, the resulting olefins contain C—C bonds conjugated with a double bond, which has a bond-breaking energy of about 60–62 kcal/mole, besides reaction inhibition, acceleration arises which was not taken into account until now (*11, 12*). Taking into account the initiation of the chains by the reaction of the type $CH_2{=}CH{-}CH_2{-}CH_3 \rightarrow C_3H_5\cdot + CH_3\cdot$ makes it possible to obtain the following interesting calculations. In the cracking of n-paraffins of C_6 and higher, the chains are initiated at less than 1% cracking, owing to the decomposition of the C_4 olefins and higher which have the C—C bond conjugated with the double bond. A simplified scheme of the cracking of n-paraffins with longer chain lengths can be expressed as follows:

$$M \rightarrow R_2\cdot + R\cdot, \quad K_1 \approx 10^{13} e^{-62000/RT}$$
$$R\cdot + A \rightarrow RH + A\cdot, \quad K_2 \approx 10^{-13} e^{-8000/RT}$$
$$A\cdot \rightarrow M^x + R\cdot,$$
$$R\cdot + M^x \rightarrow RH + R_2\cdot, \quad K_4 \approx 10^{-14} e^{-5000/RT}$$

$2R_2\cdot \rightarrow$ the products of recombination or disproportion, where
$M^x = C_3$ and higher olefins
$M =$ olefins with the C—C bond conjugated with the double bond
$R_2\cdot =$ allyl-type radicals
$R\cdot =$ active radicals
$A =$ starting paraffin
From the scheme it follows:

$$-\frac{d[A]}{d\tau} = K_2 \times \frac{K_1[M]}{K_4[M^x]}[A],$$

and considering that $[M]/[M^x] = $ constant ≈ 0.5, we get

$$-\frac{d[A]}{d\tau} \approx 10^{14} e^{-65000/RT}[A].$$

The resulting equation was found empirically by E. B. Burk (*3*) and M. D. Tilicheev (*22*). The above-studied mechanism of the cracking of n-paraffins explains the absence of dependence of the cracking rate on reaction extent. This differs from the case of the low molecular weight paraffins experimentally determined by Kasanskaya (*6*) and Panchenkov and Baranov (*19*) for n-octane and n-hexadecane.

Self-acceleration in cracking was found for propene (*17*), 2-butene (*16*), isobutene (*21*), methylcyclopentane, and cyclohexane (*8*).

These facts are difficult to explain. Taking into account the initiating influence of olefins—the cracking products having the C—C bond conjugated with the double bond—makes it easy to explain the self-acceleration of cracking. The weakest bond in propene, 2-butene, and isobutene has a bond-breaking energy of about 77 kcal/mole. The formation of radicals initiating the process in methylcyclopentane and cyclohexane is possible at the bond breakage of C—H (about 94 kcal/mole). Chain initiation is considerably accelerated by breaking this bond (only 60–62 kcal/mole) when forming the olefins along the C—C bond conjugated with the double bond.

Literature Cited

(1) Arbusov, J. A., *Uth. Sap. MGU* **1945**, 89, 1.
(2) Bass, C. J., Vlugter, J. C., *Brennstoff-Chem.* **1964**, 45, 321.
(3) Burk, E. B., *J. Phys. Chem.* **1931**, 35, 2446.
(4) Darwent, B. de B., Roberts, R., *Discussions Faraday Soc.* **1953**, 14, 55.
(5) "Handbook of Chemistry," Vol. 1, 2nd ed., Goskhimizdat, Moscow, 1963 (in Russian).
(6) Hurd, C. D., Spense, L. U., *J. Am. Chem. Soc.* **1929**, 51, 356.
(7) Kazanskaja, A. S., Ph.D. Thesis, 1956.
(8) Küchler, L., *Trans. Faraday Soc.* **1939**, 35, 874.
(9) Kudrjavtseva, J. I., Vedeneev, V. I., *Dokl. Akad. Nauk SSSR* **1960**, 134, 828.
(10) Ljadova, J. I., Vedeneev, V. I., Voevodsky, V. V., *Dokl. Akad. Nauk SSSR* **1957**, 114, 1269.
(11) Magaril, R. Z., *Khim. i Tekhnol. Topliv i Masel* **1967**, 1, 7.
(12) Magaril, R. Z., *Neft. i Gaz. Prom., Inform. Nauchn. Tekhn. Sb.* **1968**, 6, 61.
(13) Melville, H. W., Robb, J. G., *Proc. Roy. Soc. (London)* **1950**, A202, 181.
(14) Minkoff, G. I., "Frozen Free Radicals," Interscience, New York, 1960.
(15) Moin, F. B., *Usp. Khim.* **1967**, 36, 7, 1223.
(16) Moor, V. G., Frost, A. V., Shiljaeva, L. V., *Zh. Obshch. Khim.* **1937**, 7, 828.
(17) Moor, V. G., Strigaleva, N. V., Frost, A. V., *Zh. Obshch. Khim.* **1937**, 7, 860.
(18) Panfilov, V. N., Voevodsky, V. V., *Kinetika i Katiliz* **1965**, 6, 577.
(19) Pantchenkov, G. M., Baranov, V. J., *Neft. i Gaz. Prom., Inform. Nauchn. Tekhn. Sb.* **1958**, 1, 203.
(20) Semenov, N. N., "On Some Problems in Chemical Kinetics and Reactivity," Izv. AN SSSR, 1958 (in Russian).
(21) Stepuhovitch, A. D., Mitenkov, F. M., "Collection of Statistics on General Chemistry," Izv. AN SSSR, 1953, p. 234 (in Russian).
(22) Tilicheev, M. D., "Chemical Cracking," Gostoptekhizdat, Moscow, 1941 (in Russian).
(23) Trotman-Dickenson, A. F., Steacie, E. W. R., *J. Chem. Phys.* **1951**, 19, 1, 329.
(24) Vedeneev, V. I., Gurvitch, L. V., Kondratjev, V. N., Medvedev, V. A., Frankevitch, E. L., "Energy Discontinuities in Chemical Bonding. Handbook of Ionization Potentials and Electron Affinity," Izv. AN SSSR, Moscow, 1962 (in Russian).
(25) Voevodsky, V. V., "Problems in Chemical Kinetics, Catalysis, and Reactivity," AN SSSR, Otd. Khim. Nauk 1955, 150 (in Russian).
(26) Voevodsky, V. V., Kondratjev, V. N., *Progr. Reaction Kinetics* **1961**, 1, 41.

RECEIVED January 12, 1970.

The Chemical Refinery in Perspective

MATTHEW F. STEWART and JAMES T. JENSEN

Arthur D. Little, Inc., Cambridge, Mass. 02140

> *The rapid growth in petrochemical demand relative to fuels demand, improvements in the technology of handling petrochemical feedstocks, and possible shortages of traditional feedstocks have focused attention on the "petrochemical refinery." While the production of chemicals from crude oil is technically possible, economic feasibility depends heavily on the pricing levels and price stability of coproducts produced from crude. Within the present petrochemical processing environment many coproduct prices are influenced heavily by the buffering effect of large internal supply–demand balances within the fuels refinery. With the growth in importance of chemical refineries, coproduct supply, demand, and pricing will become more heavily influenced by the chemical markets themselves.*

Within the past several years discussion of the petrochemical refinery concept has been growing steadily. The term itself is somewhat vague and seems to mean different things to different people. Technically it is quite possible to charge crude oil into a processing plant which produces nothing but chemicals and internal fuel, making no saleable gasoline, middle distillates, or fuel oils. While in a strict sense this might define the petrochemical refinery, many use the term more loosely to apply to a refinery in which chemical products are maximized economically, but limited quantities of saleable fuel products are also produced. This would distinguish it from the conventional fuels refinery which usually includes some chemical manufacture as ancillary activities.

The reason for the growing interest in the petrochemical refinery seems to be two-fold. From an oil company point of view, the sheer size of some of the newer petrochemical plants is so large that their feedstock requirements and by-product return streams tend to interfere with the usual processing to conventional fuel products in all but the largest refineries. For example, a modern billion pound per year ethylene plant which is designed to run on full-range naphtha would pre-empt all of a virgin naphtha cut of a 120,000-barrels per day refinery, leaving no virgin naphtha for direct gasoline blending or for catalytic reforming into high octane gasoline blending stock.

On the other hand the chemical companies view the petrochemical refinery as a way out of their increased dependence on refiners for large volumes of feedstreams and simultaneous reliance on the refiners for disposal of by-products. The petrochemical refinery which can convert crude oil directly into saleable chemicals represents a potential remedy for chemical companies who are concerned about their potential vulnerability in having to rely so heavily on their rival oil company competitors.

Configurations of Petrochemical and Fuels Refineries

There is little question that the petrochemical refinery can perform its function technically. Whether it can also do so economically and commercially is a complex question which is discussed later. It is possible to consider direct cracking of whole crude into olefins, but such processes —pioneered principally in Germany, France, and Japan—are as yet comparatively new and untried commercially. A more likely route to the petrochemical refinery involves integrating both conventional coil cracking for olefin manufacture and extraction of aromatics from reformed and cracked naphthas with conventional refinery units. Such a refinery would produce primarily ethylene, propylene, butadiene, and aromatics. Other variations and other chemicals are possible, but the steam cracker and the aromatics extraction unit are the backbone of most schemes.

Figures 1 and 2 show comparative flow diagrams for a petrochemical refinery and a conventional fuel refinery. The difference between the 2 approaches is immediately apparent in the treatment of the naphtha fractions. Whereas, in the fuels refinery they are destined for the gasoline pool, directly as light straight run gasoline and *via* catalytic reforming as high octane reformate, all of the naphtha cut in the petrochemical refinery goes to steam cracking with the exception of the 160°–300°F naphtha heart cut which contains the benzene–toluene–xylene precursors. These go directly to catalytic reforming. The middle distillates are also sent to the steam cracker in the petrochemical refinery. In the fuels refinery these streams are used primarily for kerosene, jet fuel, diesel fuel, and distillate heating oils; the heavier distillates are usually used as catalytic cracking or hydrocracking feed. The handling of atmospheric bottoms in the fuels refinery often is a critical part of refinery economics, particularly in the United States. The common processing route involves vacuum distillation to provide additional cat cracker feed coupled with further fuel oil reduction processing such as coking to yield additional

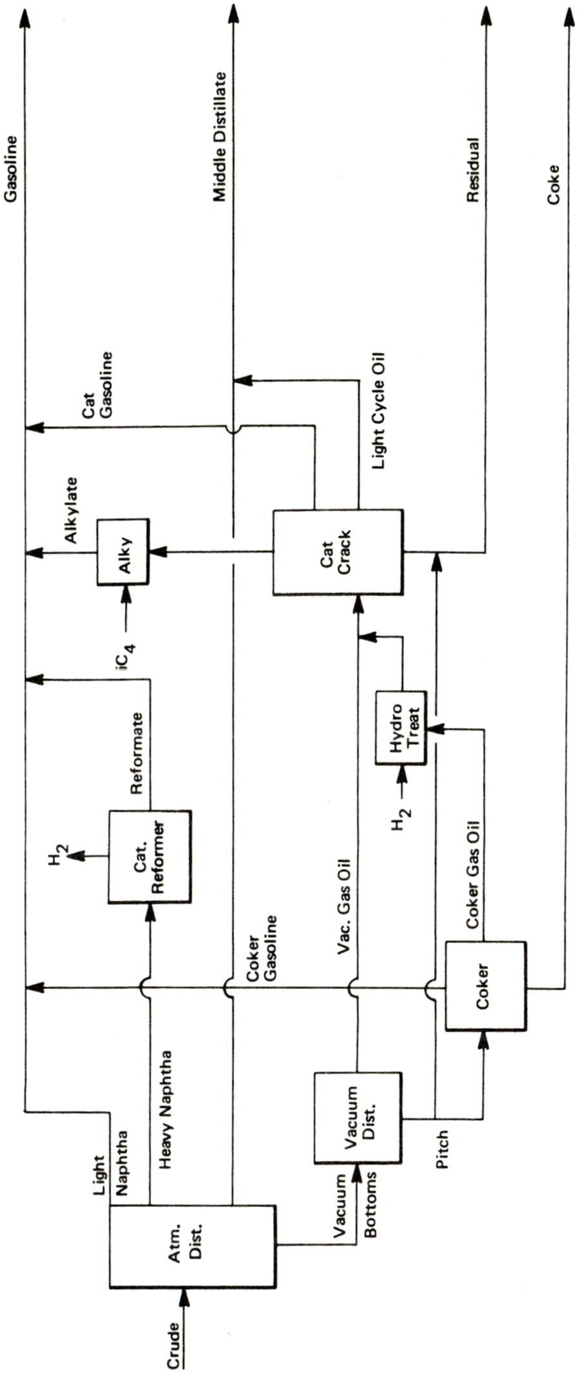

Figure 1. Fuels refinery

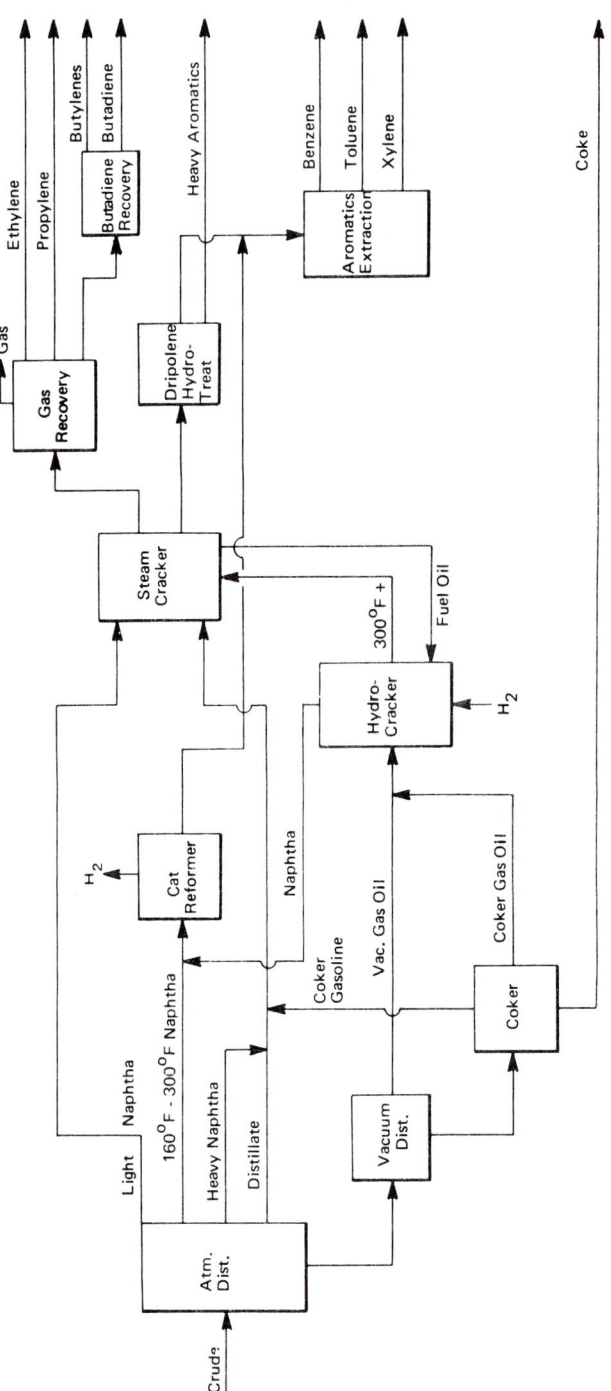

Figure 2. Petrochemical refinery

gasoline and more cracker feedstock. In the petrochemical refinery vacuum distillation may also be used to provide a gas oil for steam cracker feed while the vacuum bottoms can be processed further, using fuels refinery processing, to provide additional steam cracker and aromatics reformer feedstock.

The end result of the fuels refinery scheme shown in Figure 1 (a typical U.S. high gasoline yield refinery) is that the straight run, cracked, and reformed naphthas are blended to make a yield of about 50–55% gasoline, while the straight run distillates together with cracked distillates are blended to add another 25–30% of premium distillate yield for sale. Residual fuels and internal fuel make up the remainder. Gasoline and middle distillates, however, dominate fuel refinery yields and economics.

The petrochemical refinery of Figure 2, on the other hand, is carried largely by the premium olefins, ethylene and propylene (about 40% yield), by high value butadiene (about 4% yield), and by aromatics (about 16% yield). Another 10% of miscellaneous products are produced leaving about 30% gaseous or residual fuel yield. The comparable yield of internal and external fuel value streams from the usual U.S. fuels refinery is commonly nearer 10–12%.

Relative Economics of Petrochemical and Fuels Refineries

Two important measures of refinery economics are (1) refinery complexity, which is reflected largely in the capital investment and required capital charges to cover depreciation, return on investment, and taxes on that investment; and (2) the financial burden imposed on premium product margins as a result of degradation of raw materials to a low fuel value.

Refinery Complexity. The fuels refinery of Figure 1 is designed for U.S. conditions since it makes a maximum yield of gasoline and distillates through extensive cracking and fuel oil reduction processing. In Europe, Japan, and the Caribbean, where the market demands a higher residual fuel oil yield and the spread between product pricing and crude oil reflects these market conditions, the optimum refinery is much simpler. The favored refinery design has been the hydroskimming type; cracking, where practiced at all, has been comparatively mild. A European refiner who has to sell residual fuel oil at 30¢ to 50¢ per barrel below his crude cost can still live with an over-all refinery yield of 40% of fuel oil and fuel gas since the refinery investment cost is only $400–500 per daily barrel of processing capacity. A U.S. refiner who faces a residual fuel oil sales differential of $1.70 per barrel below crude cost can afford to install a complex refinery costing $1200 or more per daily barrel in order to reduce fuel oil and gas yields to a level of 10–12%.

Raw Material Degradation to Fuel Value. A feature of refineries which is often overlooked is that although a fuels refinery is in business to sell fuel, it is the premium fuels such as gasoline and middle distillates which bear most of the burden of raw material and processing costs, for the internal fuel gases and residual fuel oil are generally valued below the average raw material cost. The value at which residual fuel is sold, is usually less than the cost of the crude oil (whether compared per pound, per barrel, or per million Btu) because of the competitive price of coal and natural gas. This loss of value requires an added margin for the premium products to carry.

A U.S. petrochemical refinery must face the same market environment of high crude cost/low fuel value and therefore is also interested in fuel oil reduction to make additional petrochemical feedstocks. An Eastern Hemisphere petrochemical refinery could explore higher fuel oil production alternatives without the extreme economic penalties existing in the U.S. The result might be more nearly a combined chemicals/fuels refinery as the marketing of the heavy fuel oil presents comparatively simple marketing problems relative to selling gasoline.

A major reason for the high incentive in the United States to reduce fuel oil is the large spread between high U.S. crude oil prices and the prices at which residual fuel oil must be sold. The U.S. oil import program is an important factor in this spread since restrictions on crude oil imports have raised the domestic price about $1.40 per barrel above land in costs of comparable international crude oils, while the possibility of reasonably free import of residual fuel oil from the Caribbean have kept U.S. East Coast and international prices at closely related levels. Any significant easing of the oil imports program such as recommended to the President in the spring of 1970 by the Task Force could reduce this spread. This could reverse the U.S. trend to more complex refining, shifting us to a more European style than has been the case in the past. Such a trend would also result from markedly higher prices for low sulfur fuel oil produced from sweet U.S. crudes. While the crude oil residual price spread would have to drop sharply to effect a major change in U.S. refinery complexity, if it were to occur, the outlook for the petrochemical refinery would presumably be improved.

Interaction. The effect of the trade-off between refinery complexity and the loss in value of raw material degraded to fuel value for fuels refining is illustrated in Table I. For a European refinery, approximately 0.67 lb of fuel gas and residual fuel oil is made per 1.0 lb of products selling at premium over crude oil. Where crude oil cost to the refiner is $1.80/bbl and fuel value is $1.50/bbl (Table II), the yield of fuels poses a value loss of only 0.09¢/lb of fuel. Because this value loss is so low, the high fuel yield is not a significant penalty and only requires that each

Table I. Comparison of Fuel Value Losses with Investment Charges

	Lbs Degraded to Fuel/lb Premium Products	Fuel Value Loss ¢/lb of Fuel	Fuel Value Loss ¢/lb of Premium Products	Capital Charge @ 20%/lb of Products	Total Fuel Loss plus Capital Charge/lb of Product
European refinery (European crude cost)	0.67	0.09¢	0.06¢	0.14	0.20
European refinery (U.S. crude cost)	0.67	0.58¢	0.39¢	0.14	0.53
U.S. refinery (U.S. crude cost)	0.14	0.58¢	0.08¢	0.25	0.33
U.S. ethylene plant (ethane/propane)	0.29	0.39¢	0.12¢	0.66	0.78
U.S. ethylene plant (naphtha)	0.27	0.98¢	0.27¢	0.48	0.75
U.S. ethylene plant (gas oil)	0.46	0.65¢	0.30¢	0.48	0.78
U.S. petrochemical refinery (U.S. crude cost)	0.43	0.58¢	0.25¢	0.88	1.13

Table II. Values of Raw Materials, Fuel, and Investment Used in Table I

	¢/lb	$/bbl	Investment, ¢/Annual lb Premium Products
U.S. crude oil	1.06¢	$3.20	—
European crude oil	0.59¢	$1.80	—
Fuel value	0.48¢	$1.50	—
U.S. ethane/propane	0.87¢	$1.33	—
U.S. naphtha	1.46¢	$3.78	—
U.S. gas oil	1.13¢	$3.36	—
Investment			
U.S. refining	—	—	1.24¢
European refining	—	—	0.68¢
ethylene (ethane/propane)	—	—	3.32¢
(naphtha)	—	—	2.42¢
(gas oil)	—	—	2.41¢
U.S. petrochemical refining	—	—	4.43¢

pound of premium refinery products realize an added 0.06¢/lb to cover the loss in fuel oil value. The component of capital charges for the comparatively low investment European refinery would add another 0.14¢/lb of premium products to the required margin resulting in a total of 0.15¢/lb.

Were the same refinery to charge crude at U.S. prices of $3.20/bbl, the loss in fuel value would be 0.58¢/lb of fuel. This would require an added revenue of 0.39¢/lb from sales of premium products. The combination of low capital charge with high fuel value loss would bring the required margin to 0.53¢/lb.

Were the refinery to use U.S. crude but at U.S. complexity, it could minimize the fuel value loss per pound of premium products (0.08¢/lb) by investment in fuel oil reduction processes. Though the capital charge per pound of premium products would then be more than twice that of the European refinery the combination of the two cost allowances would be only 0.33¢/lb rather than the 0.53¢/lb margin for the simpler high fuel yield refinery, but this is still above the European level of 0.15¢/lb.

The same relationship between plant complexity and loss in raw material value to fuel also exists in petrochemical processing. The three common raw materials for ethylene manufacture in the U.S.—ethane/propane, naphtha, and gas oil—exhibit similar trade-off relationships between processing complexity and fuel value loss (*see* Table I).

Degradation to fuel is minimized with ethane/propane or naphtha, while gas oil involves a far greater loss of raw material value to the fuel pool. Naphtha as a raw material, however, is so expensive relative to fuel value that its comparatively low degradation level of 0.29 lb/lb of premium products still requires 0.26¢/lb added margin on premium products to cover it compared with ethane/propane requirement of only 0.12¢/lb. Even though gas oil degrades 0.46 lb/lb, its low cost relative to fuel value requires an added margin on premium products of 0.30¢/lb; this is not significantly different from that of naphtha.

The petrochemical refinery shown in Figure 1 also degrades a substantial quantity of raw material to fuel (despite the fuel oil reduction processing characteristic of U.S. fuel refineries). Because of its complexity, however, both in heavy fuel oil reduction and petrochemical processing, it has a very high capital charge per pound of premium products sold, and the total required additional margin on premium products is 1.13¢/lb of premium products (Table I). Its economics, therefore, depend heavily on the pricing levels which it realizes for the olefin and aromatics products.

Product Value Realization. Since only 70% of the original crude oil feedstock in the petrochemical refinery emerges as premium products and since the capital investment per pound of products is so high, the feasibility analysis of any petrochemical refinery scheme is very sensitive to the sale realizations of the olefin and aromatic products. This makes it extremely important both that the product realizations can be achieved for the individual plant in question and that there is some assurance of product price stability so that petrochemical refining margins do not deteriorate and make the proposed plant uneconomic at some future date.

The first requirement—that the realizations be available to the individual project in question—is a potential source of difficulty. While a petrochemical refinery project will have some flexibility to shift yields between ethylene and propylene and aromatics, the range of that flexibility is not great. Thus for each premium product to carry its share of project economics, an outlet must be available for it in approximately the same yield relationship which the refinery produces. For products such as benzene, for example, transportation costs are low enough that the petrochemical refinery output can be moved to other locations without a serious loss of netback. For others such as ethylene, transportation is more costly, and unless an adequate market outlet is available nearby, project economics could deteriorate. This problem is likely to be most acute with products produced in comparatively small yields. There may be little incentive to upgrade them by adding a derivatives plant for their production alone, and yet they may not be transportable to some place where they can be used at premium values. Such products in a fuels refinery (ethylene from a cat cracker, for example) are often left in the refinery fuel pool to be taken out only after some use develops for them, often in connection with another project. However, in a petrochemical refinery where the degradation to fuel is already high and profit margins are thin, these small losses may be a burden to the project.

For the petrochemical scheme of Figure 1, two such potentially troublesome streams are the mixed butylenes from the steam cracker (after butadiene extraction) and the heavy aromatics from the dripolene (pyrolysis gasoline) stream. Within a fuels refinery each of these streams can be disposed of easily. The butylene would go to butylene alkylation for producing a high octane gasoline blending stock. The heavy aromatics, after hydrogenation, could be blended directly to the gasoline pool. Disposal of these products at premium values outside a fuels refinery is more difficult. Although Figure 1 assumes they can be disposed of at reasonable prices, chemical markets for butylenes and heavy aromatics are limited. Three recent petrochemical refinery papers (2, 3, 6) suggested petrochemical refinery designs which were similar in major respects though differing in detail. Two proposals simply assumed sale of these 2 streams as is (presumably to a refiner). The other proposed scheme called for disposal of the butylene stream in a Wulff furnace to make ethylene and acetylene and the dealkylation of heavy aromatics to produce naphthalene. Whether the latter two processing alternatives would provide a better netback on these products than their use for gasoline in a fuels refinery is questionable.

Product Price Influences. The second problem facing a petrochemical refinery is that of stability of product prices, particularly where internal flexibility to shift yields between products is limited either by direct process considerations or by the construction of fixed size petrochemical derivative plants as part of a petrochemical complex. The economics of the petrochemical refinery will be carried primarily by the ethylene, propylene, butadiene, and aromatics. Benzene will be the principal aromatic produced particularly if the market situation makes it appropriate to dealkylate toluene to produce more benzene. The prices of these products are far from stable. Most of these products have been undergoing price decline since the early 1960's. The factors which influence prices are complex. They include not only the usual considerations of supply and demand but the effect of improving economics for new facilities arising from technological advances. A good analysis of these relationships for each of the products is contained in two recent volumes (5). Where scale economies, technological improvements in processes, or improving feedstock or by-product credit relationships are in effect for new plants, the pressures on prices tend to be downwards. However, petrochemical refinery feedstock, by-product, and even final product values are often heavily involved in the economics of fuels refining and may therefore run counter to what would otherwise seem to be the trend. Although a petrochemical refinery may be able to operate independently of a fuels refinery in a physical sense, it cannot do so in an economic sense since it depends so heavily on petrochemical price trends, and these in turn are influenced by fuels refinery economics.

The price relationships of the principal petrochemical products from a petrochemical refinery such as that of Figure 1 tend to fall into four broad categories. First are those chemical products which are not fuels or fuel intermediates themselves. While their pricing may be affected to some extent by fuels refinery product prices, their prices are primarily set by chemical competition and the effects of alternative process routes to the same product. Butadiene clearly falls in this class. Ethylene does also if ethylene alkylation to gasoline is regarded as a special situation having only a limited long term influence on ethylene prices.

Second are those products which while having chemical uses are primarily fuels (used in an impure state) and are therefore usually priced in a direct relationship with their value as a premium fuel. Toluene and xylene both fall in this category, and their pricing tends to depend on their gasoline blending values.

Third are those petrochemicals whose value is set primarily by chemical competition, but their direct interaction with the gasoline pool has a buffering effect on their price changes. Benzene, particularly in its relationship to toluene through toluene dealkylation, falls in this category.

Fourth are those chemicals whose pricing is usually set in a fuels refinery context because chemical uses are small relative to fuels use. These chemicals are subject to pricing fluctuations because of a complex relationship to premium fuel values. Propylene in the United States is clearly in this category.

In Europe and Japan, where naphtha cracker by-products are much larger relative to the gasoline pool than in the U.S., the relationship between by-product supply (which might result from normal ethylene growth) and the chemical demand for by-products influences by-product price. In the U.S., however, the buffering effect tends to obscure these chemical supply-demand relationships and make price forecasting more complex. For a petrochemical refinery, which has relatively high capital investment and a high level of value degradation to fuel, an understanding of the trends in price pressures for the main chemicals may be the key to deciding whether to go with a petrochemical refinery scheme or to try some other route. For a chemical company which wants to avoid relying directly on premium fuel outlets for products the price–pressure trend analysis is doubly important. An inflexible petrochemical refinery scheme which depends heavily on realizing maximum coproduct credits for all chemicals may not be able to withstand price erosion on some of its by-products.

Recently most of the large oil companies announced their intention to market low-lead or lead-free gasoline to reduce automobile exhaust gas pollution. This gasoline, although having an octane number lower than leaded regular gasoline, will be sold at a premium price, which is considered necessary to cover increased manufacturing costs as well as the substantial investment in new distribution and marketing facilities. Almost all the new automobiles made in 1971 will require low-lead or lead-free gasoline. During the next decade the gasoline clear (unleaded) octane pool will gradually increase, and refiners will operate reformers at a higher severity and make more alkylate to meet the rising requirement for high octane gasoline blending components. The demand for high octane aromatics will increase more rapidly than before, and the value of these aromatics will rise. Benzene will not increase as much in value as a gasoline blending component as the higher octane toluene and xylenes. However, because toluene demand will rise, dealkylation plants will tend to be shut as the toluene is diverted to the gasoline market. As a result, the price of benzene, as well as toluene and xylenes, is likely to increase over the next few years. This increase in value of aromatics will raise the value of the by-product pyrolysis gasoline from cracking gas oil and naphtha in the manufacture of ethylene.

Although propylene alkylate and butylene alkylate have a lower clear octane number than the heavy aromatics, their sensitivity (research minus motor octane number) is much lower than the aromatics. Since automobile engines require a high motor octane number to run efficiently, the value of alkylate in the octane pool will increase. This will have a positive effect on the price of propylene and butylene in chemical markets. The quantity of alkylate that will be produced will be limited by the economics of making olefins and making or buying isobutane. Again, refiners will have the advantage over chemical companies in using aromatics and olefins as valuable gasoline blending stocks.

Petrochemical Refinery Products

Ethylene. Ethylene is the most important product from the petrochemical refinery shown in Figure 1, constituting nearly 40% of the value of premium products from the refinery (*see* Table III). Ethylene prices have been drifting down for a number of years, resulting in comparatively low experienced rates of return for relatively new steam cracking plants. This downward drift has been a result of economics of scale from larger plant sizes, improvements in cracking technology, and a broadening of the base of feedstocks which can be used, but overcapacity and increased competition have assured that these benefits would be passed on to the customer.

The future outlook for ethylene prices depends heavily on feedstocks. Natural gas liquids—primarily ethane/propane—constitute about

Table III. Percent of Value from Petrochemical Refinery

	Yields			
	Percent Total	*Percent Premium Value*	*Price*	*Percent Premium Products*
Ethylene	26.0	37.2	2.8¢	37.3
Propylene	14.0	20.0	2.7¢	19.4
Butadiene	4.0	5.7	7.0¢	14.3
Benzene	8.6	12.3	2.8¢	12.4
Toluene	4.3	6.1	2.1¢	4.6
Xylene	3.1	4.4	2.5¢	3.9
Butylene	5.0	7.2	2.0¢	5.0
Heavy gasoline	2.6	3.7	1.6¢	2.1
Butane	2.4	3.4	0.8¢	1.0
Fuel oil, gas, and coke	30.0	—	—	—
	100.0%	100.0%		100.0%

60% of the raw material for U.S. ethylene manufacture. These feedstocks were regarded widely as in tight supply two years ago, particularly with the imposition of heating value limitations on natural gas by the Federal Power Commission. Natural gas liquids then went through a period where they were long and prices were weak. Now it appears that the shortage is resuming. Alternative raw materials include naphtha, gas oil, as well as crude oil should the idea of the petrochemical refinery prove itself commercially. The impact of U.S. oil import controls on petrochemical feedstocks, particularly for ethylene, has been an extremely important part of the potential pricing outlook (4, 5).

Propylene. Propylene has been growing rapidly for chemical uses although probably no more than a third of the propylene which is actually produced in the United States reaches chemical markets. The largest single source of propylene is refinery operations, particularly the off-gases from catalytic cracking. The largest single use of propylene is for alkylation utilizing isobutane to produce a high octane gasoline blending stock. Because the bulk of supply and demand is internal to refineries, propylene prices are determined largely by refinery gasoline economics. This has been particularly true during 1969 when the surplus of the light hydrocarbons from natural gasoline plants coupled with changes in internal refinery light hydrocarbon balances caused a sharp drop in the price of normal and isobutane (1). Low cost isobutane increases the value of propylene to a refiner since the resulting propylene alkylate blending value in gasoline normally remains unchanged. Thus, propylene for chemical uses assumes a higher price since refiners are unwilling to sell it as long as they can use it to make gasoline, taking advantage of low priced isobutane. The chemical refinery, like naphtha and gas oil cracking units, makes a high yield of propylene by-product relative to ethylene and is thus favored in times of strong propylene prices. Obviously the prospective petrochemical refinery investor must consider how long isobutane price levels remain depressed in judging the sensitivity of project economics to current propylene pricing.

Butadiene. Butadiene pricing seems to have reflected increased supply from both ethylene by-product sources and old plants partially modified to increase yields. The favored route for new direct capacity is probably butane dehydrogenation but the growth of more U.S. ethylene capacity based on gas oil would substantially increase by-product supply. Since butadiene growth is slower than ethylene growth with a heavy shift toward refinery liquids, cracking for ethylene would tend to depress butadiene supply/demand balances and prices.

Aromatics. The aromatics have peculiar supply/demand problems. With the gradual reduction in importance of supply from coal coking (which could be considered an inflexible yield coal refinery) more than half of U.S. benzene is now provided directly by naphtha reforming and another quarter is derived from hydrodealkylation of toluene. Both toluene and xylenes are produced substantially in excess of chemical requirements from catalytic reforming and are usually left in the gasoline pool at premium blending values. The petrochemical refinery then faces the prospect of competitive sales of toluene and xylene at fuel-based prices or the inclusion of process capacity to convert toluene to benzene and to separate xylene into its isomers. A fuels refinery could retain its gasoline blending alternatives.

Conclusion

Since most of the petrochemicals from the petrochemical refinery scheme have values in the U.S. which are so intimately involved with premium fuel values and have price trends which are "buffered" by fuel refinery requirements and economics, the petrochemical refinery must be considered in an oil industry context.

For the key chemical products which must support the economics of a petrochemical refinery pricing uncertainties appear inevitable. However, as long as chemical feedstock demands in the U.S. continue to grow more rapidly than fuels and as long as chemical companies remain uneasy over their growing reliance on refiners, the petrochemical refinery concept will receive active attention from chemical companies.

We feel that in the near future, the independent petrochemical refinery will be difficult to realize. Instead oil companies will extend their basic fuels refineries into more petrochemicals production taking advantage of the flexibility and product value optimization which a fuels/petrochemical refinery allows.

Literature Cited

(1) Bryson, A. E., *Natural Gas Processors Assoc. Meetg., New Orleans*, **March 1968.**
(2) Gambro, A. John, Caspers, J., *World Petrol.* **April 1967.**
(3) Gambro, A. John, Caspers, J., Newman, J., *World Petrol.* **Dec. 1968.**
(4) Stewart, Matthew F., *A.I.M.E. Meetg.,* **Feb. 1969.**
(5) Stobaugh, Robert B., Jr., "Petrochemical Manufacturing and Marketing Guide," 2 vols., Gulf Publishing Co., Houston, 1968.
(6) Struth, B. W., List, H. L., *A.I.Ch.E. Meetg., New Orleans,* **March 1969.**

RECEIVED January 12, 1970.

Product Optimization in the Petrochemical Refinery

HARRY M. WALKER

Monsanto Co., Alvin, Tex.

>*Economic study of various configurations of the petrochemical refinery by computer simulation reveals that profit gains, sufficient to render such a plant attractive in the U. S. economic environment can be achieved by co-producing premium liquid fuels. Without liquid fuel production the best return on investment calculated was 9.8% (before federal income tax), increasing to a more acceptable 17% with maximum gasoline and kerosene production. The basic chemical unit is better (20.5%) under European economic conditions but is not improved greatly by fuels production. Fuels production improves the operating divisor and permits the sale of the higher valued crude fractions in their best markets. The chemical raw material cost is reduced by cracking only lower valued fractions (U. S. economic environment).*

The term "petrochemical refinery" first appeared in the technical literature in the early 1960's (2). Since that time numerous authors have used the concept in discussing proposed integrated plant complexes designed to convert crude oils into chemical products in high over-all yield (1, 3, 4, 6, 8). In most cases the design objective seems to have been the production of chemicals with a minimum of by-product liquid fuel products. The configurations proposed have all involved thermal pyrolysis for ethylene and propylene production as the basic core of the integrated operation. Units for recovering BTX (benzene, toluene, xylenes) and butadiene are standard blocks. Catalytic reforming for the production of additional BTX aromatics is frequently suggested (10) as are units for toluene dealkylation, naphthalene production, the production of ethylene, propylene, or butylene alkylate, the dehydrogenation of butylenes to butadiene, isobutylene recovery, ethylbenzene distillation, delayed coking (9), or other possible processes. In one case a Wulff unit for acetylene manufacture was suggested (5).

One cannot question the technical feasibility of the schemes advanced. In nearly all cases the building blocks are known—*i.e.*, available technologies. The question of economic feasibility is quite another matter. This paper deals with some aspects of the latter question.

While the development of a proposal involving many closely integrated units may represent great technical ingenuity, economic factors must control. Every unit, every capital item must meet the test of providing an adequate return on the investment. Perhaps more importantly all products must be competitive. Is the product obtained at a cost fully competitive with the cost of the same product obtained in other ways, from other sources, or by other producers? The latter question is the more difficult to answer and the one which has received little attention in the past. In a refinery where numerous products arise from a common feedstock processed, in part, through common units it is extraordinarily difficult to determine a cost for individual products, let alone provide answers to searching questions regarding "competitiveness." Nevertheless, this paper will attempt to provide some insights into this problem.

This discussion will be concerned with integrated hydrocarbon processing, starting with crude oil as feedstock. Plants charging pre-processed fractions such as LPG, naphtha, gas oil, natural gasoline, etc. are definitely outside this concept since they imply a reliance upon the by-products of other hydrocarbon processing operations. The proper selection of the products to be produced from this crude oil input and the process steps used to obtain them and their scale will determine the economic feasibility of the scheme.

Products of Thermal Cracking

First, let us consider in some detail the products which can be obtained *via* thermal cracking of liquid hydrocarbons. Table I includes all of the products known or believed to be marketed from commercial steam-cracking operations. Certainly the lower molecular weight materials are well known. Since the liquid products of cracking are less familiar, they are discussed in some detail.

The C_5 fraction varies widely both in quantity and in component distribution. This variation is both a function of pyrolysis severity and of feedstock, though as the feed becomes heavier, its nature becomes less important. Typical C_5 fraction analyses are found in Table II. The partial or complete hydrogenation of some of these fractions yields a gasoline blending stock in which the concentration of the preferred

Table I. Products from Pyrolysis

Hydrogen	Butane/butene mix	Isoprene monomer
Acetylene	2-Butylene	Methylacetylene–allene conc.
Ethylene	Mixed C_5's	Dicyclopentadiene
Ethane	Isoprene conc.	Pyrolysis gasoline
Propylene	Piperylene conc.	Hydrotr. C_9 aromatics
Propane	Benzene	Alkyl naphthalene conc.
Mixed C_4's	BTX raffinate	Mixed xylenes
Butadiene	Naphthalene	Ethylbenzene
Toluene	Carbon black oil	Aromatic oils (500°–700° F.)
o-Xylene	m-, p-Xylenes	Pyrolysis tar
p-Xylene	Cracked C_9 aromatics	Mixed C_{8-9} aromatics
Resin oil	Coke	Pyrolysis pitch (200 ° F.S.P.)

Table II. Typical C_5 Fraction Analyses

Component	Wt. %			
C_5H_6	19	14	8	26
Isoprene	30	17	14	11
Piperylene	14	20	9	12
C_5H_{10}	15	14	16	13
Isopentane	6		24	12
n Pentane	5		26	15
Cyclopentane	2			
Miscellaneous	9	35	3	11
% Cyclics	25		9	31

branched and cyclic compounds is high. Leaded octane values in the 100 range are attainable.

The BTX fraction shows a similar variability arising from pyrolysis severity and feedstock (Table III). In general, the aromatic content ranges upwards of 50% and may reach nearly 100% in high severity operation. BTX ratios vary widely, depending on feedstock. LPG-type feeds yield benzene predominantly among the aromatics, while the heavier feeds show increased yields of toluene and C_8 aromatics. In all cases benzene appears to be synthesized predominantly from paraffin or naphthene chain fragments by condensation reactions. Toluene and the C_8 aromatics appear to derive principally from aromatics in the feed with some pyrolytic side-chain dealkylation. In all cases styrene predominates in the C_8 fraction (7).

The commercial processing of this cut invariably involves hydrogenation. For gasoline end use partial reduction to saturate diolefins generally is adequate, the chief target being to satisfy gum formation specifications in the final gasoline blend. Gasolines having blending values of over 130 RON (research octane number) are obtainable.

Table III. Typical C_6–C_8 Fractions

Components	Wt. %				
C_5's	.8	6.7	14.8		1
Benzene	38.8	43.6	42.8	43.8	36
Toluene	29.5	27.4	10.9	27.8	33
Styrene	7.5	1.0	.9	5.0	7
Ethylbenzene	4.3	.9		1.9	3
Xylene	16.4	3.7	.6	9.7	13
C_9 aromatics	1.0	2.7			1
C_{6-8} unsaturates	1.7	14.0	4.6	9.3	4
Dicyclopentadiene			11.1		
Methycylopentadiene			2.0		
Other			12.3	2.5	2
% Aromatics	97.5	80.0	55.2	88.2	93

Table IV. Typical Resin Oil

Component	Wt. %
Vinyltoluene	16.0
Indene	14.0
Methylindene	7.0
C_2 styrene	6.0
α-Methylstyrene	2.5
β-Methylstyrene	1.5
Divinylbenzene	1.5
Styrene	1.5
Dicyclopentadiene	11.0
Methyldicyclopentadiene	9.5
C_3 alkylbenzene	22.5
C_4 alkylbenzene	7.5
Naphthalene	3.5
Xylene	1.0

When the aromatic chemicals are the desired products, hydrogenation is more severe to yield an olefin-free product suitable for extracting the aromatics by one of the conventional recovery processes.

Attention is called to the fact that the nonaromatics are predominantly cyclic in nature—methylcyclopentadiene being the non-aromatic moiety normally present in the greatest concentration. Reduction of the cyclic olefins and diolefins yields naphthenes rather than paraffins. As a result, sharp separations in the extraction unit becomes more difficult. This is particularly a problem in the C_8 range when the individual isomers are to be separated for high grade chemical markets.

The C_9 range contains an interesting mixture of indenes, styrenes, alkylbenzenes, and dimers of cyclopentadiene and methylcyclopentadiene. Commercially this fraction is sold for polymerization to hydrocarbon resins. Table IV gives the analysis of a typical "resin oil." Since this fraction lies in the gasoline boiling range, it can be converted to a valuable gasoline blending component by hydrogenation for reducing the styrenes and the dimers. RON values of 105 are attainable.

The C_{10}–C_{13} or "alkylnaphthalene concentrate" fraction boiling from 400°–550°F contains principally bicyclic aromatics of the naphthalene series. Also present are methylindene and higher indenes plus biphenyls, higher alkylstyrenes, and alkylbenzenes. Nonaromatics are present in low concentrations.

The attainable yield of naphthalene *via* hydrodealkylation of this fraction ranges from less than 50% to greater than 70%, depending upon the basic pyrolysis raw material, pyrolysis severity, and fractionation. By close cutting in the 400°–450°F range it is possible in some cases to produce concentrates from which solid naphthalene can be obtained by crystallization. In general, the size of this fraction is greater as heavier stocks are fed to pyrolysis, but the concentration of naphthalene ring compounds in the fraction is less.

With moderate or high severity cracking the density of the 400°–550°F fraction will approximate 1.00 with all higher fractions being heavier. The presence of these "heavy oils" which sink in water complicates the design of quench systems in pyrolysis units.

The 550°F+ pyrolysis oil fraction will normally be a solid tar at room temperature. However, distillation of this material yields additional heavy oils in the 550°–1000°F boiling range (7). Analysis of these oils indicates the full spectrum of higher homologs of the aromatics present in the lower fractions plus additional series such as phenanthrenes, fluorenes, and others. Such oils prepared from pyrolysis tars are essentially totally aromatic, and in many respects they approximate the corresponding materials obtained from coal tar distillation. However, they are more completely hydrocarbon in nature, being essentially free of oxygen and nitrogen compounds and quite low in sulfur. These oils are totally unlike the heavy oils, bright stocks, etc. normally prepared from virgin petroleum cuts, being essentially free of paraffins, naphthenes, or asphaltenes.

The ultimate pitch product is a black tar best characterized by its softening point. This can vary from below room temperature to above 200°F by control of front-end distillation. Its specific gravity is normally 1.15–1.20, and it is incompatible with asphalt or residual fuel oil. It finds markets as a raw material for petroleum coke or carbon black and for various specialty end uses. It is a preferred feed for needle coke manufacture. It can also be used as a low sulfur fuel oil.

Economic Aspects

The economic health and profitability of the petrochemical refinery depends upon the revenue from all products produced sufficiently exceeding the cost of all raw materials plus other expenses to yield an adequate profit. This is true of all economic enterprises. Focusing on this aspect, we examine Figure 1 where U. S. plant net market values of the various products under consideration are shown in an ascending scale from the lowest valued product—fuel gas—to the highest—isoprene monomer. Product columns are shown for the petrochemical refinery and for a typical fuels refinery. The third column positions the basic raw material crude oil and its direct distillation fractions. The scale is logarithmic, purely for convenience. A typical petrochemical refinery unit producing olefins and aromatics *via* pyrolysis or pyrolysis plus naphtha reforming yields only about 65% of products having a value greater than the crude oil fed. For the modern fuels refinery this figure exceeds 80%. The burden of large yields of low value by-products weighs heavily upon the economics of any processing scheme.

To examine more completely economic aspects of the petrochemical refinery, evaluations were made of integrated grass-roots units varying from a simple basic olefins producer to a rather completely developed "all-chemical" petrochemical refinery. These are diagrammed in Figures

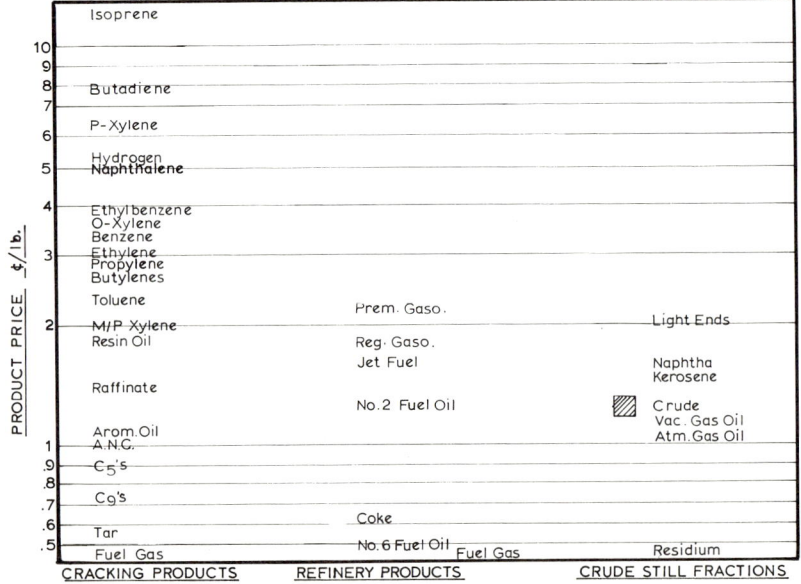

Figure 1. Relative values for U.S. (Gulf Coast) petrochemical and fuel refineries

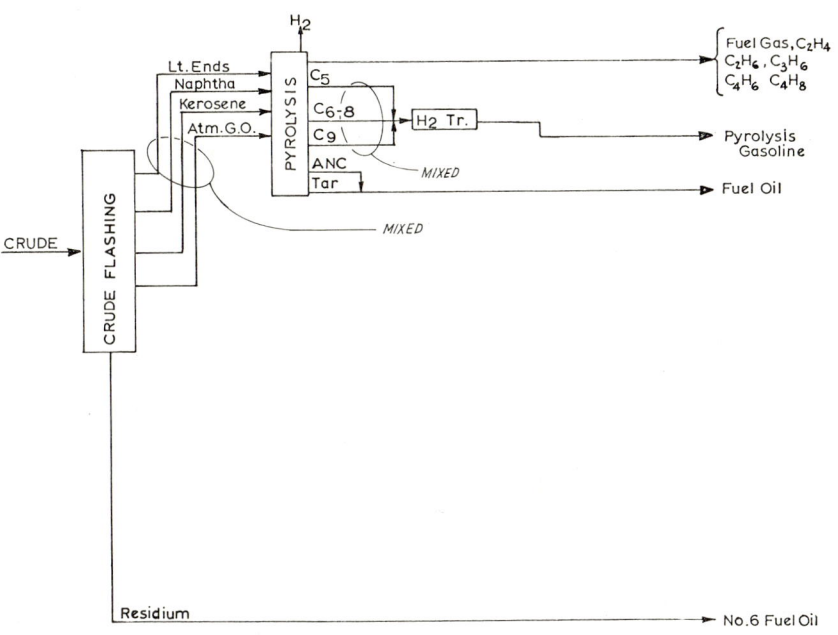

Figure 2. Basic petrochemical refinery (Case 01)

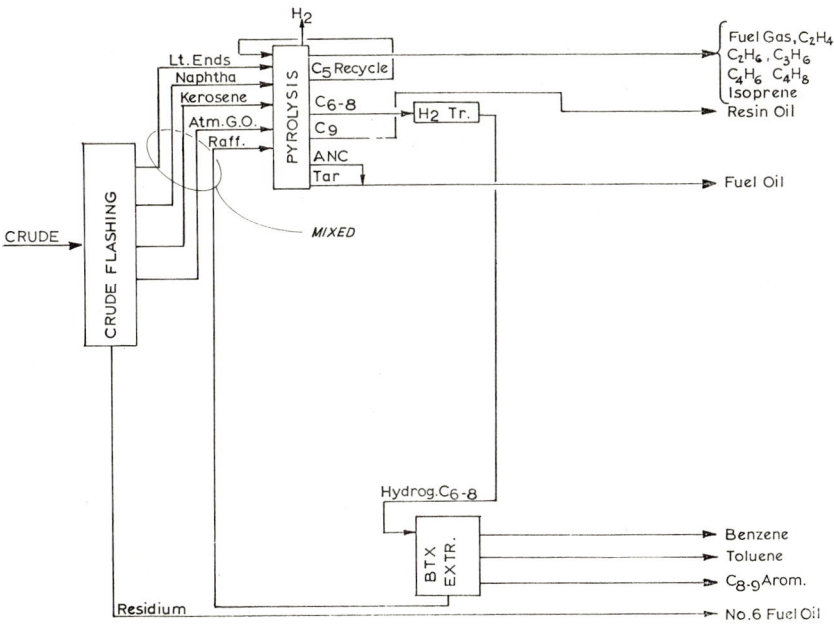

Figure 3. Flow sheet for adding BTX (Case 02)

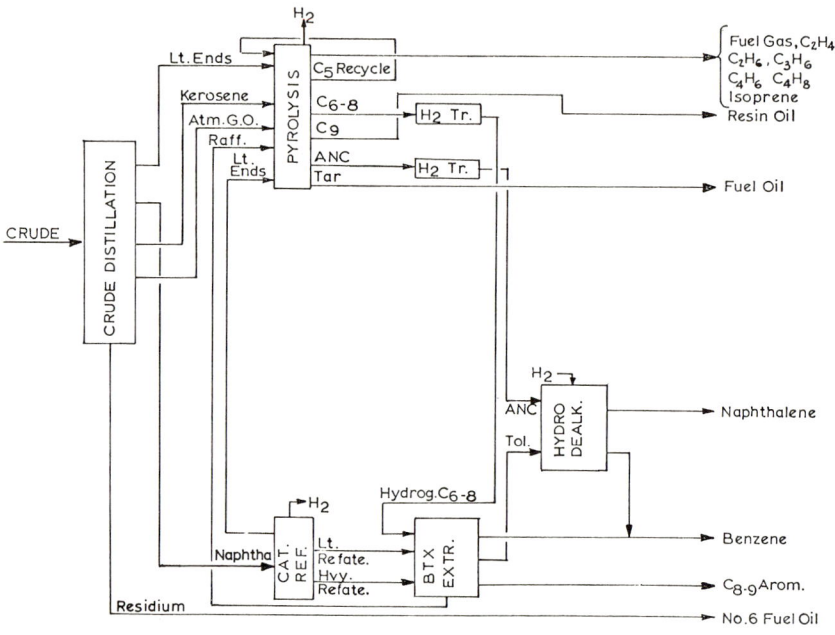

Figure 4. Flow sheet for adding HDA and reforming (Case 04).

2, 3, and 4. The study was extended further (discussed later). From these cases several generalizations regarding petrochemical economics can be drawn.

In general, the weighted average product value for the petrochemical refinery product slate lies between 2.1 and 2.6 cents/lb. Typical fuels refineries might show 1.6–1.7 cents/lb. The comparative margins over crude oil cost might be 0.8–1.2 cents for the petrochemical unit and perhaps 0.4 cent for the fuels unit.

While these figures would seem to indicate a comfortable advantage for the chemical unit, there are offsetting factors. Investment and operating cost factors favor the fuels unit. That these factors combine to render the chemical unit marginal in the U. S. economy is borne out by the fact that no crude based all-chemical petrochemical refinery has yet been constructed in the United States.

Figure 5 is a repeat of Figure 1 for European conditions and reveals a very different situation. The European petrochemical refinery product slate is only a little less valuable than in the United States largely because of the high value of ethylene and gaseous fuels. Since the cost of crude oil is appreciably less, the indicated petrochemical margin is greater than in the U. S. economy. In one specific case to be discussed the margin for the same product slate increased from 0.8 to 1.1 cents/lb under European price/feed cost conditions. This was caused almost entirely by the lower European crude oil cost.

In the United States the preferred product slate favors large by-product yields; in Europe maximum ethylene production with low by-product make is favored. Had the European case been modified to provide for maximum ethylene yields, the indicated margin improvement would have been considerably greater.

With the much greater operating margins attainable under foreign economic conditions, the petrochemical refinery in its basic all-chemical configuration will inherently be much more desirable than in the United States. I expect to see the construction of a number of such units in various foreign countries.

Having added support to the generally accepted belief that the petrochemical refinery is desirable under foreign but not under U. S. economic conditions, let us examine some of the reasons for this difference.

The first and perhaps most obvious reason is the oil imports program which has served to create an artificial barrier maintaining the price of crude oil in the United States about $1.50/bbl above the prevailing price in much of the rest of the world. It is no coincidence that this difference amounts to about 0.5 cent/lb exceeding the gain in margin we saw for the petrochemical refinery under foreign conditions. Since the imports program protects the premium fuels markets as well as the U. S. crude

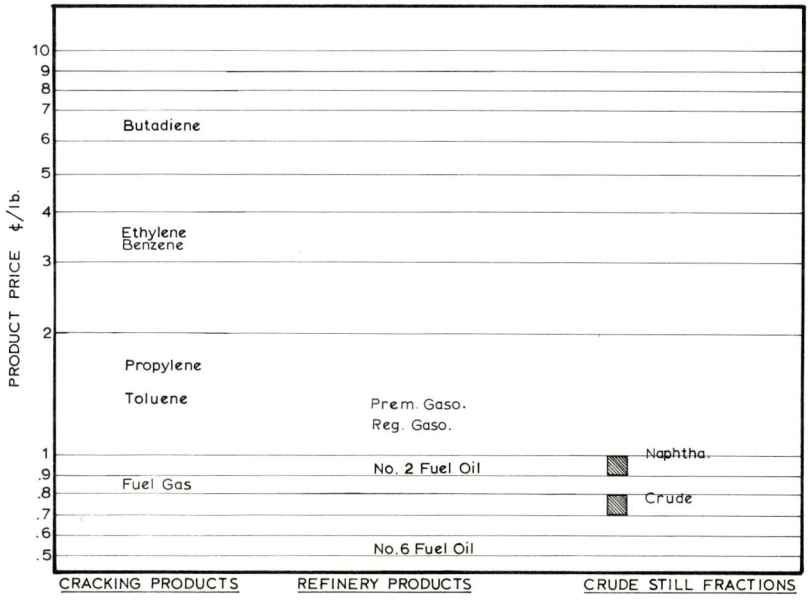

Figure 5. Relative values for European petrochemical and fuel refineries

market, the profitability of fuels refineries has been preserved. The U. S. chemical producer finds no such protection.

Other factors operate to the advantage of the fuels refinery and to the disadvantage of the petrochemical unit. The first is simply the question of divisor. While a typical fuels refinery might run 150,000 bbls/day, the typical chemical refineries considered later run 30,000–50,000 bbls/day. Yet the two contain the same functional units—notably crude distillation and catalytic reforming, though on a very different scale. Clearly the petrochemical refinery must attempt to improve its divisors if it is to become more competitive.

Consideration of the nature of the petrochemical refinery itself gives some clues as to another source of its profit problem. In the simple, basic unit depicted in Figure 2 thermal cracking dominates the operation. Over 90% of the crude input is consumed without regard to relative values. Thus, it is an indiscriminate cracker of butanes, light naphtha, heavy naphtha, kerosene, distillate, atmospheric gas oil, and vacuum gas oil. Since acceptably similar product slates can be obtained from many of these fractions, it is obvious that the economics suffer when the high valued naphtha and kerosene fractions are thermally cracked.

The addition of BTX operations as in Figure 3 does nothing to alleviate this situation. Even the rather completely developed petrochemical refinery of Figure 4 (Case 04), still suffers considerably from the problem.

Recapping the problems of the simple petrochemical refinery in the United States, they are:

(1) The artificially high cost of crude oil owing to the imports program.

(2) Small divisors on many units compared with the fuels refinery.

(3) Inadequate chemical markets for some of the products.

(4) Indiscriminate cracking of crude fractions without regard to alternate use values.

Recognizing these problem areas, I suggest these approaches to improve the economics of the petrochemical refinery under U. S. economic conditions:

(1) Build only units of competitive divisor as judged by fuels refinery standards.

(2) Convert all problem by-products and other available recycle or intermediate streams to high value, large volume, protected market, premium fuels—*i.e.*, gasoline or jet fuel.

(3) Thermally crack only those hydrocarbon fractions having the lowest value in the U. S. economy.

This can be accomplished if we accept one fundamental change in our concept—*i.e.*, permit liquid fuels as major products. Most of the difficultly marketable products are either blendable into gasoline or convertable to gasoline by standard processing steps. At the same time all of the more valuable crude fractions owe their value to their gasoline or jet fuel potential. Hence, marketing these fractions in their logical markets would raise profits. The volume production of premium fuels will also alleviate much of the divisor problem.

Model Cases

The economic evaluation cases previously mentioned examine the points suggested with respect to several configurations of the petrochemical refinery. Case 01 is the basic olefins unit (Figure 2) which disposes of its C_5–C_9 fractions as a semihydrogenated "pyrolysis gasoline." The higher boiling fractions are sold as fuel oil. Butadiene separation is considered to be a part of the basic olefins operation.

Case 02 adds BTX recovery (Figure 3). This requires separating the C_5 fraction; isoprene monomer concentrate is added as a product, and the remaining C_5's are recycled to the furnaces. A C_9 fraction (resin oil) is also separated and marketed.

Case 04 (Figure 4) adds hydrodealkylation of both toluene and the alkylnaphthalene concentrate fraction. To produce the necessary hydrogen as well as to increase BTX production catalytic reforming of all naphtha in the crude is added. Case 04 represents a relatively complete development of the "all-chemical" configuration of the petrochemical refinery.

These cases were solved on the computer using our nonlinear simulation routine which is capable of modeling plants of this type extremely precisely. No attempt was made to optimize cracking conditions; all cases used basic low severity, low ethylene-yielding high by-product ratio furnace conditions. However, each furnace feed was given a separate cracking pattern permitting measurement of the economic effect of the shifting furnace feed slate. The program computes utilities and other elements of conversion cost as a function of throughput section by section. The capital investment in each unit is also computed using the conventional power function of capacity.

Basic cost and price data from Figures 1 and 5 were used in the computations. The results of six pertinent cases for the "all-chemical" refinery are contained in Table V.

The capital investment figures assume the development of a new site with all necessary auxiliary facilities such as loading docks, storage tanks, utilities generation, cross-country pipelines, underground storage, etc. In addition, an allowance for working capital at 10% sales is in-

Table V. Investment/Return Data for All-Chemical Refineries

Case	Total Capital, $M	Pre-Tax Return, %
	U.S. Gulf Coast	
01[a]	84.1	3.5
02[b]	93.3	6.5
04[c]	114.4	9.8
	Europe	
01E[a]	75.8	12.8
02E[b]	84.1	16.5
04E[c]	103.3	20.5

[a] Base
[b] Add BTX
[c] Plus dealkylation and refining

cluded. The indicated return-on-investment figures are simple investors' rate of return and are calculated prior to federal income taxes.

It must be stressed that this study was designed to examine the specific economic points under discussion. It was not practical to optimize the cases in other respects. In particular, no attempt was made to optimize cracking severity, cracking furnace conditions, or energy recovery. Thus, the cases are definitely conservative in their reported earnings potential.

The data of Table V confirm my previous statements about the relative economics of the petrochemical refinery under U. S. and foreign conditions.

All cases are basically oriented toward the U. S. economic situation. The European values were obtained by substituting European product pricing, crude oil cost, and utilities costs in the computer inputs. Plant capital was taken as 10% less than the U. S. figure. A cost adjustment was also made to reflect this capital difference. However, no effort was made to optimize the cracking situation for that environment. By using high severity cracking and maximum recycle, the European cases could be improved considerably.

Seeking the effect of changes in the basic orientation which I have suggested, I ran a series of cases involving the production of gasoline as a major co-product and finally also the production of jet/turbine fuel.

Case 05 (Figure 6) represents a minimal gasoline production. Here all of the raffinate, crude still light ends, and reformer light ends were diverted to form the nucleus of the gasoline pool. In addition, the recycle C_5's and the excess resin oil fractions were hydrotreated and added. For balance, part of the xylenes–C_9 aromatics mixture was also included. Case 06 adds the entire heavy reformate fraction to the gasoline pool without extraction. Figure 7 shows the flow pattern. Case 07 (Figure 8)

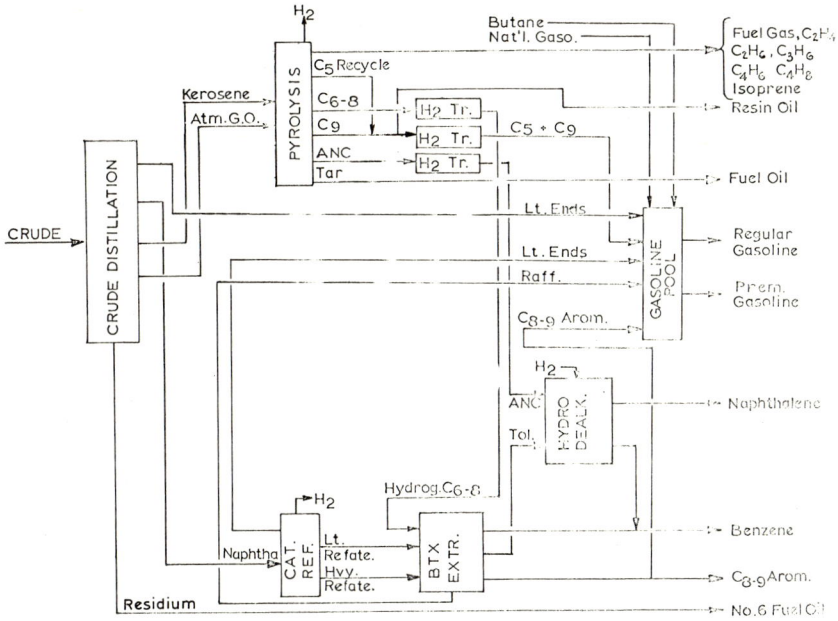

Figure 6. Flow sheet for minimum gasoline case (Case 05)

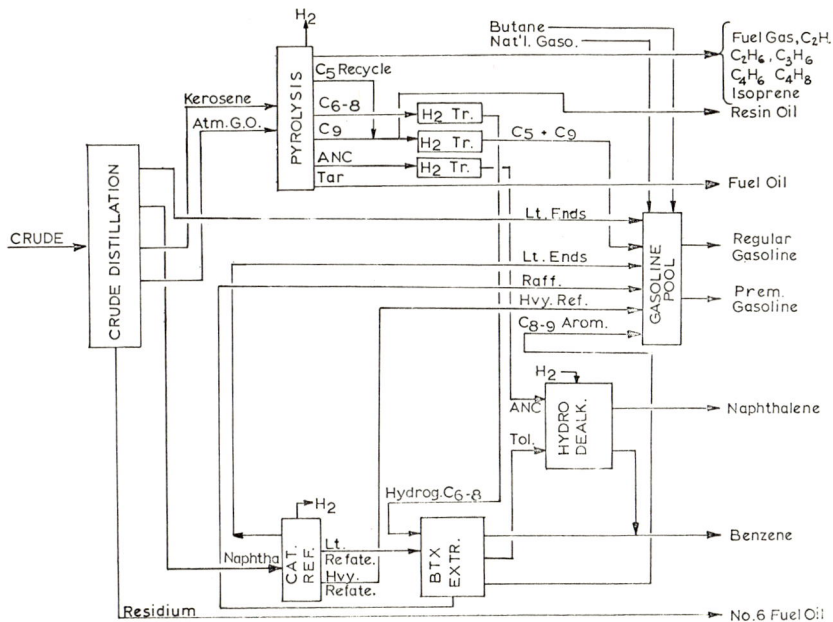

Figure 7. Flow sheet for maximum gasoline case (Case 06)

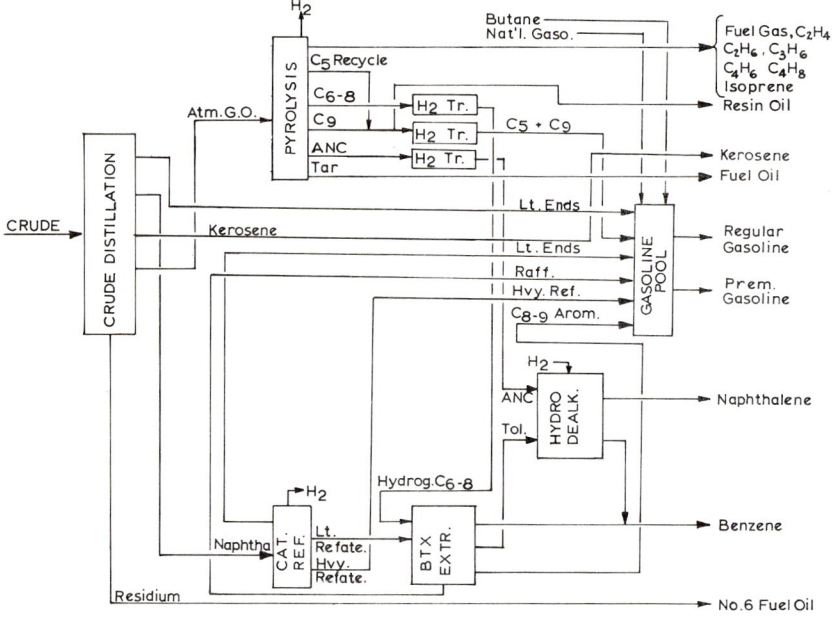

Figure 8. Flow sheet incorporating kerosene for sale (Case 07)

Table VI. Fuel Co-Product Cases

Case	Gasoline bbls/day	Total Capital, M	Pre-Tax Return, %
		U.S. Gulf Coast	
04	0	114.4	9.8
05[a]	12000	126.5	12.8
06[b]	20000	127.8	15.2
07[c]	30000	159.1	17.0
		Europe	
04E	0	103.3	20.5
05E[a]	12000	113.4	21.1
06E[b]	20000	114.6	23.5
07E[c]	30000	141.6	27.5

[a] Minimum gasoline
[b] Maximum gasoline
[c] Plus 33000 bbls/day kerosene sold

diverts the kerosene fraction from furnace feed to jet fuel sales, but otherwise the flow plan is identical with Case 06 (Figure 7). These cases are summarized in Table VI. For comparison the best all-chemical case (04) is also included.

Clearly the co-production of the premium liquid fuels has greatly improved the economics of the petrochemical refinery in the U. S. environment. The gains for the foreign situation are much less dramatic—in one case negligible.

Tables VII and VIII analyze rather precisely what has been achieved to change the profit so sharply. Table VII gives the divisor effect. Note that the divisors on the crude still and reformer in Case 04 are marginal; in Case 07 they are fully competitive by fuels refinery standards. The production of co-product benzene has also increased tremendously.

Table VII. Unit Divisors

Case	Ethylene Unit[a]	Crude Still[b]	Reformer[b]	Benzene[c]
01	750	36500		
02	750	43500		38
04	750	51500	14000	102
05	750	62500	17000	108
06	750	65000	17600	112
07	750	115000	31000	146

[a] $\overline{M}$ lbs/yr
[b] bbl/day
[c] $\overline{M}$ gal/yr

Table VIII. Pyrolysis Furnace Feeds

	Bbl/Day	Wt %
Case 04		
Crude light ends	1490	3
Reformer light ends	2060	4
C_5 recycles	1260	2
Raffinate	6300	13
Kerosene	14900	34
Atmospheric gas oil	19100	44
Case 06		
Kerosene	17150	43
Atmospheric gas oil	23200	57
Case 07		
Atmospheric gas oil	41000	100

Table VIII shows the change in cracking furnace feed composition as we step up the series. In Case 04 this feed was composed of high value fuels components to the extent of 56%. In Case 07 no premium fuel fractions were cracked.

While these cases were based upon a light crude feed, other cases which have been run using heavier crudes show much the same result.

For crudes yielding appreciable fractions of vacuum gas oil a hydrocracker was used to process this material to lighter fractions which were blended with the corresponding virgin crude streams. In general, such cases are somewhat poorer where chemicals only are produced and somewhat better where gasoline is co-produced.

The fully developed petrochemical refinery as modeled in Cases 06 and 07 represents a complete melding of the fuels refinery with olefins–aromatics processing where the efficient scale of the fuels operation is retained while the chemical units provide the best possible end use for the lower valued hydrocarbon streams. At the same time they return certain products whose most valuable use is as components of the gasoline pool.

Summary

The petrochemical refinery without premium liquid fuels is attractive in the foreign economic environment. To achieve an attractive configuration in the U. S. economy, such fuels are essential products. They help provide competitive unit divisors and take advantage of the protected markets created by the oil imports program. I believe that this is the direction the U. S. industry will take. If the chemical producers don't take the lead, the fuels producers certainly will.

Literature Cited

(1) Beavon, P. K. *et al.*, *Chem. Eng. Prog.* **July 1969**, 65, 23.
(2) *Chem. Eng. News* **March 11, 1963,** 21.
(3) Frank, S. M., *Oil Gas J.* **Nov. 22, 1965,** 76.
(4) Freiling, J. G. *et al.*, *Hydrocarbon Processing* **Nov. 1968,** 145.
(5) Gambro, A. J. *et al.*, *World Petrol.* **April 1967,** 88.
(6) Gambro, A. J. *et al.*, *World Petrol.* **Dec. 1968,** 41.
(7) Monsanto Co., unpublished data.
(8) Rowell, R. L., AIChE Meeting, New Orleans, La., March 1969.
(9) Struth, B. W., AIChE Meeting, New Orleans, La., March 1969.
(10) Spritz, P. H., *Oil Gas J.* **Sept. 2, 1968,** 108.

RECEIVED December 16, 1969.

9

The Manufacture of Propylene

ALVIN H. WEISS

Department of Chemical Engineering, Worcester Polytechnic Institute, Worcester, Mass. 01609

> *Propylene demand will grow to the 11-billion lb level by 1973. Propylene from either heavier ethylene feed stocks or European imports will not alleviate the shortage completely. On the other hand, it is not expected that price will exceed 3.1 cents/lb. In spite of decreasing propylene availability, refiners will consider release of alkylate stocks at this level. Development of an economic process for direct propylene production is in the future. Dehydrogenation or iodinative partial oxidation processes for propylene from propane are neither commercially proved nor have they been demonstrated to have economic promise. Dehydrogenation in the presence of sulfur may bypass propane dehydrogenation equilibrium limits, and preliminary experimental data are presented.*

A decade ago there was little belief that a propylene demand would arise. There was no doubt that its use for petrochemicals and plastics would grow, but propylene production was thought to be entirely adequate.

There are two main sources of propylene production—"refinery" and chemical." The former is derived from catalytic cracking and is used mainly for refinery purposes—*i.e.*, polymer gasoline and alkylate. The latter is derived from ethylene plants and is marketed mainly for petrochemical usage. In both cases, the propylene is a by-product and not a directly manufactured product. In 1963 Davis (*19*) described propylene as the "bargain olefin" at 2.25 cents/lb and predicted no shortage in sight. He pointed out that 1962 total domestic refinery derived propylene capacity was 17.1 billion lbs annually, and chemically derived propylene was 3.4 billion lbs annually. All of the propylene cannot be recovered economically. Davis estimated that "available" propylene amounted to 17 billion lbs, of which chemical uses constituted

3 billion lbs. He predicted the growth in catalytic cracking would add a billion lbs/year to the pool, while chemical uses would only increase from 3.0 billion lbs in 1962 to 3.6 billion lbs in 1965.

However, by 1965, *Chemical and Engineering News* (*C & EN*) pointed out that domestically available propylene was 18 billion lbs and that chemical propylene consumption had reached 4.6 billion lbs (*10*). A severe shortage was in sight. Besides burgeoning propylene use, the shortage was also a result of the unexpected changeover to zeolitic cracking catalysts, which by their very nature produce less propylene than do silica alumina catalysts. In addition, hydrocracking entered the scene, and this growth represents a situation where no propylene at all is produced from gas oil. Octane demands were resulting in propylene being used for alkylate; and the only balancing factor was the phasing out of a certain amount of "poly" gasoline capacity because of the product sensitivity (difference between research and motor octane numbers).

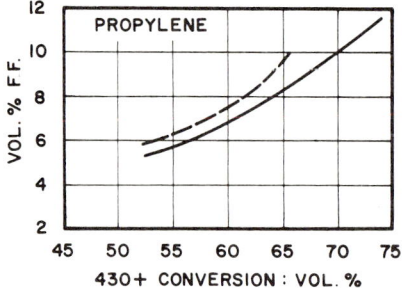

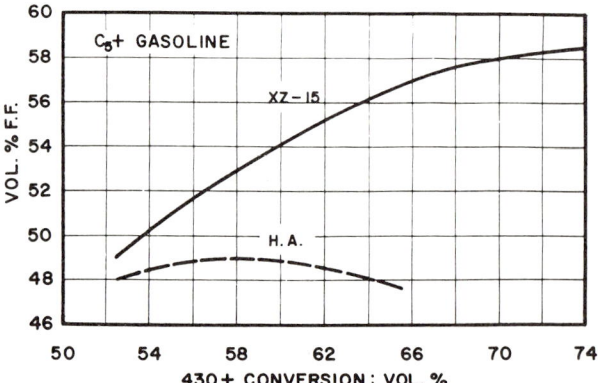

Figure 1. Comparison of Davison fluid catalysts for cracking west Texas gas oil. Lower propylene yields and higher gasoline yields result with molecular sieve XZ-15 vs. high activity silica alumina (5).

Table I. Comparison of Molecular Sieve and Kaolin Catalyst Cracking Evaluations (39)

Charge stock	Heavy Sweet Gas Oil Plus Recycle	
Reactor temperature, °F	915	
Recycle in feed, vol %	21	
Catalyst	Mol. Sieve	Kaolin
Yields (fresh feed basis)		
Conversion, vol %	70.0	70.0
Gasoline (385°F at 90%), vol %	56.7	47.3
C_4 cut, vol %	13.3	16.3
Dry gas, wt %	8.5	10.6
Coke, wt. %	2.2	6.7
C_4 cut breakdown, vol % of C_4's		
n-Butane	14	12
Isobutane	41	41
Total Butenes	45	47
Dry gas breakdown, wt. % of Gas		
Propane	21	21
Propylene	31	35
C_2 and lighter	47	44
Gasoline research octanes		
Clear	89.5	92.0
+3 cc TEL	97.0	97.7

The effect of the molecular sieve catalysts on the propylene picture can be seen quantitatively by examining comparative data for both conventional and molecular sieve catalysts. Baker *et al.* (5) presented an in-depth comparison of performance of Davison fluid catalysts for West Texas gas oil cracking. Figure 1 shows their plots of gasoline yield and propylene yield for their molecular sieve catalyst XZ15 and their conventional high alumina fluid cracking catalyst. Not only does the molecular sieve intrinsically produce less propylene at a given conversion, but the effect is magnified at a given gasoline production.

Mills, Ashwill, and Gresham (39) compared pelleted Houdry kaolin and sieve catalysts for cracking heavy sweet gas oil. Table I shows their results obtained at constant conversion of 70 vol %. Results correspond to those for fluid catalyst: 9.4 vol % additional gasoline results, and propylene yield is reduced from 3.7 to 2.6 wt %.

Because of the impact of the sieves and of hydrocracking Ockerbloom and Stuart (43) of Sun Oil have projected, on the basis of the gasoline market, that economically recoverable propylene from refineries will remain relatively constant—in the range of 14 to 16 billion lbs/year through 1975. Figure 2 shows their prediction; the upper limit results from operations to maximize conversion to gasoline, and the lower limit results from operations to minimize feed capacity. They also predict that

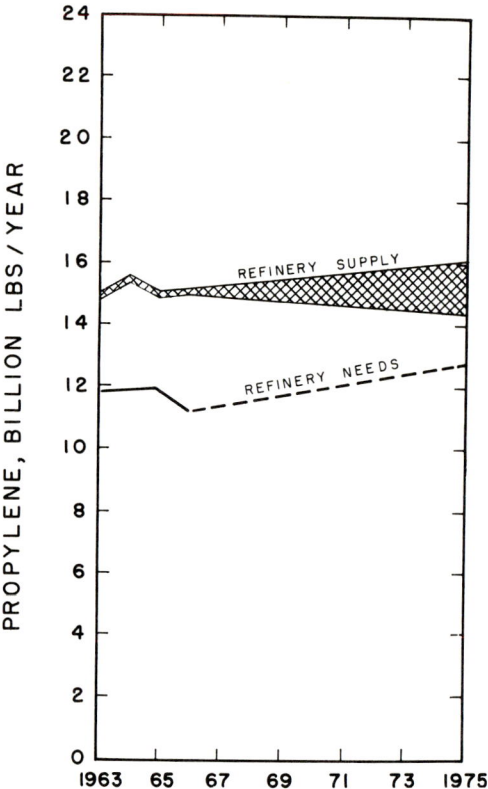

Figure 2. Propylene available from refineries will remain at the 2 billion lb/year level (43).

refinery needs for poly gasoline and propylene alkylation will remain relatively constant at the 12 billion lbs/year level.

Attention to the propylene shortage had crystallized with the publication of a major study by Stobaugh (56) in 1967, in which he analyzed thoroughly the sources and markets for propylene. He suggested that by 1970 over-all propylene production from both refining and chemical sources would probably not exceed 19 billion lbs, and chemical demand would have increased to 7.5 billion lbs at a value of 2.5 cents/lb. Stobaugh lists both U.S. propylene producers and their plant capacities as well as propylene consumers.

Ockerbloom and Mitchell (11, 43) pointed out that the shortage predicted by 1970 could be aggravated by factors such as growth of propylene chemical demand even beyond forecast levels and by legislation against lead alkyls.

$C \& EN$ (12) has listed propylene markets which show that by 1968 Stobaugh's 7 billion lb prediction had been met and that price had risen past 2.5 cents to 3.0 cents/lb for chemical grade propylene (90–95% purity). Table II shows the use breakdown for propylene and also a prediction of 11 billion lbs chemical demand by 1973.

At present there is no question that the propylene shortage is a reality, and there is need to assess the potential as well as the actual sources of propylene. In this respect, the price that propylene should seek will provide a frame of reference. Consider that the gasoline value of propylene equals approximately 2.8 cents/lb (12.5 cents/gal ÷ 4.35 lb/gal) and that propylene recovery costs should not exceed 0.2 cents/lb. The total—3.0 cents/lb—is a break-even area for a refiner. He will be sorely tempted to release propylene above this price and satisfy gasoline commitments by other means. The exact price, of course, will vary for each refinery and will be a function not only of its raw materials *vs.* sales picture but also of its accounting techniques. The 3.0 cents/lb figure could vary by as much as ±0.5 cents/lb, and will depend on surplus of refining capacity (pool octane), isobutane, refining processes in place, etc.

Even so, 3.1 cents/lb to 3.2 cents/lb appears to be the area (at 1969 gasoline price levels) for eventual propylene stabilization which will be affected by supply-demand response. Justification of propylene expansion on the basis of selling propylene at an eventual 3.5 cents/lb to 3.9 cents/lb market price will be very risky.

Unless restrictions are eased, imports will probably play only a small role in satisfying propylene demand. European propylene (12) can be

Table II. Propylene Demand (12)

Product	Propylene Demand (millions of pounds)	
	1968	1973
2-Propanol	1550	2000
Acrylonitrile	1300	2000
Polypropylene	900	1800
Propylene oxide	800	1550
Heptene	750	750
Cumene	500	830
Butyraldehyde	450	700
Dodecene	350	250
Nonene	300	350
Glycerine	130	170
Epichlorohydrin	100	200
Isoprene	100	150
EPDM rubber	30	100
Acrylic acids	10	200
Miscellaneous	50	100
Totals	7320	11,150

Table III. Effect of Increasing Ethylene Yield

Feedstock	Change Feed	
	Propane[a] →	Ethane
Coil temperature, °F	1460 →	1515
Coil outlet pressure, psig	10.0 →	10.5
Steam-to-feed mole ratio	0.4	1.1
C_3H_6/C_2H_4 mole ratio	0.32 →	0.02

[a] Ethane recycled to extinction

purchased for 4.5 to 9 cents/gal, shipping costs are 1 to 3 cents/gal, and the import ticket costs 3 cents/gal. Consequently, depending on the situation, foreign material can be had from 2 cents/lb to 3.5 cents/lb. The lower figure is not as likely as the higher, and the importer's naphtha quota will be affected adversely. This latter will, in effect, reduce the propylene produced from pyrolysis units in this country.

Besides catalytic cracking, the only other major source of propylene in the near future will continue to be as a pyrolysis by-product. Even an all-chemical coal refiner will make little propylene. OCR estimates (*13*) that a 100,000 bbl plant will only produce 28MM lbs/years of propylene. The areas of significance that must be considered for now are the pyrolysis by-product processes.

Pyrolysis Processes for Ethylene

There is a wealth of information available on ethylene, and the reader can obtain quick technical entree to the field from the review by Davenport (*18*). Flow sheets and descriptions of commercially licensed processes are published in the "Petrochemical Handbook" issues of *Hydrocarbon Processing*. Processes are licensed by C. F. Braun, M. W. Kellog, Lummus, Selas, Petrocarbon Developments, Stone and Webster, Lurgi, and Badische Anilin- & Soda-Fabrik. These are all pyrolysis processes which operate in the presence of steam to suppress coking. Cracking furnaces are used in all except the last two processes, which are fluid bed. The Lurgi process uses sand as a heat carrier, and Badische uses coke formed from the charge stock. The advantage of the fluid beds is that both processes are capable of using crude oil as feed. It is doubtful if crude oil pyrolysis will be economic in the United States in the near future.

Burke and Miller (*8*) published in 1965 a two-part report on ethylene which provides an overview of the world ethylene producers, plants, capacities, feedstocks, and economics. Much of their information is augmented and updated by an economic study published by Freiling, Huson, and Summerville (*22*) of Lummus. Growth in ethylene demand is ex-

on Propylene Yield Furnace Operations (54)

Increase Severity

n-Butane			73.2° API S.R. Distillate		36–38° API Gas Oil	
1415 →	1435 →	1448	1365 →	1410	1335 →	1375
6.8	6.8	6.8	12.5	15.3	8.5	6.5
0.28	0.32	0.35	3.05	7.5	4.0	7.9
.77 →	.53 →	.47	0.57 →	0.32	0.52 →	0.35

pected to increase by 10%/year—from 16 billion lbs/year in 1970 to 25 in 1975 to 36 in 1980. This 10% rate is confirmed by Sawyer (52) from U. S. Tariff Commission data and by Prescott (46) from predictions by Union Carbide, Shell, and Enjay.

Freiling *et al.* estimated that 30 billion lbs/year new capacity would have to be built to satisfy the 20 billion lbs/year demand increase from 1970 to 1980. This accommodates retiring plants more than 15 years old (8 billion lbs capacity) and a 2 billion lb installed capacity excess. The industry is presently 2 to 5 billion lbs over capacity. The retirement of old plants is disputed by Prescott, who cites examples of operational 30-year-old plants.

To relate these ethylene statistics to propylene, consider the position of the older and therefore smaller ethylene producer. As his commitments for ethylene grow, with a fixed capacity unit he has no choice but to shift from propane to ethane feed or to operate at higher and higher severity with a heavier stock. All things being equal, this reduces propylene yield for either a given unit or for a given feedstock but does increase ethylene production by at least 10% in a plant.

Table III shows the effect of shifting furnace operation from propane fresh feed to ethane. Data are from Schutt and Zdonik (54). The reduction of propylene yield from ethane to negligible levels in favor of increased ethylene production cannot be done if a plant has propylene commitments. Because propylene requirements cannot be satisfied with ethane feed, Ericsson (14) has concluded that propane will continue to be the preferred feedstock to make ethylene. Actually, 85% of the U.S. ethylene plants are located in the Gulf Coast area so that they can obtain and operate on economical ethane and propane feeds. The need for propane pyrolysis has resulted in a renewal of experimental interest in this area, and in-depth studies have been made by Crynes and Albright (17) and by Buekens and Froment (7).

Table III also shows how propylene yield is reduced if ethylene production of a furnace is maximized by increasing severity for a given stock. Examples chosen are *n*-butane, straight run distillate, and gas oil.

There is the possibility of simultaneously increasing ethylene and propylene yield in a unit by judicious switching to a lighter feed stock.

Table IV. Effect of Lighter Feedstock on Ethylene Production Fluid Bed Unit (28)

Feedstock	Cracking Temperature, °F	C_3H_6/C_2H_4 Mole Ratio	Capacity of One Unit, metric tons/year	
			Maximum Ethylene	Propylene[a]
Iraq Crude Oil	1346	.43		
	1400	.37	40,000	22,000
Kuwait Heavy naphtha	1517	.29		
	1562	.17	60,000	15,000
Kuwait Light naphtha	1427	.40		
	1526	.28	70,000	29,000
Propane	1418	.39		
	1436	.20	100,000	31,000
Ethane	1562	.02		
	1508	.00	110,000	4,000

[a] At maximum ethylene capacity.

Table IV shows such data for a Lurgi sand cracker (28). Even though propylene-to-ethylene mole ratio is reduced by switching from naphtha to propane feed, propylene production from a unit increases. Unfortunately, most processors would not have this feedstock flexibility.

The general effect, then, as a refiner's ethylene demands increase, is a severity increase on a given feedstock with a resulting decrease in propylene production. At some point he will have to build new facilities. When the new plant is built, the design capacity will be undoubtedly greater than his current needs, and his new unit will be placed on stream —possibly at 70–75% of design ethylene capacity. The old unit would presumably be moth-balled until such time as the new facilities cannot meet production demands.

Consider what has happened—we must have just gone from famine to feast in propylene. An over-capacity unit can operate at low severity to optimize propylene yield. Eventually, the new unit will operate at high severity, and then the old unit can be returned to operation. However, by-product demand is such that it is not unreasonable to contemplate the possibility of adapting in the interim some older units to *n*-butane feed. *n*-Butane is presently in oversupply (15), and more propylene can be made from it than from propane (see Table III). Table V presents data of Freiling et al. (22) for various feed stock requirements and by-product productions per pound of ethylene. If any existing ethylene units are not retired but are switched from ethane or propane to *n*-butane feed, one can immediately prorate feed rates and calculate that for every billion pounds ethylene capacity, an additional 111 MM lbs of propylene can be made from former propane units and 236 MM lbs additional propylene from former ethane units. Some fraction of Freilings' estimated 8 billion lbs ethylene capacity will be retired by 1980 but not all. Additional propylene quantities can be counted upon from switching these existing units to n-butane feed.

The main answer to the propylene situation lies in this demand-over-capacity picture for ethylene and in the nature of the new facilities that will be built. Consider that the size for economic ethylene plants has grown from the 75 million lb level in the 1950's to the present 500 billion lb level. Even 1–5 billion lb annual ethylene capacity units are predicted. Lambrix, Morris, and Rosenfeld (35) of Kellogg estimate that the investment for billion lb/year plants operating from heavy oil feedstocks is in the range of \$60 to \$69 MM. Investment in ethylene production is entering the range for building a medium sized refinery, and indeed these plants are regarded as "chemical" refineries, whose economic justification is based on by-product recovery.

Table V shows demand ratios of by-products to ethylene that have been projected by Freiling through 1980. Propylene demand will not grow as fast as ethylene, and the present level of 0.59 will reduce to 0.49 lb propylene/lb ethylene. n-Butane, naphtha, or gas oil feed can satisfy this, and thus, in all probability, much of the 30 billion lbs of new ethylene capacity will be from heavier feeds. Naphthas and gas oils are presently economical only in Europe.

n-Butane will be a feedstock for some new construction. Naphtha will probably continue in short supply, but the gas oil picture is receiving considerable attention, and Lambrix et. al. (35) showed that a billion lb/year of ethylene gas oil "chemical refinery" can be a very profitable venture even at Gulf Coast prices. For the right feedstock, a return on investment as high as is 36% calculated. Cijfer, Kitzer, and Wall (16) of Selas state that even plants now in current construction for operation on naphtha feed are being rated on gas oil in anticipation of future feed shortages. They provide extensive yield data from gas oil feeds and project that at the 450 million lb/year ethylene production level, the balance point for gas oil vs. full range naphtha feedstock occurs when gas oil cost is approximately 70% of naphtha cost (on a weight basis).

Isobutane Cracking

In the preceding section it was noted that an ethylene producer operating to maximize ethylene production would produce significantly smaller quantities of propylene. If he, or even a refiner without ethylene facilities, were to have simultaneous commitments for propylene and

Table V. Effect of By-Product Production

Feedstocks	Ethane	Propane	n-Butane
Feed requirements	1.31	2.38	2.50
Propylene	.04	.38	.52
Butadiene	.02	.08	.07
Aromatics	.00	.08	.06

[a] All figures in lb/lb ethylene produced.

isobutylene for alkylate production or other use, it should be advantageous to consider mild steam cracking of isobutane to isobutylene and propylene. Isobutane is often available for this purpose owing to the higher ratio of isobutane to n-butane produced by hydrocrackers and by catalytic cracking units using molecular sieve type catalysts. Several months ago Coastal States Petrochemical Co. started operating an isobutane steam cracker, designed and constructed by Foster Wheeler Corp. It is reported that Foster Wheeler is licensing the Coastal States' isobutane cracking process.

The following five independent reactions describe the production of gaseous products from pyrolysis of isobutane:

$$i\text{-}C_4H_{10} \rightleftarrows i\text{-}C_4H_8 + H_2$$

$$i\text{-}C_4H_{10} \rightleftarrows CH_4 + C_3H_6$$

$$H_2 + C_3H_6 \rightleftarrows C_3H_8$$

$$C_3H_8 \rightleftarrows C_2H_4 + CH_4$$

$$H_2 + C_2H_4 \rightleftarrows C_2H_6$$

Extrapolation of thermal data of Egloff, Thomas, and Linn (20) suggests that $i\text{-}C_4H_{10} \rightarrow C_2H_4 + C_2H_6$ is not a primary reaction of significance (at 1202°F and both 14.2 and 99.5 psia). Also a reaction suggested by Tropsch, Thomas, and Egloff (57) $i\text{-}C_4H_{10} \rightarrow i\text{-}C_4H_8 + C_3H_8 + CH_4$ is not necessary to describe the production of gaseous products.

The primary products produced in isobutane cracking are propylene and isobutylene. Propane, ethylene, and ethane and methane are produced in subsequent secondary reactions as well as significant amounts of liquid products (dripolenes).

The isobutane pyrolysis studies by Egloff, Thomas, and Linn (18) established gaseous product selectivity relationships for operation at 14.2 and 99.5 psia. Unfortunately, liquid products were not reported, and their propylene and isobutylene selectivity data are probably high. Nevertheless, their data are useful for establishing shapes of curves and trends.

Demands on Ethylene Feedstocks[a] **(22)**

Ratios			Demand Ratios		
Full Range Naphtha	Light Gas Oil	Heavy Gas Oil	1970	1975	1980
3.20	3.80	4.29			
.52	.57	.61	.59	.53	.49
.14	.15	.17	.22	.17	.14
.39	.29	.30	.60	.53	.48

To avoid confusion, these data are not included on the figures of this study, but it should be understood that they were used to develop the figures.

More reliable are atmospheric pressure pyrolysis data of Marek and Neuhaus (36), who did account for liquid products. Marek and Neuhaus' data show no significant selectivity differences between 1112°F (+) and 1202°F (▽) (see Figures 3, 4, and 5). Hepp (25) has patented an isobutane cracking process operating at 1200°–1350°F, 130–150 psig.

Figures 3, 4, and 5 present, respectively, mole % i-C_4H_8, mole % C_3H_6, wt % total i-C_4H_8 + C_3H_6 selectivity–conversion plots of Marek and Neuhaus' data. Also included are thermal data of Tropsch, Thomas, and Egloff (57) at 1031°F, 725 psia and of Hepp and Fry (26) at 1022°F, 2500 psia. Not included are data at 1300°F, 215 psia and 1250°F, 515 psia, Hepp's patent, which are roughly the same selectivity levels as Marek's atmospheric data.

Consider the case of 50% isobutane conversion. This is a "magic number," for which product from an isobutane cracker is equimolar in olefins and isobutane (except for a minor total selectivity correction). This mixture could be fed directly to an alkylation unit. Total propylene plus isobutylene weight percent selectivity at atmospheric pressure is 65 wt %. The price spread of isobutane at approximately $1.89/bbl to alkylate value olefins at $4.62/bbl or better (these prices are equivalent to 2.20 cents/lb isobutylene and 2.53 cents/lb propylene) suggests that isobutane cracking may be significant in the near future by alleviating olefin demands for alkylate, in effect, releasing more propylene for chemical consumption.

More to the point of this study is that propylene selectivity at atmospheric pressure is almost constant at approximately 38 mole % across the entire isobutane conversion range (Figure 4). Isobutane cracking could be used to produce chemical propylene directly, and the justification to do this instead of producing propylene for alkylate is only a function of price. The price that propylene would have to command to justify its chemical sale, rather than alkylate use, is basically its gasoline value, which is the 3.0 cents/lb figure developed earlier in this paper.

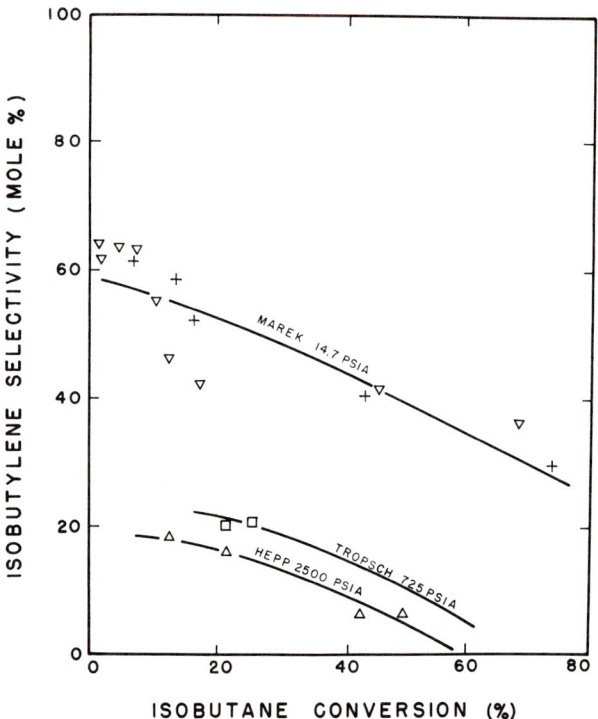

Figure 3. Pyrolysis of isobutane. Isobutylene selectivity decreases with conversion and pressure.

It is recommended that refiners scrutinize isobutane cracking economics to see if chemical propylene manufacture by this route is justified for their particular situation. In all probability, some significant portion of chemical propylene needs can and will be met by special situation isobutane cracking.

The Houdry Catadiene Process

When one contemplates the shortage of propylene and the availability of propane, it is natural to postulate whether catalytic dehydrogenation of propane rather than pyrolysis of propane would be an inherently economic process. Unwanted side reactions are minimized in catalytic processes, and the reaction:

$$C_3H_8 \rightleftharpoons C_3H_6 + H_2$$

appears at first glance to be disarmingly simple. However, it will be shown that equilibrium considerations of this endothermic reaction mitigate against high conversions at temperatures low enough to prevent

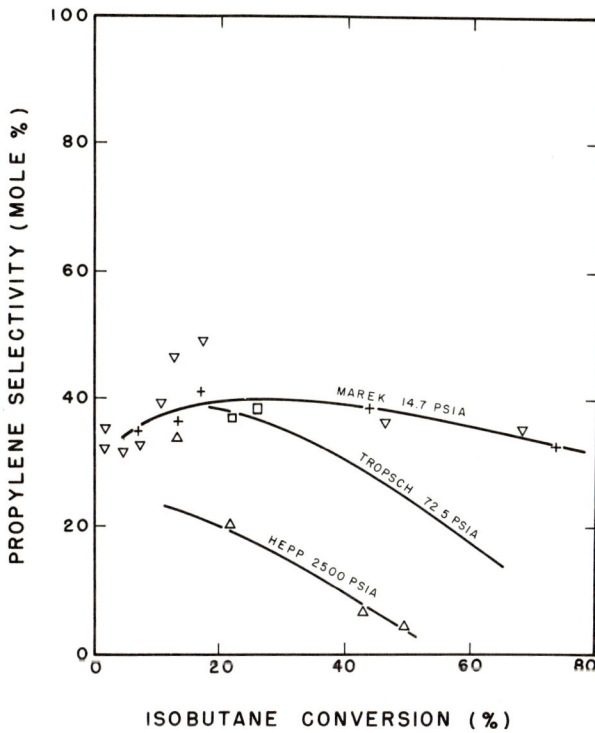

Figure 4. Pyrolysis of isobutane. Propylene selectivity is almost constant at atmospheric pressure.

cracking by the following reactions, which produce fuel gas and deposit coke on the catalyst:

$$C_3H_8 \rightarrow C_2H_4 + CH_4$$
$$C_3H_8 \rightarrow Coke + nH_2$$
$$C_3H_6 \rightarrow Coke + mH_2$$

Hydrogen produced from dehydrogenation and coking can also back-hydrogenate some of the ethylene formed:

$$H_2 + C_2H_4 \rightleftarrows C_2H_6$$

In a dehydrogenation process ethylene would not be recovered for sale, and the above reaction would actually be an advantage as a means of adding heat and removing hydrogen.

It is well known that, in addition to butadiene from butane, the Houdry Catadiene process can produce propylene from propane. This process has been in successful operation to produce butadiene and butene

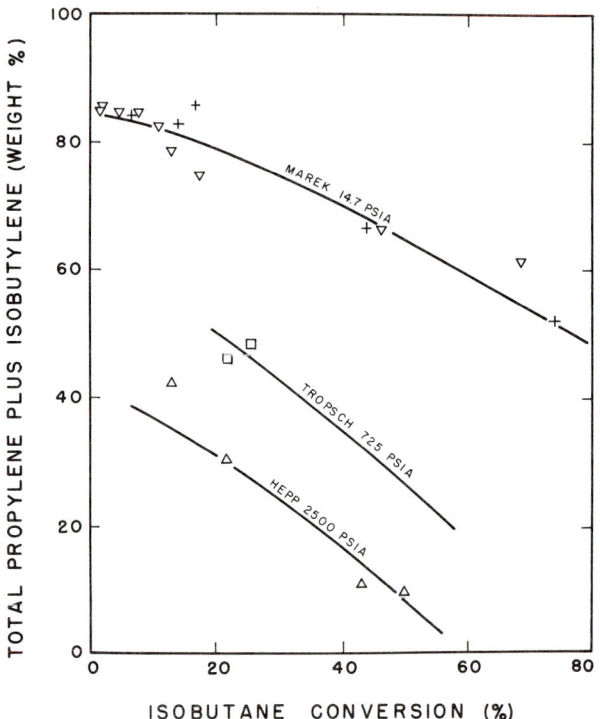

Figure 5. Pyrolysis of isobutane. High total selectivities at low pressures can be had.

from *n*-butane since World War II. It is the only commercially proved process for dehydrogenating *n*-butane, and 13 plants with an aggregate capacity of a billion lbs of butadiene per year have been built (*30*). The Catadiene process is available for license, but because of the large quantities of butadiene that are now available, both by import and as a by-product from ethylene units, there is little expectation that additional Catadiene capacity will be built to produce butadiene in the United States (*29*).

The major paper on the Houdry Catadiene process is that published by Hornaday, Ferrell, and Mills in 1961 (*27*). Their main concern is butadiene, but yield data are provided on dehydrogenation of other stocks, specifically, for present purposes, propane.

Based on the information provided by Hornaday *et al.* it is possible to conceptualize a flow scheme for propane dehydrogenation and to describe the Catadiene process in terms of propane. Figure 6 is the flow sheet for the process, which is cyclic. Fresh propane is combined with unconverted recycled propane and preheated to *ca.* 1150°F. Higher feed temperatures could result in excessive charge stock pyrolysis. The propane then passes over a shallow bed of catalyst mixed with inert material. Heat stored in the catalyst bed both from burning off coke and from the sensible heat of large quantities of regeneration air (5 to 10 lbs/lb feed) satisfies the endothermic requirements of the dehydrogenation reaction. Catalyst temperatures vary not only from the beginning to the end of an on-stream period as a consequence of this endothermicity but also from the inlet to the outlet of the bed. Consequently, in stating "temperature" of the commercial adiabatic reaction, one must realize that this is a nominal figure, indicative of level only—roughly 1175° ±25°F.

Figure 6 has been drawn with five reactors. Two reactors are dehydrogenating propane, two are having the coke deposited on catalyst burnt off with air, and one is being evacuated, purged, and valve-changed, at any given time. Other combinations of reactions are possible, and Hornaday *et al.* present the timing cycle for this 2-2-1 case.

The cited article does not give operating conditions for propylene. The catalyst is chromia-alumina. Because of equilibrium, vacuum operation and hence short catalyst beds are required, as suggested by the compressors shown in Figure 6. Reactor effluents, after being quenched quickly, are compressed and separated conventionally.

Propane conversion is only 54%, and a propylene selectivity and recovery of 76 and 93 mole %, respectively, result in a propylene yield of 38%. Neither methylacetylene nor allene is listed among the reaction products, and coke production amounts to 4% of propane converted.

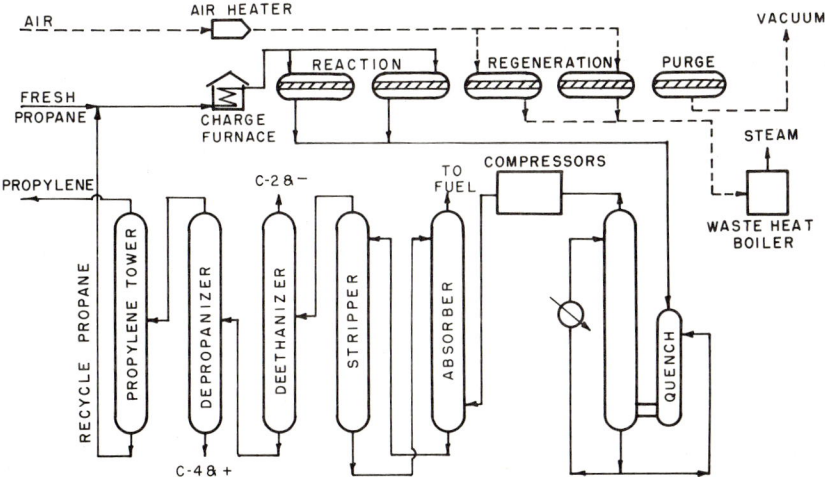

Figure 6. The Houdry Catadiene process. Flow sheet for propane dehydrogenation.

The most important point—low conversion—can be seen from the following equilibrium conversion tabulation at 0.23 and 1.00 atm total pressure

Temp., °F	K_p atm	Equilibrium Conversion, %	
		0.23 atm	1.00 atm
1100	.289	75	47
1150	.575	85	60
1200	.835	89	68

Equilibrium constants for propane dehydrogenation were obtained from API 44 data, and conversions X_e were calculated from the simple formula

$$X_e = \frac{K_p}{\pi + K_p}$$

where π = total pressure in atmospheres.

Those who are interested in chromia-alumina catalyst are referred to the detailed review article by Poole and MacIver (*44*). These authors also studied the sintering of chromia-alumina (*45*) and found that chromia in the catalyst sinters faster than alumina. Their data suggest that catalyst activity loss may be a serious problem at the temperature levels required for high propane conversions.

The kinetics of propane dehydrogenation over chromia-alumina have been determined by Tyurgaev *et al.* (*58, 59, 60*). Empirical expressions for coking and cracking, in conjunction with a Langmuir-Hinshelwood dehydrogenation model, described performance over their chromia-alumina catalyst. They found that their catalyst activity was low—only twice that of thermal. This is a good generalization since in these temperature ranges catalytic activity of almost any catalyst begins to approach thermal activity. Despite its modest activity for propane, chromia-alumina catalyst is the only commercially proved catalyst that could dehydrogenate propane. A review of experimental studies on catalytic propane dehydrogenation (mainly Soviet work) was published by Weiss (*61*), and the reader is referred to this for detailed operating conditions.

The important point, for present purposes, is that in terms of propane dehydrogenation the Houdry Catadiene process must cope with conversions that are limited by equilibrium considerations as well as high temperatures that are adverse both to selectivity and activity. Vacuum operation and multireactor cyclic operation combine to require both high compression costs and high plant investments. No Catadiene plant for propane dehydrogenation has yet been built. A catalyst or process improvement that can accommodate the problems peculiar to propane and yet be profitable enough to stay within the modest product-charge price differential of propylene and propane has not been announced by Houdry, as of this writing.

It should also be noted that any existing Houdry butane dehydrogenation unit (with minor modification) can operate at either full or partial capacity on propane feed or on mixed propane–butane feeds. The high butadiene-to-ethylene ratios that can be anticipated as ethylene operations are eventually switched to gas oil feeds (see Table V) suggest that propane dehydrogenation may be an unhappy prospect in store for some Houdry butane dehydrogenation units. However, the ease of n-butane and isobutane dehydrogenation, relative to propane, would probably influence a Houdry operator to consider C-4 alkylate olefins as products in preference to propylene. This will, of course, result indirectly in release of propylene from alkylate commitments but not its direct manufacture. For the present Houdry units will probably continue to produce butadiene and only design quantities of butylene (as at Petrotex). Barring special situations, no new Houdry construction for propane feed or propane feed to existing units will be economically justifiable in the forseeable future.

The economic and therefore commercially suitable processes for obtaining propylene are as a by-product of catalytic cracking, as a by-product of ethylene manufacture, and as a co-product of an isobutane cracker.

Oxidative Dehydrogenation of Propane

The limitations resulting from the unfavorable propane dehydrogenation equilibrium were discussed above. Any *selective* oxidant that could remove the H_2 formed in the propane dehydrogenation reaction would eliminate the equilibrium problem.

Much attention in the literature has been devoted to oxidative dehydrogenation processes for n-butane since equilibrium is a serious problem in butadiene manufacture also. However, much of the zest for the economic future of oxidative dehydrogenation was lost when Shell's Berre plant, based on iodinative dehydrogenation, was not placed on stream. Whether or not this was caused solely by European butadiene oversupply is speculative, but the net effect is that as yet there is no proved commercial oxidative dehydrogenation process to produce butadiene that could be adapted to propylene.

The state of the art in partial oxidation of propane to propylene is assessed below. Unlike the preceding sections which dealt with commercially practical processes this section will concern itself with evaluating the promise of the partial oxidants—oxygen, halogens, and sulfur—that have been proposed for effecting the propane to propylene reaction.

Oxygen. There has been a significant amount of research reported on the partial oxidation reaction of propane with propylene (as distinct from complete oxidation to CO and CO_2, which can be found in the combustion literature). The goal of much of the partial oxidation research has been to produce oxygenated compounds, such as aldehydes and alcohols. These are the predominant products at temperatures below 750°F (*24*). Olefin production at significant conversion levels is never highly selective, but there is a reasonable range of temperature—750° to 950°F—in which the desired reaction proceeds as:

$$\tfrac{1}{2}O_2 + C_3H_8 \rightarrow C_3H_6 + H_2O$$

Maximum yield of propylene, which occurs near the higher temperature, according to Calderbank *et al.* (*9*), is only 8.5 wt %, and this is accompanied by 5 wt % ethylene as a co-product. Above 950°F, oxidative cracking is the main reaction, and while yields of light paraffins and unsaturates result, almost no propylene and large amounts of carbon monoxide become the product distribution.

The interested reader is referred to surveys of research in the thermal reaction by Medley and Cooly (*38*) and by Rovertson (*49*), who cover the field of propane partial oxidation before 1960. Process variables are reported by Satterfield *et al.* (*50, 51*), Mitchell (*40*), and Newitt (*42*). Mechanism, surface effects, and theoretical considerations are given in papers by Albright *et al.* (*1, 2*), Ferguson and Yokley (*21*), and Knox *et al.* (*33, 34*).

The preceding discussion of thermal partial oxidation encompasses work in empty reactors and in fixed and fluid beds of inert materials, such as quartz or sand. A new type of reactor has been developed by Fenske and his associates (*31*), which they call a "raining" reactor. Inert solids are permitted to rain down through a reaction zone. Tests in the range 600°–1050°F were run that showed that solids rate was a critical variable, and the results of a major process variable study were presented. Residence times were varied from 0.02–0.5 minute, pressure from 1–8 atm, and propane to oxygen mole ratios from 2–20.

Two size ranges of particles were used: 80–100 mesh and 35–40 mesh; the optimum raining solids rate was about 0.6 lbs solids/lb oxygen. Yields of propylene possible are higher than in conventional reactors: 10.1 wt % propylene with coproduct of 4.4 wt % ethylene. The improvement in propylene yield is attributed to the raining solids, which stop free radical chains and inhibit further reaction.

There have also been attempts to use catalysts such as Al, Al_2O_3, Cr_2O_3–Al_2O_3, and Cr_2O_3–bauxite, and the reader is referred to partial oxidation studies reported by Baccaredda and Nencetti (*4*). The last catalyst specified gave the best performance—17% C_3H_6 and 10% C_2H_4 in the product gases. Yields are not specified, but one would expect extremely low selectivity on porous catalysts.

The partial oxidation of propane with oxygen appears, in the absence of a homogeneous catalyst, to be far from having the potential of commercialization as a result of inherent poor selectivity to propylene. This necessitates that conversions be kept low. The advantages of not only supplying heat for the reaction but also of shifting the propane dehydrogenation equilibrium are outweighed by the intrinsic non-selectivity. One should not compare the partial oxidation of propane to dehydrogenation but rather to pyrolysis, to assess its potential merit, which lies mainly in a desirable shift in the ratio of propylene to ethylene that is produced from propane:

	Catalytic Dehydrogenation	Oxidative Dehydrogenation		Pyrolysis
		Raining Reactor	Al_2O_3– Bauxite	
Conversion, %	54	25	—	90
C_3H_6/C_2H_4 Mole ratio	∞	1.55	1.70	0.32

A method has been claimed for improving partial oxidation yield using a homogeneous catalyst with air or O_2; processes using catalytic amounts of halogens have been patented by Petrotex (6) and by El Paso Natural Gas Products Co. (55). The latter claims that if reactants having catalyst:oxidant:propane ratio of 0.15 lbs I_2: 3.0 lbs O_2: 1 lb C_3H_8 are processed at 1200°F, 1 atm, and 0.05 minutes contact time, propane conversion of 72% and selectivity of 59 mole % are achieved. For reference, catadiene conversion and selectivity are 54% and 75.6 mole %, respectively, and research efforts to increase El Paso's partial oxidation yields past this level will be needed to effect an economic process in which iodine values are recovered and recycled.

Halogens. The following discussion, evaluating halogens as partial oxidants, is quoted directly from Weiss (61), who has reviewed the utility of this type of partial oxidation for production of propylene.

These are processes using halogen (chlorine, bromine, or iodine) as the oxidant; and the reader is recommended to reviews by King (32) and by Wolf et al. (62), which evaluate the patented processes, mainly with respect to the production of butadiene.

Basically, the iodinative processes function in the following manner, using propane as an example:

I_2 reacts with propane to produce propylene and hydrogen iodide

$$CH_3CH_2CH_3 + I_2 \rightleftarrows CH_3CH = CH_2 + 2HI$$

Oxygen present in the reacting system then reacts with the HI produced faster than with hydrocarbons to regenerate the I_2

$$4HI + O_2 \rightarrow 2I_2 + 2H_2O$$

Halogens other than I_2 are generally not useful because of the large number of substitution products that result. Also, the process may be operated without added oxygen, and then the HI produced is converted to I_2 in an external Deacon-type process. The Shell patent literature for this method to produce butadiene is provided by King. Bataafse Petroleum Maatschappig N.V. (47) has patented a vapor phase non-catalytic process for the propylene reaction, which takes place at atmospheric pressure, 0.86 I_2/C_3H_8 mole ratio, 36 seconds residence time, and 550°C. Conversion of 51 mole %, and selectivity to propylene of 93 mole % are claimed. The tremendous amounts of I_2 that require handling as well as the fact that 2% propyl iodide is formed (from which I_2 must be recovered) suggest that this process may not be economically applicable for propylene manufacture. Any losses of expensive iodine would be intolerable in this process.

Both the thermodynamics and the kinetics of the dehydrogenation of propane by iodine vapor

$$C_3H_8 + I_2 \rightleftarrows C_3H_6 + HI$$

were characterized by Nangia and Benson (41).

Another modification of the iodinative-type process for the purpose of reducing I_2 usage is a series of Shell patents (see King) that describe the use of a metal oxide or hydroxide acceptor that is capable of reacting with the HI formed during dehydrogenation in the range of 500°C—e.g.

$$2HI + NiO \rightarrow NiI_2 + H_2O$$
$$\text{solid} \quad \text{solid}$$

$$HI + LiOH \rightarrow LiI + H_2O$$
$$\text{molten} \quad \text{molten}$$

I_2 is recovered by oxidizing the acceptor in a cyclic reaction system, both acceptor and I_2 are returned to the dehydrogenation

$$2NiI_2 + O_2 \rightarrow 2NiO + 2I_2$$

$$4LiI + O_2 + H_2O \rightarrow 4LiOH + 2I_2$$

Even in the improvements, the complete recovery of expensive iodine values and decomposition of iodinated compounds require elaborate facilities. Substitution of either Br_2 (48) or Cl_2 (3) for I_2 so greatly increases the level of halogenated byproducts that the utility of the processes is obviated. It is doubtful that the iodinative process will find any commercial application for propylene manufacture.

Organic iodine recovery is less of a problem in a propylene process than in a butadiene process. Diiodobutene polymerizes in the quenching section, and methyl iodide–butadiene separation is difficult.

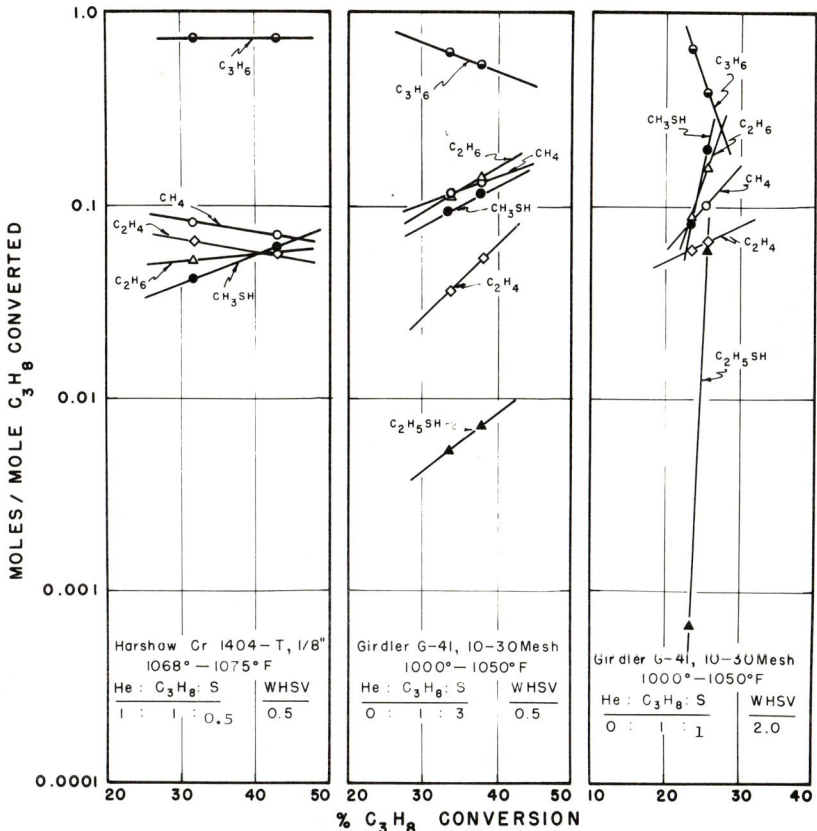

Figure 7. Dehydrogenation of propane in the presence of sulfur. Ratios of helium, propane, and sulfur are approximate.

Sulfur. To assess whether sulfur can be used as a partial oxidant for propane, exploratory experiments have been made at Worcester Polytechnic Institute in which propane and propane–helium mixtures were saturated partially with sulfur at atmospheric pressure and then passed over commercially available chromia-alumina catalysts. No methylacetylene was detected by thermal conductivity gas chromatography using a 20-foot squalane column, but significant amounts of methyl and ethyl mercaptans were found. Figure 7 illustrates the nature of the gaseous products obtained. Continued experimentation will establish the system more concretely. No coking data or results from sustained operation are available as yet. The results shown in Figure 7, while preliminary, show that the reaction

$$C_3H_8 + \tfrac{1}{2}S_2 \rightleftarrows C_3H_6 + H_2S$$

is indeed feasible and that it is readily catalyzed by chromia-alumina. The operation shown is, of course, far from the optimum needed for a commercial propylene process.

Equilibrium data for the above reaction have been calculated by Mauras (37). The slightly exothermic reaction (rather than seriously endothermic as in conventional dehydrogenation) has practically complete equilibrium conversion over the temperature range that would be of practical interest.

Temperature, °F	*Equilibrium Constant*	*Equilibrium Conversion,* %
842	45	93
1022	110	96
1202	156	97

Actually, the system is not the simple reaction indicated. Besides the cracked products CH_4, C_2H_4, C_2H_6, the experimental data of Figure 7 show that methyl and ethyl mercaptan form. It has not been decided yet whether two distinct reactions are actually taking place over the chromia-alumina:

$$C_3H_8 \rightleftarrows C_3H_6 + H_2$$

$$H_2 + \tfrac{1}{2}S_2 \rightleftarrows H_2S$$

Furthermore, sulfur has been written in the diatomic form S_2, although there is actually a distribution of polyatomicity at the temperatures of interest. Mauras has also calculated the equilibria between the species S_2, S_4, S_6, and S_8: At 1 atm total pressure and temperatures above 1000°F, he showed that equilibria calculated using the polymeric distribution of sulfur do not differ from those based on diatomic sulfur.

$$C_3H_8 + \frac{1}{n}S_n \rightleftarrows C_3H_6 + H_2S$$

It is significant that, unlike oxidative dehydrogenation, there has been no serious amount of attention given to sulfoxidation as a process to produce four carbon or higher olefins and diolefins. The reason is that thiophene formation occurs. For example:

$$C_4H_{10} + \tfrac{1}{2}S_2 \rightleftarrows \underset{S}{\square} + H_2S$$

Chromia-alumina is a catalyst for the above reaction.

There is no reason to believe that chromia-alumina is the only catalyst suitable for sulfoxidation of propane. Garwood et al. (23) showed that significant amounts of H_2S are produced from propane in the presence of molecular sieves. His procedure was to pass propane through a layer of sulfur on top of a catalyst bed while raising the temperature.

Studies were made from 650° to 813°F. Some of the gas analyses resulting from his work are shown in Table VI, which lists the catalysts studied as well as a comparative run on inert 96% silica glass. The miniscule yields of propylene and benzene reported do not account for the large quantities of H_2S measured.

Table VI. Reaction of Propane with Sulfur[a]

| Catalyst | Gas Analysis, wt % | | | Propane |
	Propylene	Benzene	H_2S	
Vycor	—	—	1.4	98.2
Rare earth exchanged X	0.5	—	26.0	71.8
Na-zeolite X	1.2	1.0	33.4	63.6
Presulfided Na-zeolite X	nil	1.3	34.8	62.8

[a] Examples from Garwood (23).

Table VII. Reaction of Propane with Sulfur[a]

Catalyst	None	Cr-Bi-P [a] Ni-Al [b]	None
Av. reaction temp., °F	1150	1090(a) 1015(b)	1410
Pressure, psia	45	80	20
Total feed, lbs			
Propane	1140	1910	842
Propylene	12	642	104
C_4+	78	192	95
Sulfur	1866	3929	1204
H_2S	—	85	—
Steam	—	449	—
N_2, Ar	—	21	—
Reactor Effluent, lbs			
Propane	180	950	32
Propylene	540	1329	152
Methylacetylene	328	196	509
C_1, C_2	72	39	227
Tar	20	42	70
C_4+	38	152	9
Sulfur	773	3113	191
H_2S	1151	935	1055
Steam	—	449	—
N_2, Ar	—	21	—

[a] Examples from Schuman (53).

Schuman (53), in a recent patent, claims that methylacetylene and propylene are the products obtained when sulfur and propane react either thermally or in the presence of catalysts. Examples reproduced directly from his patent are given in Table VII; note that no mercaptans are reported.

Obviously, the technology for oxidative dehydrogenation of propane by sulfur is now in its infancy. The advantage of using sulfur as an oxidant would be its low cost; elaborate recovery systems that are part of iodine processes would not be required. However, mercaptans that form would have to be recycled to extinction.

Conclusions

We have seen that by 1973 catalytic cracking will only satisfy 2 to 4 billion lbs/year of a projected 11 billion lb/year propylene demand. Most of the balance will be produced as a by-product of ethylene manufacture. Shifting from ethane and propane to heavier stocks such as n-butane and gas oil will satisfy propylene needs. Some propylene will also be produced from isobutane steam crackers as an isobutylene co-product.

New construction of propylene capacity on the basis of a propylene market price significantly above gasoline value plus recovery cost will be questionable, economically. At about the 3.1¢/lb level, refiners will consider releasing a portion of the 12 billion lbs/year propylene committed for alkylation, and they will satisfy their octane needs by other means. The future may hold the promise of a direct process for propylene, but economic and technological breakthroughs are needed. Available catalytic and iodinative dehydrogenation technology cannot be applied economically to propane. Oxidative dehydrogenation processes using either oxygen or sulfur, are conceptually of interest but undeveloped technically.

Acknowledgment

The author thanks the Esso Research and Engineering Corp. for their support of propane dehydrogenation experimentation at Worcester Polytechnic Institute. The isobutane cracking study was funded by Chem Systems, Inc.; the assistance and expertise obtained from Bronek Dutkiewicz, Bert Struth, and Martin Sherwin of Chem Systems and from John Johns on technical and economic points is deeply appreciated.

Literature Cited

(1) Albright, L. F., *Chem. Eng.* **1967**, (15), 197.
(2) Albright, L. F., Locke, S. A., McFarlane, D. R., Glahn, G. L., *Ind. Eng. Chem.* **1960**, 52, 221.
(3) Arganbright, R. P., U. S. Patent **2,971,995** (1961).

(4) Baccaredda, M., Nencetti, G., *World Petrol. Congr.*, **5th, 1959**.
(5) Baker, R. W., Blazek, J. J., Maher, P. K., Ciapetta, F. G., Evans, R. E., *Ann. Meetg. 3rd, Natl. Petrol. Refining Assoc., San Antonio, Tex., 1964*.
(6) Belgium Patents **617,890-617,893**.
(7) Buekens, A. G., Froment, G. F., *Ind. Eng. Chem. Process Design Develop.* **1968**, 7, 3, 435.
(8) Burke, D. P., Miller, R., *Chem. Week*, **Oct. 23, 1965**, 63; **Nov. 13, 1965**, 69.
(9) Calderbank, P. H., Havanian, V. P., Talbot, F. D. F., *J. Appl. Chem.* **1967**, 7, 425.
(10) *Chem. Eng. News*, **June 13, 1966**, 34.
(11) *Ibid.*, **April 22, 1967**, 135.
(12) *Ibid.*, **Aug. 5, 1968**, p. 26.
(13) *Ibid.*, **Jan. 6, 1969**, p. 40.
(14) *Ibid.*, **March 24, 1969**, 18.
(15) *Ibid.*, **June 10, 1968**, 38.
(16) Cijfer, H. J., Kitzen, M. R., Wall, F. M., *AIChE Natl. Meetg., 64th, New Orleans, La., 1969*.
(17) Crynes, B. L., Albright, L. F., *Ind. Eng. Chem., Process Design and Develop.* **1969**, 8, (1), 25.
(18) Davenport, C. H., *Petrol. Refiner* **March, 1960**, 39, 125.
(19) Davis, W. H., *Chem. Eng. Progr.* **1963**, 59 (2), 28.
(20) Egloff, G., Thomas, C. L., Linn, C. B., *Ind. Eng. Chem.* **1936**, 28, 1283.
(21) Ferguson, R. E., Yokley, C. R., *Intern. Symp. Combust.*, **7th, 1959**, 113.
(22) Freiling, J. G., Hirson, B. L., Summerville, R. N., *Hydrocarbon Process Petrol. Refiner* **1968**, 47 (11), 145.
(23) Garwood, W. E., Hamilton, W. A., Kerr, G. T., Myers, C. G., U. S. Patent **3,247,278** (1968).
(24) Hatch, L. F., *Hydrocarbon Process, Petrol. Refiner* **1964**, 43 (11), 171.
(25) Hepp, H. J., U. S. Patent **2,816,150** (Dec. 10, 1957).
(26) Hepp, H. J., Frey, F. E., *Ind. Eng. Chem.* **1953**, 45, 410.
(27) Hornaday, G. F., Ferrell, F. M., Mills, G. A., *Advan. Petrol. Chem. Refining* **1961**, 4, 451.
(28) *Hydrocarbon Process. Petrol. Refiner* **1965**, 44 (11), 205.
(29) *Ibid.*, **1968**, 47 (9), 21.
(30) *Ibid.*, p. 230.
(31) Jones, J. H., Danbert, T. E., Fenske, M. R., *Ind. Eng. Chem., Process Design Develop.* **1969**, 8 (1), 17.
(32) King, R. W., *Hydrocarbon Process. Petrol. Refiner* **1966**, 45 (11), 189.
(33) Knox, J. H., *Trans. Faraday Soc.* **1960**, 56, 1225.
(34) Knox, J. H., Wells, C. H. J., *Trans. Faraday Soc.* **1963**, 59, 2786.
(35) Lambrix, J. R., Morris, C. S., Rosenfeld, H. J., *AIChE Natl. Meetg., 64th, New Orleans, La., 1969*.
(36) Marek, L. F., Neuhaus, H., *Ind. Eng. Chem.* **1933**, 25, 516-519.
(37) Mauras, H. J., *Chim. Phys.* **1964**, 61 (9), 1343.
(38) Medley, J. D., Cooley, S. D., *Advan. Petrol. Chem. Refining* **1960**, 3, 309.
(39) Mills, G. A., Ashwill, R. E., Gresham, T. L., *Hydrocarbon Process. Petrol. Refiner* **1967**, 46 (9), 121.
(40) Mitchell, R. L., *Petrol. Refiner* **1956**, 35 (7), 179.
(41) Nangia, P. S., Benson, S. W., *J. Am. Chem. Soc.* **1964**, 86, 2770.
(42) Newitt, D. M., *Chem. Rev.* **1937**, 21, 299.
(43) Ockerbloom, N. E., Stuart, A. P., *Hydrocarbon Process. Petrol. Refiner* **1967**, 46 (5), 25.
(44) Poole, C. P., Jr., MacIver, D. S., *Advan. Catalysis* **1967**, 17, 223.
(45) Poole, C. P., Jr., Kehl, W. L., MacIver, D. S., *J. Catalysis* **1962**, 1, 407.
(46) Prescott, J. H., *Chem. Eng.* **Jan. 13, 1969**, 28.
(47) Raley, J. H., *German Patent* **1,137,430** (1962).

(48) Raley, J. H., Mullineaux, R. D., *German Patent* **1,194,400** (1965).
(49) Robertson, N. C., "The Chemistry of Petroleum Hydrocarbons," Vol. II, p. 365, Reinhold, New York, 1955.
(50) Satterfield, C. N., Reid, R. C., *J. Chem. Eng. Data* **1961**, 6, 302.
(51) Satterfield, C. N., Wilson, R. E., *Ind. Eng. Chem.* **1954**, 46, 1001.
(52) Sawyer, F. G., *Hydrocarbon Process. Petrol. Refiner* **1969**, 48, 1, 122.
(53) Schuman, S. C., U. S. Patent **3,344,203** (1967).
(54) Schutt, H. C., Zdonik, S. B., *Oil Gas J.* **Feb. 13, 1956**, p. 98.
(55) Steinmetz, W. E., *French Patent* **1,346.343** (1963).
(56) Stobaugh, R. S., *Hydrocarbon Process. Petrol. Refiner* **1967**, 46 (1), 177.
(57) Tropsch, H., Thomas, C. L., Egloff, G., *Ind. Eng. Chem.* **1936**, 28, 324.
(58) Tyuryaev, I. Ya., *Izv. Vysshikh. Uchebn. Zavdenii, Khim. i Khim. Tekhnol.* **1959**, 2, 797.
(59) Tyuryaev, I. Ya., Mukhina, T. N., Bushin, A. N., Gurina, P. S., *Khim. Prom.* **1959**, (3), 201.
(60) Tyuryaev, I. Ya., Mukhina, T. N., Pavlova, V. B., Kolyaskina, G. M., *Khim. Tekhnol., Topliv Masel* **1958**, 3 (12), 90.
(61) Weiss, A. H., *Hydrocarbon Process. Petrol. Refiner* **1968**, 47, 4, 123.
(62) Wolf, C. N., Bergman, R. I., Sittig, M., *Chem. Week* **May 28, 1966**, 113.

RECEIVED January 12, 1970.

10

Kinetics of the Demethylation of Methylcyclohexane

M. M. JOHNSON and H. J. HEPP

Phillips Petroleum Co., Research and Development Division, Bartlesville, Okla. 74003

> *The reaction of hydrogen and methylcyclohexane over supported nickel catalysts was studied at conditions where methane and cyclohexane were principal products. With high nickel content catalysts, activated at 700°F, the reaction was difficult to control in fixed bed reactors, and conversion was limited to 30% by runaway reaction to methane. Reducing the nickel content to 25% and activating the catalyst for longer periods at demethylation conditions gave improved results. In a fluidized bed reactor, conversion of 55% with yields of 66 lbs of cyclohexane/100 lbs of methylcyclohexane reacted were obtained at 575°F, 150 psig, and H_2/methylcyclohexane molar ratios of 6. The conversion flow rate data agree with an integrated rate expression arrived at by assuming the methylcyclohexane reacts on catalyst sites not occupied by hydrogen.*

Cyclohexane is a large volume petrochemical used mainly in nylon manufacture. In 1965 the free-world consumption was 1.1 million tons, and this is expected to double substantially by 1970 (5). Methods for manufacturing cyclohexane are thus of considerable interest. While presently most cyclohexane is made by hydrogenating benzene (1), some is derived from natural sources. A potential method for manufacturing cyclohexane is by the selective hydrodealkylation of methylcyclohexane. Such a process is considerably more direct than the demethanation of toluene, followed by hydrogenation of the resulting benzene.

The general reaction was studied by Haensel and Ipatieff (2) in the early 1940's. They found that paraffin hydrocarbons containing a methyl group attached to a secondary carbon atom could be demethylated selectively when passed, along with hydrogen, over a properly activated nickel or cobalt catalyst at carefully controlled conditions of time, temperature, pressure, and hydrogen ratio. For example, neohexane or triptane could be produced in good yield from 2,2-dimethyl- or 2,2,3-trimethylpentane by hydrogenolysis over a nickel catalyst at temperatures in the range 518°–662°F. The reaction is quite exothermic—about 13 kcal/mole of methane formed—and temperature control is difficult. Haensel found that tempering the activity of his catalyst by pretreating at 1000°F was useful in reducing the tendency for temperature runaways. Methyl groups attached to tertiary or quaternary carbons are less active than those attached to secondary carbons, thus making selective removal of the latter possible.

The products obtained from hydrogenolytic cleavage of alkylcyclohexanes show that the ring is quite stable, and successive loss of methyl groups is the principal reaction. Kochloefl and Bazant (4) studied the demethylation of alkylcyclohexanes at low conversions and listed cyclohexane as the sole product of demethylation of methylcyclohexane at 428°F using a nickel–alumina catalyst with hydrogen at 1 atm. The hydrogenolysis of alkylcyclohexanes is apparently somewhat more complicated than that of the isoparaffin since secondary butylcyclohexane hydrogenolysis yields normal propylcyclohexane. Isopropylcyclohexane would be the expected product on the basis of the reactivity rules cited above.

The demethylation of methylcyclohexane over a variety of supported nickel catalysts with hydrogen at 300 psi was examined by Shuikin and Tien (6). Over the temperature range 626°–680°F employed, dehydrogenation to aromatics and isomerization of the cyclohexane formed to methylcyclopentane complicated the reaction. A catalyst consisting of nickel on alumina which had been treated with HF gave the highest yield of cyclohexane—30.1% per pass based on methylcyclohexane. However, isomerization to methylcyclopentane was extensive and was undoubtedly catalyzed by the HF treatment. The formation of lower alkanes was attributed to opening of the cyclopentane ring.

Since the demethylation of methylcyclohexane with hydrogen is a potential route to cyclohexane, this reaction was studied under conditions where methane and cyclohexane were the principal products, and methylcyclopentane formation and dehydrogenation to aromatics were substantially avoided.

Experimental

Apparatus and Procedure. Fixed Bed. The reactor consisted of a vertically supported, 1 inch diameter stainless steel tube heated in a 28 inch, 3-sectional electric furnace. The top section of the reactor, which served as a reactant vaporizer and preheater, was packed with 70 ml of

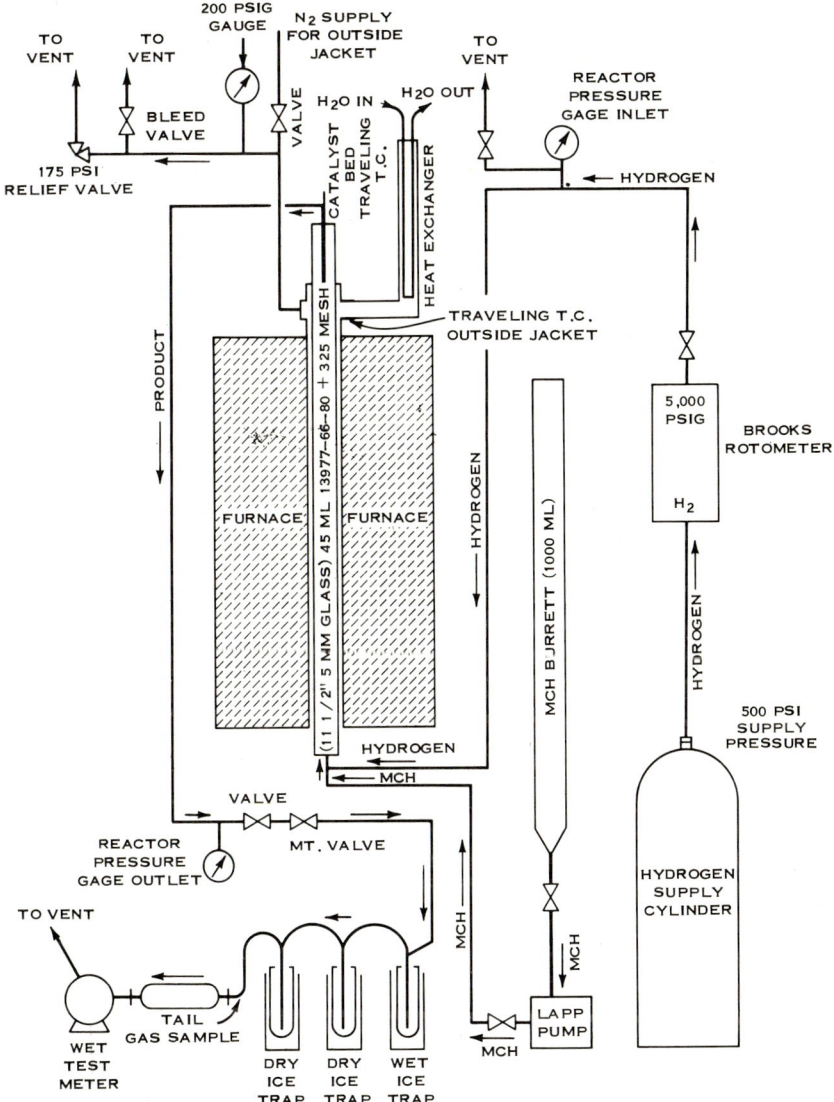

Figure 1. Schematic of fluidized bed equipment

5 mm glass beads. The middle 10-inch reaction zone was packed with 100 ml of $-6/+20$ mesh catalyst and the bottom section with 75 ml of 5 mm glass beads. A thermowell located coaxially in the reactor permitted temperature measurements in all sections. Except for the most active catalyst tested, nickel on kieselguhr, this arrangement permitted approximately isothermal control of the catalyst bed.

The methylcyclohexane and hydrogen entered at the top of the reactor. The total effluent passed through a motor valve at the exit of the

Table I. Demethylation of

Period	1	2	3
Activation temp., °F (hrs)		700 (16)	
Average temp., °F	450	535	535
Pressure, psig	152	152	150
LHSV (vol/vol cat/hr)	0.40	0.39	0.87
H_2/MCH mole ratio	6.67	6.52	2.84
Conversion, % MCH reacted			
Total	2.2	30.0	17.9
By Hydrogenolysis	2.2	30.0	17.9
k'', atm/hr[b]	0.086	1.15	1.29
Yield of products, lb/100 lb MCH reacted			
CH_4	15.9	18.6	15.6
C_2H_6–C_4H_{10}	—	4.2	—
C_5H_{12}	3.3	4.4	4.1
C_6H_{14}	3.8	6.9	11.3
C_7H_{16}	7.1	6.4	9.6
Methylcyclopentane	14.7	5.7	5.1
Cyclohexane	57.4	56.5	56.6
Toluene	—	—	—
	102.2	102.7	102.3

[a] Catalyst used was 70% nickel–kieselghur.
[b] Rate constants are for the hydrogenolysis reaction only. Dehydrogenation was not included.

reactor, which reduced the pressure to atmospheric. A wet ice trap followed by two dry ice–acetone traps was used to collect the liquid product. Samples of the effluent gas from the last trap were taken at regular intervals and analyzed by mass spectrometry. The liquid product was analyzed by gas chromatography. Some of the liquid product was carried over from the last trap as a mist, but material balances usually accounted for over 94% of the hydrocarbon charged.

FLUIDIZED BED REACTOR. As some difficulty was encountered in operating the fixed bed reactor at predetermined temperatures, a reactor heated with boiling biphenyl was built to improve heat transfer. The reactor was a ¾-inch od stainless steel tube jacketed with a 1-inch schedule 80 steel pipe. The annulus between the two concentric tubes was filled with 70 grams of biphenyl. Vigorous boiling and refluxing of the biphenyl ensured isothermal operation of the reactor. This reactor was operated as a fluidized bed with −80/+325 mesh catalyst. Details of this reactor and the related equipment are given in Figure 1. The same sample train and procedure described for the fixed bed reactor was used in this part of the study.

Catalysts. The nickel-on-kieselguhr catalyst used in initial work was the commercially available Harshaw-0101 T. This catalyst was crushed and sieved to −6/+20 mesh for fixed bed studies. Three nickel–alumina

Methylcyclohexane in Fixed Bed Reactor[a]

4	5	7	8	9
		900 (4)		
535	535	538	575	600
300	75	75	75	75
0.84	0.76	0.90	0.82	0.84
3.06	6.10	1.24	5.30	3.30
9.1	30.0	11.6	10.0	28.1
9.1	30.0	4.5	7.7	14.6
1.23	1.20	0.136	0.298	0.575
14.8	20.1	6.0	17.0	14.0
—				
4.2	3.9	0.6	—	0.8
6.8	10.3	0.7	2.0	1.8
15.2	8.0	4.1	8.0	3.1
3.0	2.9	2.0	4.1	2.6
58.3	55.8	25.8	48.4	32.9
—	—	59.0	22.6	46.6
102.3	101.0	98.2	102.1	102.6

catalysts of varying nickel content were prepared from the nitrate salt by precipitating a basic nickel carbonate in a slurry of Alon-C alumina using a potassium carbonate solution. The supernatant solution was decanted, and the remaining solution removed by vacuum filtration. After washing with distilled water and drying at 295°F for 3 hours, the dry cake was crushed to $-6/+20$ mesh for use in fixed bed reactors.

Discussion of Results

A nickel-on-kieselguhr catalyst containing about 70% of nickel in the reduced form, similar to the catalyst used by Haensel, was employed in the initial work. Since the removal of the methyl group from methylcyclohexane is considerably more difficult than removal of an unbranched methyl from a paraffin, the catalyst was pretreated for maximum activity by reducing in a stream of hydrogen for 15 hours at 700°F.

The data obtained are listed in Table I. This catalyst had demethylation activity at temperatures as low as 450°F. As the temperature was increased, other factors held essentially constant, the conversion increased, and a methylcyclohexane conversion of 30% was obtained at 535°F. All attempts to increase conversion beyond this point resulted in an uncontrollable temperature rise. Methane and other light paraffins were the sole products. As shown in the table, cyclohexane yields based on MCH reacted were but little affected by operating conditions, being in the range of 56–58% by weight. Some methylcyclopentane was also present, probably arising from isomerization of cyclohexane caused either by the nickel or by some residual acidity of the support. This isomerization activity decreased as the catalyst aged.

Accompanying the demethylation reaction were various ring-opening reactions to give 2- and 3-methylhexane and some n-heptane. The dehydrogenation of MCH to toluene also occurred. Ring-opening reactions are closely related to the demethylation reaction and, as might be expected, parallel the desired reaction. Thus, it is probably impossible to avoid completely the formation of paraffins. Successive demethylation to form lower molecular weight paraffins occurs rapidly. However, dehydrogenation to toluene can be made small by increasing reactor pressure and hydrogen concentration and limiting reactor temperature to 600°F as a maximum.

In an effort to increase conversion and selectivity, an attempt was made to moderate the activity of a second portion of this catalyst by pretreating at 900°F for 4 hours in hydrogen at 1 atm. The data are listed in periods 7, 8, and 9 of Table I. The activity of the catalyst was reduced sharply. Attempts to increase conversion by lowering operating pressure, reducing hydrogen flow rate, or by increasing reactor temperature resulted in toluene formation. At conditions where toluene formation was limited, the selectivity to cyclohexane was reduced by the high temperature treatment. Similarly, the use of a cyclohexane diluent to control reactor temperature using the 70% Ni catalyst activated at 700°F was unsatisfactory because 1.6% of the cyclohexane itself was converted, which reduced selectivity greatly.

The data obtained with this catalyst show the same general trends as noted by previous workers. In addition to the three major competing reactions taking place—*i.e.* demethylation, ring opening, and dehydrogenation—the kinetics show the same features. MCH conversion is increased as temperature is increased, or pressure lowered, or hydrogen concentration reduced.

In an effort to obtain a less temperature-sensitive system, lower nickel content catalysts were prepared on an alumina support and tested for demethylation activity. The first, Preparation A, with a nominal nickel content of 50 wt % was activated at 700°F in a slow stream of hydrogen at atmospheric pressure for 16 hours. This catalyst was tested at conditions similar to those employed with the nickel–kieselguhr catalyst reported above. The results are given in Table II.

As may be seen from examination of these data, results are very similar to those obtained with the nickel–kieselguhr catalyst. The catalyst

Table II. Hydrogenolysis with 50% Nickel–Alumina Catalyst

Run No.	1	2	3	4	5
Activation temp., °F (hrs)	700 (16)				600 (4)
Average temp., °F	457	482	517	519	588
Pressure, psig	150	150	300	150	150
LHSV (vol/vol cat/hr)	0.42	0.46	0.42	0.40	0.60
H_2/MCH mole ratio	6.1	6.3	6.6	7.1	7.3
Hours of catalyst use	1.5	3.0	6.0	6.5	14
Conversion, % MCH reacted	2.7	5.2	7.2	14.9	49.6
k'', atm/hr	0.109	0.232	0.560	0.576	2.83
Composition of liquid product, wt %[a]					
C_5H_{12}		0.05	0.06	0.14	1.41
C_6H_{14}	0.15	0.27	0.37	0.66	5.56
C_7H_{16}	0.32	0.65	0.95	1.48	3.28
Methylcyclopentane	0.86	0.54	0.28	0.42	1.49
Cyclohexane	1.06	3.24	4.53	9.68	33.70
Methylcyclohexane	97.60	95.23	93.78	87.59	55.25

[a] The gas from the dry ice trap was hydrogen containing 0.35–2.63 mole % methane, except in Run 5, where methane was 9.15%.

was active at a temperature of 457°F. However, nearly equal amounts of cyclohexane and its isomer methylcyclopentane were formed at this lower temperature. As the duration of the test was extended, the amount of methylcyclopentane decreased. With increasing temperature, the conversion to cyclohexane increased, and the same general trends noted in the initial work were obtained—i.e., increased conversion at lower operating pressure, and the ring cleavage reaction closely paralleling the demethylation reaction. In an effort to increase the conversion, the pressure was reduced to 105 psi at a temperature of 535°F. This resulted in a runaway reaction to light paraffins, and the catalyst was deactivated. Rate constants calculated for this catalyst, as discussed later, are almost identical to those obtained with the nickel–kieselguhr catalyst, as shown in Figure 2. In view of the differences in composition and the influence of the catalyst pretreatment used to produce the nickel metal and an active catalyst, this similarity must be considered a coincidence.

In a subsequent test the remainder of Preparation A was used to explore the effects of catalyst activation temperature and test duration. After a short 4-hour catalyst activation period at 600°F with hydrogen at atmospheric pressure, testing was started at 532°F, and the temperature was raised over a period of 8 hours to give a catalyst bed temperature of 579°F. The cyclohexane content of the liquid effluent increased from 7 to 28% during this period. The system was maintained under a slow stream of hydrogen overnight, and testing resumed the following morn-

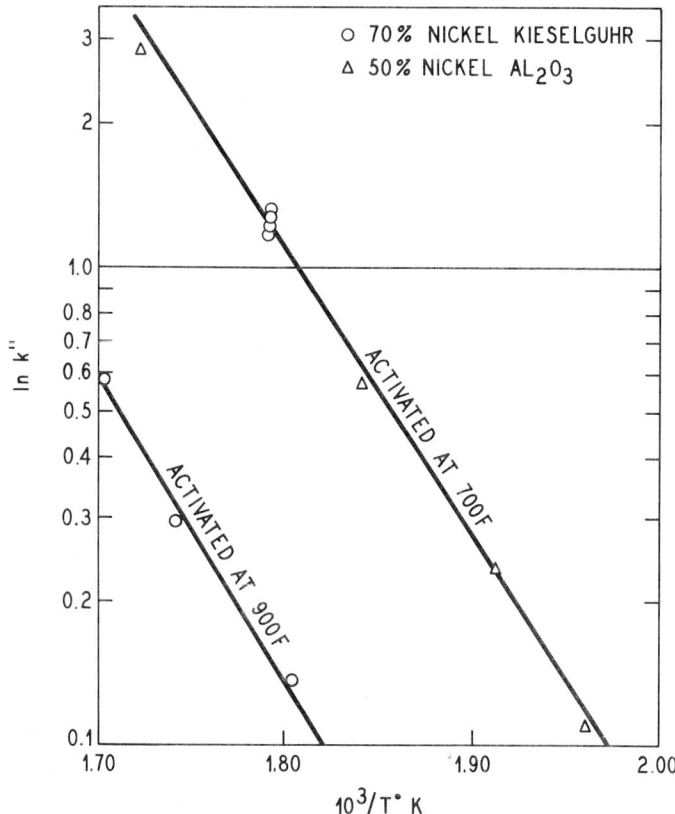

Figure 2. Rate constant vs. temperature

ing at 150 psig, 0.6 LHSV, 565°F, and H_2/MCH molar ratio of 6. The liquid effluent contained 14% cyclohexane after the first hour, and this rose to 34% after 6 hours. Table III shows the concentration–time dependence. The results after 14 hours are shown in Table II, Run 5.

Table III. Cyclohexane Content vs. Time

Catalyst Age, hours	8	9	10	11	12	14
Temperature, midpoint of bed, °F	565	582	581	585	586	588
Cyclohexane Concentration, wt %	14	16	23	25	27	34

As indicated in Table III, an increase in catalyst bed temperature accompanied increased conversion, and part of this increase in conversion undoubtedly resulted from the change in temperature rather than an increase in catalyst activity. The results in Table II show that improved selectivities and higher conversions could be obtained from supported nickel catalysts activated at lower temperatures for longer times.

As it was difficult to maintain predetermined catalyst bed temperatures in a fixed bed reactor, a fluidized bed reactor was built, and a 26 wt % nickel catalyst was tested, Preparation B. The catalyst was activated or reduced at the conditions of the run, and results are summarized in Table IV.

As these data show, the catalyst increased in activity with time as the nickel was reduced. After 6.5 hours, the catalyst reached 69% steady-state activity and after 15 hours, 90%. The increase in activity appears to follow a first-order rate law at temperatures near 575°F.

$$\ln \frac{1}{1-x} = 0.16\ t$$

where x = fraction of steady state activity

t = test duration, hours

A gradual improvement in activity and selectivity was noted up to 36 hours, and slow continued improvement is quite probable beyond this point. The amount of neohexane and toluene in the liquid effluent continued to decrease with time, indicating some change in catalyst properties. During the test, temperature traverses were made throughout the fluidized catalyst bed. Temperatures from top to bottom of the bed varied by 2°F.

Once near steady-state activity had been reached (19 hours), the temperature was decreased 15°F, and the effect of decreased temperature on conversion and selectivity established, Run 4. An estimated activation energy for the conversion of methylcyclohexane is 28 kcal/mole, only 2 kcal less than the approximate value for the nickel–kieselguhr catalyst. The effect of decreased operating pressure is shown by the data of Run 5. Conversion increased, and efficiency to cyclohexane decreased slightly. The same effect was noted previously in fixed bed tests with Preparation A.

No plugging of the small inlets in the control valve downstream of the reactor was noted, and fines formation and carryover of catalyst were not a problem at the low fluidization velocities used in this test. Bed expansion was not measured directly, but it was calculated to be about 7–10%.

The combined liquid effluent from the fluidized-bed run was collected, and a complete analysis was made. The data are shown in Table V. The extensive isomerization taking place is of some interest; all of the C_6 paraffins except diisopropyl are present in the effluent, along with cyclohexane–methylcyclopentane, MCH–dimethylcyclopentane, etc. Nickel is, of course, known to have isomerizing activity, but in addition the catalyst may be functioning as a dual-function isomerization catalyst.

Table IV. Hydrogenolysis with 26% Ni

Period	1	2	3	4
Temperature, °F	582	577	578	562
Pressure, psig		150		
LHSV (vol/vol/hr)	0.82	0.91	0.82	0.84
H_2/MCH mole ratio	6.0	5.6	6.0	6.2
Hours of catalyst use	6	15	20	24
Conversion, % of MCH	38	50	54	38
k'', atm/hr	2.90	4.32	4.24	3.07
Composition of liquid products, wt %				
C_4H_{10}	—	0.21	0.28	0.19
C_5H_{12}	0.34	2.16	2.73	1.72
C_6H_{14}	1.48	4.76	5.32	3.87
C_7H_{16}	1.12	2.78	2.42	2.49
Methylcyclopentane	0.50	1.09	1.04	0.77
Cyclohexane	21.62	34.42	37.37	25.63
Methylcyclohexane	74.30	54.46	50.80	65.29
Toluene	0.40	0.09	—	—
	99.76	99.97	99.96	99.96
Yield of products, lb/100 lb MCH reacted				
CH_4–C_4H_{10}		19.52	20.87	15.81
Methylcyclopentane		1.98	1.72	1.91
Cyclohexane		62.68	62.07	63.57
C_5–C_7 Paraffins		16.06	17.75	20.62
Toluene		2.19	0.17	—
		102.43	102.58	101.91

[a] No prior treatment was used. Reduction occured during run.

The effect of nickel content and catalyst preparation on demethylation activity was pursued further. A catalyst of 10.5 wt % nickel, Preparation C, was tested in the fixed bed reactor, after activation in place at 540°F for 2 hours, with the feed at 0.8 LHSV, 150 psig, and a H_2-to-methylcyclohexane mole ratio of 6. The temperature was raised over a period of 17 hours to 580°F, other factors held constant, and the effluent analyzed for cyclohexane at various temperatures. The conversion to cyclohexane was low, and only 1.5 wt % cyclohexane was found in the liquid effluent at 580°F. Under somewhat similar conditions, a conversion of over 50% of the methylcyclohexane was obtained with the 26 wt % nickel catalyst. These results suggest that interaction of nickel with the support occurs, and the resulting nickel-aluminate is not active for the demethylation reaction.

Two additional catalysts were tested that confirm that interaction of nickel with the support or substrate leads to loss of demethylation on Al_2O_3 in a Fluidized Bed Reactor[a]

5	6	7	8
567	572	574	575
105		150	
1.00	0.86	0.84	0.90
5.0	6.1	6.0	6.2
27	30	34	36
56	49	53	55
3.80	4.04	4.27	4.24
0.58	0.18	0.23	0.17
3.48	1.60	2.10	1.94
5.56	4.31	4.81	4.07
2.47	3.26	2.88	2.81
1.16	1.03	0.73	0.99
36.85	33.30	36.90	38.52
49.85	56.29	52.30	51.45
—	—	—	—
99.95	99.97	99.95	99.95
24.98	22.60	22.30	19.14
1.80	1.82	1.23	1.69
57.38	60.87	62.20	65.80
18.87	17.17	16.97	15.44
—	—	—	—
103.03	102.46	102.70	102.34

activity. Each of these catalysts contained 20% nickel. The first was a coprecipitated nickel–silica from nickel nitrate and water glass, and the second was a coprecipitated nickel–alumina from nickel nitrate–aluminum nitrate. X-ray and magnetic susceptibility measurements indicated excellent dispersion of the nickel throughout the support in both preparations. These catalysts were activated in hydrogen at 1 atm and 700°F for 16 hours. They were tested for demethylation activity at 0.4 LHSV, 560°F, and a H_2-to-methylcyclohexane feed ratio of 6. No activity was noted for either catalyst, suggesting that nickel metal is the active entity for demethylation activity.

Kinetics

The data obtained in this study, together with literature data, show that for a given catalyst, activity varies with both the composition and the conditions employed in the reduction to nickel. However, for a given catalyst which has been activated in a definite manner, the conversion of

Table V. Analysis of Liquid Composite Run in a Fluidized Bed Reactor

Component	Wt %
C_4's	.16
$i\text{-}C_5$	.55
$n\text{-}C_5$	.85
neo-C_6	.13
2-MeC_5	1.31
3-MeC_5	1.12
$n\text{-}C_6$	1.02
CyC_5	.26
DMeC_5's	.15
MeCyC_5	.79
2-MeC_6	.72
3-MeC_6	1.69
$n\text{-}C_7$	.25
CyC_6	34.64
MeCyC_6	55.80
Benzene	.12
Undetermined	.20
Toluene	.23
	100.00

Gas Analysis

	Mole %
H_2	90.3
CH_4	9.6
C_4	.1
	100.0

MCH increases with temperature, decreases as space velocity is increased, decreases as pressure is increased, and decreases as hydrogen mole ratio is increased. The latter observations suggest that MCH and hydrogen are competing for sites. The conversion of MCH increases as more sites become available owing to lowered hydrogen activity, and decreases as hydrogen crowds it out by occupying more sites. The rate of reaction on this basis may be expressed as follows:

$$r = k\,\theta_v \tag{1}$$

Following Hougen and Watson (3), the fraction of vacant sites, θ_v is given by

$$\theta_v = \frac{1}{\left(1 + \dfrac{N_H}{N_T} K_1\pi + \dfrac{N_M}{N_T} K_2\pi\right)} \tag{2}$$

which reduces to the following when it is assumed that the surface coverage by hydrogen is large.

$$\theta_v = \frac{1}{\frac{N_H}{N_T} K_1 \pi} \tag{3}$$

Expressing the moles of hydrogen, N_H, and the moles of methylcyclohexane, N_M, in terms of the fraction reacted, f, and the initial concentration in the feed gives for the rate equation

$$\frac{dW}{FN_0} = \frac{df}{\frac{k \cdot N_T}{K_1 \pi N_H}} = \frac{K_1 \pi (B - f) \, df}{k (B + 1)} \tag{4}$$

Collecting constants and integrating gives for $Bf > \dfrac{f^2}{2}$

$$f = \frac{(W)}{FN_0} \frac{k'}{\pi} \frac{(B + 1)}{B} \tag{5}$$

The following equivalent form of this equation is more convenient and was used to calculate the rate constants listed in the tables and plotted in Figure 2.

$$k'' = (\text{LHSV}) \frac{(B)}{B + 1} \pi f \tag{6}$$

A plot of ln k'' vs. $1/T$ for the nickel-kieselguhr catalysts is shown in Figure 2. The activation energy for the reaction is estimated at 30 kcal/mole. As mentioned earlier, this value was confirmed substantially by data taken during Period 4 of the fluidized-bed run (Table IV) where temperature control was better. In this run 28 kcal/mole was obtained despite the difference in catalyst composition.

Nomenclature

B = Molar ratio of H_2 to MCH in feed, moles/mole
f = Fraction of MCH reacted
FN_0 = Molar feed rate of MCH, moles/hr
k = Rate constant, moles/hr/gram of catalyst
k'' = Rate constant, atm/hr
K_1, K_2 = Adsorption equilibrium constants, atm^{-1}
LHSV = Volume of MCH feed/hr/volume of catalyst, hr^{-1}
N_H = Moles of H_2 per unit mass of feed, moles/gram
N_M = Moles of MCH per unit mass of feed, moles/gram
N_T = Total moles of feed per unit mass of feed, moles/gram

t = Time, hrs
x = Fraction of steady-state activity
W = Mass of catalyst, grams
θ_v = Fraction of total sites that are vacant
π = Reactor pressure, atm

Literature Cited

(1) *Chem. Week*, **Nov. 20, 1965**, 41.
(2) Haensel, V., Ipatieff, V. N., U. S. Patents **2,422,670–2,422,675** (1947).
(3) Hougen, O. A., Watson, K. M., "Chemical Processes Principles," Chap. 19, Wiley, New York, 1947.
(4) Kochloefl, K., Bazant, V., *J. Catalysis* **1967**, 8, 250–260.
(5) *Oil Gas J.*, **May 22, 1967**, 86.
(6) Shuikin, N. I., Tien, Hsing-Hua, *Izv. Akad. Nauk SSSR, Otd. Khim. Nauk* **1960**, 2014–2018.

RECEIVED January 12, 1970.

11

Catalytic Dehydrogenation of Higher Normal Paraffins to Linear Olefins

JAMES F. ROTH, JOSEPH B. ABELL, LOYD W. FANNIN, and ANDREW R. SCHAEFER

Central Research Department, Monsanto Co., St. Louis, Mo. 63166

The availability of higher (C_6^+) normal paraffins at low cost has stimulated study of the direct dehydrogenation of such paraffins to the corresponding normal monoolefins. The olefins produced in this way are mainly internal olefins, but they are nevertheless useful for industrial applications. The catalysts so far investigated require that modest levels of conversions ($\sim$ 10 to 15%) be employed to permit reasonable levels of selectivity towards monoolefin formation. Side reactions encountered in the catalytic dehydrogenation of higher normal paraffins include cracking, skeletal isomerization, formation of polyunsaturates and aromatics, and coking. Results are presented on several catalyst systems including molybdena–alumina, chromia–alumina, platinum–alumina, and carbonaceous catalysts. The highest selectivities were achieved with platinum–low acidity alumina compositions.

Recently, interest has been aroused in the possible use of higher (C_6^+) linear olefins as intermediates in the manufacture of various products, including biodegradable detergents, alcohols, homopolymers, and copolymers. Higher α-olefins have become available commercially and have been produced either by the thermal cracking of paraffins or by the controlled oligomerization of ethylene. However, the availability of higher normal paraffins at low cost has also stimulated the study of methods of converting such paraffins to linear olefins. One approach has been *via* chlorination of the normal paraffins followed by dehydrohalogenation, while another route investigated by several workers (*1, 2, 3, 5, 7, 9, 12, 13, 14, 15*) and comprising the subject of this paper, is the direct catalytic dehydrogenation of the normal paraffins to yield linear olefins of the same carbon number. The olefins produced either by the chlorination or direct dehydrogenation routes are mainly internal olefins, but they are nevertheless useful for industrial applications. This applies whenever the processes employing the olefins require reaction conditions that promote facile bond isomerization.

In a review article on catalytic dehydrogenation Kearby (8) states: "hexane and higher molecular weight hydrocarbons have a strong tendency to form aromatics or to crack and therefore have not given good yields of corresponding olefins or diolefins." To provide good dehydrogenation selectivity, it is therefore necessary to attenuate such side reactions.

The thermodynamics that prevail for dehydrogenation of higher normal paraffins require the use of relatively high reaction temperatures (400°–500°C) to achieve significant levels of conversion (16). Side reactions that accompany the desired dehydrogenation to monoolefin include cracking, skeletal isomerization, formation of polyunsaturates and aromatics, and catalyst coking. The extent of participation of these side reactions can be controlled somewhat by the choice of catalyst. However, even the best catalysts uncovered to date, in which cracking, skeletal isomerization, and coking are suppressed, tend to produce excessive polyolefins and aromatics whenever one attempts to achieve a high level of total paraffin conversion. Apparently the predominantly paraffinic character of monoolefins of high molecular weight permits sequential dehydrogenations to occur quite readily, leading to the formation of di- and triolefins, with the latter dehydrocyclizing to aromatics. To minimize the extent of sequential dehydrogenations and the accompanying formation of polyunsaturates, it is necessary to limit the amount of total paraffin conversion. Generally, good selectivities towards monoolefin formation require that total conversion be limited to the range of about 10 to 15%.

Thus, the product of direct dehydrogenation of higher normal paraffins is one diluted with a considerable amount of unreacted paraffin. The monoolefin and paraffin usually have rather close boiling points and cannot be separated readily by distillation. Separation must be effected by subjecting the olefin to some desired secondary reaction, or by some other means, and the unreacted paraffin is then ultimately recycled to the dehydrogenation process.

Several catalyst systems have been investigated in our laboratories including molybdena–alumina, chromia–alumina, platinum–alumina, and carbonaceous catalysts. Studies of dehydrogenation over these catalysts have been carried out mainly using n-dodecane as the model compound. However, similar results have also been obtained with both higher and lower molecular weight paraffins, and the results with n-dodecane are, accordingly, considered representative of the behavior of higher normal paraffins in general.

Experimental

Materials. n-Dodecane of 98% or higher purity was obtained from Humphrey-Wilkinson, and n-heptane was Phillips pure grade material. Platinum diamminodinitrite was obtained from J. Bishop and Co. and α-dodecene from Gulf. α-Dodecene was analyzed and found to be 100% olefin with a 1-dodecene content of 95%. Some of the catalysts used were obtained from commercial suppliers. Their designation, source, and manufacturer's description are given in Table I.

Table I. Catalysts Used

Material	Manufacturer	Description
Mo-0502	Harshaw	10% MoO_3 on activated alumina-1/8 inch tablets, (dehydrogenation catalyst)
Cr-0205	Harshaw	19% Cr_2O_3 on activated alumina-5/32 inch tablets, (butane dehydrogenation catalyst)
RD-150	Engelhard	0.5% Pt on activated alumina-1/16 inch extrudate (reforming catalyst)
PHF-1	American Cyanamid	0.6% Pt on activated alumina-1/8 inch tablets (reforming catalyst)
UOP-R7	Universal Oil Products	0.35% Pt on activated alumina-1/22 inch balls (reforming catalyst)

Catalyst Preparations. The preparation of molybdena–alumina catalysts has been reported (*12*). The platinum–alumina catalysts were prepared by impregnating alumina particles with an aqueous ammoniacal solution of platinum diamminodinitrite using the minimum solution technique, followed by drying of the impregnate at 150°C for at least 5 hours, and calcination of the dried composite in air at a temperature of 450°C for 5 hours. The alumina support was precalcined in air at 600°C before impregnation with the platinum-containing solution. The various alumina supports used in preparing the platinum catalysts consisted of activated aluminas obtained from commercial sources. Their methods of preparation are unknown, but they were characterized relative to physical properties. The carbon-on-alumina catalyst was prepared by carbonization of an activated alumina having a surface area of 72 M^2/gram until the carbon content was about 8 wt %. Carbonization was conducted by contacting the alumina in a tubular reactor at 540°C for 4 hours with a flowing gas stream containing a mixture of 1-butene and nitrogen (mole ratio of 1-butene to nitrogen of 1:4). The alumina support was of low acidity.

Reactor Apparatus. The reactor systems used for obtaining data on n-dodecane dehydrogenation over molybdena–alumina and platinum–alumina catalysts and the data on reaction of bare supports with α-dodocene have been described previously (*12*). The reactors used for obtaining data on n-dodecane dehydrogenation over chromia–alumina and carbon–alumina catalysts were also fixed bed reactors but differed somewhat in geometry and design. They consisted of 1-inch id 316 stainless steel pipe, 20 inches long, containing an ⅛ inch od thermowell that extended along the entire longitudinal axis of the reactor. With the thermowell in place, the volume per inch of reactor was 12.7 cc. Catalyst bed depths up to 12 inches could be accommodated. The remaining volume was filled with inert alumina balls (Alcoa T-162, ⅛ inch diameter). Heat for the reactor was provided by a series of electrical heating bands spaced along its length. The entire unit was covered with 2½ inches of thermal insulation. By means of selector switches, electrical power for the heaters could be obtained from any of four automatically controlled power supplies. The control elements for these power supplies were thermocouples placed in thermowells located longitudinally within the walls of the reactors. Thus, heating loads and control points could be shifted at will to obtain a uniform temperature profile throughout the length of the reaction zone. For example, with appreciable dehydrogenation occurring over a 10-inch catalyst bed, the axial temperature profile could be maintained at 540° ± 0.5°C. Before mixing, the liquid feed components were vaporized and preheated in packed, heated tubes, and the gaseous components were also preheated. The feed mixtures were conducted to the inlet of the reactor through heated lines. The reactor effluent was passed through a tap-water cooled, cooler-condenser before entering the sampling system. Liquid products were collected in receivers while the noncondensibles were passed through a packed condenser maintained at 10°C. Any material that condensed was returned to the liquid product receiver. The remaining gases passed through a sample-collecting chamber before going to the vent.

Analysis of Dehydrogenation Products. Products were analyzed by gas chromatographic (GC) methods. During the work reported the GC methods used were varied both with respect to instruments and details of the analysis. These variations should, however, have led to only minor differences in the analytical results. The product analysis used with molybdena–alumina and platinum–alumina catalysts is the same as that described previously (12). GC analysis of products obtained from carbon-on-alumina catalysts was performed on an F & M Model 810 chromatograph equipped using a single flame ionization detector. The column was a 7-ring polyphenyl ether (7-R PPE) coated, stainless steel capillary 0.001 inch id x 200 feet long. The operating conditions were: injector block and splitter temperature at 240°C; detector block at 285°C; after a 6-minute post injection interval at 80°C, the column oven was programmed from 80° to 170°C at 6°/minute and held at 170° for 6 minutes; carrier gas, helium; flow rate of helium 280 cc/minute split 100:1 after the injection block; hydrogen flow rate 34.5 cc/minute; air flow rate 400 cc/minute; and liquid sample size 0.3 μl. Analysis of products obtained from the chromia–alumina catalyst was performed on an F & M Model 810 chromatograph using dual flame-ionization detectors. The columns were 0.18 inch id × 10 foot long packed with 18 wt % Carbowax 20M plus 1 wt % Ag NO_3 on 60–80 mesh Chromosorb W (AW/DMCS). The operating conditions were: injection block temperature at 305°C, detector block temperature at 310°C, column oven programmed at 4°C/minute from 90° to 230°C and held at 230°C for 14 minutes, helium flow rate at 75 cc/minute, hydrogen flow rate at 35 cc/minute, air flow rate at 600 cc/minute, and liquid sample size of 0.25 μl. Weight percent composition was calculated from the integrated area and normalized to 100%.

All 3 GC methods used gave a similar sequence of eluted components. The components were grouped and designated as follows: LB, components eluted before the C_{12} paraffin consisting of lower carbon number paraffins, olefins, and aromatics resulting from cracking, and also containing iso-C_{12} paraffins; n-C_{12}, the n-dodecane; MO, the various bond isomers of C_{12} n-monoolefins; DTA, all components eluting after the C_{12} olefins consisting of diolefins, triolefins, and predominantly, substituted aromatics.

Results and Discussion

Acidity of Supports. Most of the side reactions encountered in the attempted dehydrogenation of normal paraffins to linear monoolefins are affected by the acidity of the catalyst. For example, strongly acidic catalysts would be expected to promote skeletal isomerization, cracking, dehydrocyclization, and coking. Pines and Haag (10) have shown that catalyst supports themselves may exhibit intrinsic acidity and that model reactions may be used to characterize the acidity of the support. Since all of the catalysts covered in the present work contained activated alumina as a major component, an attempt was made to characterize the acidity of several aluminas. The results are given in Table II. We decided to use the amount of LB formation (which includes cracked

Table II. Low Boiler Formation by Various Supports[a]

Support	Surface Area M^2/gram	Wt %[b] Na	Wt. % Conversion to LB	Hydrocarbon Feed
Activated Alumina A	202	0.30	0.1	n-Dodecane
			0.2	8% 1-Dodecene[c]
Activated Alumina B	72	0.19	0.0	n-Dodecane
			0.1	8% 1-Dodecene[c]
Activated Alumina C	292	0.018	0.3	n-Dodecane
			2.9	8% 1-Dodecene[c]
Activated Alumina D	83	0.32	0.1	8% 1-Dodecene[c]
Activated Alumina E	438	0.003	2.7	n-Dodecane
			6.9	8% 1-Dodecene[c]

[a] Evaluation conditions: bulk volume of support 145 cc, 8.44 grams of C_{12} hydrocarbon per minute, H_2/hydrocarbon mole ratio of 2.0, atmospheric pressure, reaction temperature of 440°C.
[b] Wt % K in each case found to be <0.01
[c] Balance n-dodecane.

products and iso-C_{12} paraffins) as the model reaction to measure support acidity. At first the interaction of *n*-dodecane with the support was investigated. However, this proved to be a relatively insensitive probe. Introduction of some 1-dodecene into the feed increased the amount of LB formation significantly and, as may be seen in Table II, permitted one to obtain more striking differences between the various activated alumina supports. We concluded that under dehydrogenation conditions most of the LB formation results from secondary reactions of the monoolefin rather than *via* direct conversion of the paraffin. This conclusion is further supported by the observation that the levels of LB formation recorded in Table II with a 1-dodecene containing feeds are similar to those obtained with platinum catalysts dispersed on the same supports and subjected to paraffin dehydrogenation catalysis under the same reaction conditions.

It appears that LB formation side reactions arise from the presence of strongly acidic sites in the alumina. Moreover, it is concluded that the sodium content of the alumina exerts a major influence on the relative population of strongly acidic sites. The results in Table II suggest that as little as a few hundredths of a per cent of sodium is sufficient to annihilate the strongly acidic sites in activated alumina. This conclusion is at least qualitatively in agreement with the findings of Pines and Haag (*10*). All of the activated aluminas, however, proved to give very facile bond isomerization of 1-dodecene so that almost all of the C_{12} monoolefin in the product consisted of various internal linear C_{12} monoolefins. In contrast, a pure silica gel (Davison Grade 59) gave under similar conditions not only very low LB formation but did not promote any significant amount of bond isomerization of 1-dodecene. It seems fitting therefore to characterize activated aluminas of low alkali metal content (<0.01%) as "strongly acidic" aluminas, activated aluminas of appreciable alkali metal content (>0.05%) as "low acidity" aluminas, and supports incapable of promoting even bond isomerization under these conditions as "non-acidic" supports (*e.g.*, pure silica gel and α-alumina). According to this classification, active sites capable of catalyzing bond isomerization of olefins are considered sites of low acidity. In the past there has been some tendency to speak of "non-acidic" activated aluminas when referring to ones that do not possess highly acidic sites, but in view of the results reported above, this description seems imprecise and should be refined in the manner proposed.

Dehydrogenation over Molybdena–Alumina. Detailed results have been reported previously (*12*) for molybdena–alumina catalysts. It appears that freshly prepared molybdena–alumina catalysts are strongly acidic and exhibit very high activity but poor selectivity towards monoolefin formation initially. After about 24 hours, these catalysts become substantially deactivated as a result of coking and then begin to afford more reasonable selectivity towards monoolefin formation. For example, after 25 hours, the following results can be obtained:

Total paraffin conversion, %	= 13.8
Conversion to MO, %	= 8.7
Conversion to DTA, %	= 3.0
Conversion to LB, %	= 2.1
Selectivity to MO, %	= 63.0

By temperature compensations, such results can be sustained for periods of a few hundred hours. However, a slow continuing deactivation occurs, and eventually the catalyst must be regenerated by combustion of the deposited coke. After regeneration, reasonable dehydrogenation results are again obtained after a rapid initial deactivation of the catalyst. Steam was found to be a particularly useful diluent with this type of catalyst. In contrast to this, the best results with molybdena–alumina for catalytic reforming are obtained over fresh catalyst and when steps such as dilution with hydrogen are employed to repress the initial rapid deactivation.

The high acidity of molybdena–alumina composites appear to be derived mainly not from the intrinsic acidity of the alumina but from active sites on a surface phase containing molybdenum atoms. The amount of LB and DTA formation that is observed initially over molybdenum–alumina is at least 10 times greater than observed over the bare support when the latter reacts with 8% 1-dodecene. In this case, the bare support was alumina B listed in Table II.

The amount of coke deposited on molybdena/alumina at the time it affords reasonable dehydrogenation selectivity is somewhat variable and depends on the surface area of the alumina support. Carbon contents of selective catalysts have been generally to be in the range of 5 to 20 wt %.

Dehydrogenation over Carbon–Alumina. A recent review of the catalytic properties of carbon (6) cites no references to catalytic dehydrogenation of hydrocarbons in the scientific literature. The patent literature, on the other hand, contains a few references to dehydrogenation of low molecular weight hydrocarbons over activated carbons. We have found recently (13) that *supported* carbon possesses interesting catalytic dehydrogenation properties and can be used to dehydrogenate paraffins, olefins, and alkyl-substituted aromatics quite generally. These supported forms of carbon can be produced in various ways but are most conveniently prepared by the pyrolysis of hydrocarbons on the surface of porous supports such as activated alumina. We have shown that the acidity of alumina has a profound effect on the state of dispersion of supported carbon (4) and on its catalytic properties towards the dehydrogenation of hydrocarbons (11, 13). For example, we have been able to prepare carbonaceous catalysts that are more active and more selective in the dehydrogenation of ethylbenzene to styrene than the presently used commercial catalysts based on iron oxide.

For the dehydrogenation of n-dodecane, a catalyst was prepared in the following way. Activated alumina B was first impregnated with sufficient sodium nitrate to incorporate 2 wt % sodium, and then it was calcined in air. The composite was carbonized with a flowing gas mixture of 1-butene and nitrogen (mole ratio of 1-butene to nitrogen of 1:3) at 540°C for about 5 hours at which time about 8 wt % carbon was deposited. This catalyst was then evaluated relative to the dehydrogenation of n-dodecane. The n-dodecane feed was diluted with steam (mole ratio of steam to n-dodecane of 5:1) and then contacted with the supported carbon catalyst at 470°C. The contact time (assuming 40% void space in the catalyst bed) was 1.25 seconds. After 14 hours of reaction, the following results were obtained:

> Total paraffin conversion, % = 10.2
> Conversion to MO, % = 7.7
> Conversion to DTA, % = 1.0
> Conversion to LB, % = 1.6
> Selectivity to MO, % = 75.0

At 19 hours of reaction, the contact time was increased to 2.5 seconds, and the following results were obtained:

> Total paraffin conversion, % = 13.0
> Conversion to MO, % = 8.9
> Conversion to DTA, % = 1.7
> Conversion to LB, % = 2.4
> Selectivity to MO, % = 70.0

The general behavior of this type of carbon catalyst supported on a low acidity alumina was very similar to that obtained with the molybdena–alumina catalyst with respect to activity, selectivity, and tendency towards slow continuous deactivation. The similarity is, in fact, so striking that it is highly suggestive that the selective dehydrogenation observed with molybdena–alumina arises not from active sites characteristic of the original fresh catalyst but rather from sites on the carbonaceous phase deposited during the first 24 hours. In short, it appears as though the molybdena–alumina catalyst exhibits selective dehydrogenation only when it is transformed to the equivalent of a supported carbon catalyst. It is believed that carbonaceous compositions, when supported or unsupported, derive their catalytic dehydrogenation properties from active sites consisting of free radicals on the surface of the carbon phase. In a sense they appear to behave like solid free radicals capable of engaging in hydrogen atom abstraction reactions similar to those well known to occur with hydrocarbon free radicals in the gaseous state. Presumably, dehydrogenation would occur as a result of sequential hydrogen atom abstractions.

Dehydrogenation over Chromia–Alumina. Chromia–alumina catalyst CR-0205 was studied relative to butane dehydrogenation and gave low conversion to butadiene under conditions of appreciable conversion to butenes. In addition, there was an exceptionally small amount of cracked products and essentially no skeletal isomerization. Accordingly, this well-known catalyst was evaluated for n-dodecane dehydrogenation. Conditions of evaluation were: temperature of 440°C, atmospheric pressure, hydrogen diluent with a hydrogen to n-dodecane mole ratio of 8 to 1. The results obtained were as follows:

Total paraffin conversion, % = 10.9
Conversion to MO, % = 7.7
Conversion to DTA, % = 2.3
Conversion to LB, % = 0.9
Selectivity to MO, % = 70.5

This catalyst gave significantly lower LB formation than molybdena–alumina or carbon–alumina but appeared to be more prone towards aromatics formation when compared at similar conversion levels.

Dehydrogenation over Platinum–Alumina. The acidity of the support is a major consideration in achieving selective dehydrogenation over platinum–alumina catalysts. The platinum–aluminas used in catalytic reforming are formulated with highly acidic aluminas by design, and these compositions give poor selectivity in the dehydrogenation of n-dodecane to linear dodecenes. This is illustrated by the following results obtained with one commercial reforming catalyst:

Total paraffin conversion, % = 14.7
Conversion to MO, % = 9.6
Conversion to DTA + LB, % = 5.1
Selectivity to MO, % = 32.0

Similar results were obtained with two other commercial platinum–alumina reforming catalysts. All three of the reforming catalysts evaluated had sodium contents of <0.001% which is consistent with the previous observation of a relationship between strong acidity and alkali metal content. The conditions used to evaluate these catalysts as well as platinum–aluminas of low acidity were as follows: flow of 8.4 grams of n-dodecane per minute, sufficient hydrogen to maintain a hydrogen–n-dodecane mole ratio of 2.0, atmospheric pressure, 145 cc of catalyst, reaction temperature of 440°C. Each catalyst was reduced in hydrogen for 1 hour at 440°C before a dehydrogenation run. Generally, analytical data were obtained over a 20-hour period of dehydrogenation catalysis.

Several platinum–alumina catalysts were prepared on low acidity aluminas such as A, B, and D listed in Table II. Practically all of the data obtained with such catalysts fell on a single selectivity curve correlating % conversion to MO with % selectivity to MO formation. Selectivity was found to be quite high, as illustrated by the following results for platinum on low acidity alumina:

$$\begin{aligned}
\text{Total paraffin conversion, \%} &= 11.6 \\
\text{Conversion to MO, \%} &= 9.5 \\
\text{Conversion to DTA + LB, \%} &= 2.1 \\
\text{Selectivity to MO, \%} &= 82.0
\end{aligned}$$

With these low acidity catalysts, selectivity varied inversely with conversion. At lower levels of conversion, it was possible to achieve selectivities to MO in excess of 90%. It was also found that the pore structure of the alumina support exerted a significant influence on both activity and selectivity. A relatively high macroporosity is desirable. The only catalysts based on low acidity aluminas that gave selectivities that departed appreciably from the general selectivity curve described above were those having a small macroporosity. Data on this macroporosity effect have been reported by Moore and Roth (9). These authors also found that a uniform platinum concentration profile gives higher catalyst activity at a fixed platinum loading than a surface activated catalyst.

A catalyst was prepared on the strongly acidic alumina E, using exactly the same preparation procedure used to make the platinum–low acidity alumina compositions. This catalyst was evaluated under the standard conditions and gave results very similar to those obtained with the commercial platinum–alumina reforming catalysts. At a conversion to MO of 8%, the selectivity to MO formation was 31%.

The dehydrogenation of n-heptane over a platinum–low acidity catalyst was also investigated. Owing to the lower molecular weight of n-heptane, somewhat higher temperatures ($\sim 20°C$) were required to achieve paraffin conversions similar to those obtained with n-dodecane. However, at these slightly higher temperatures, product distributions and selectivities were close to those found with n-dodecane.

Rapid deactivation occurs when n-dodecane is dehydrogenated over platinum–alumina without any diluent or with an inert diluent such as nitrogen. The rate of deactivation is decreased greatly when hydrogen is used as a diluent. However, even with hydrogen dilution, a slow deactivation (accompanied by carbonization of the catalyst) occurs. Eventually it is necessary to regenerate the catalyst by combustion of the coke deposits and reactivation in hydrogen.

Conclusion

Significant progress has been made in the direct catalytic dehydrogenation of higher normal paraffins to linear olefins, and further developments in this area of technology are likely to occur in the future.

Acknowledgment

The authors are indebted to Herman Pines for many stimulating discussions during this work.

Literature Cited

(1) Abell, J. B., Fannin, L. W., Roth, J. F., U. S. Patent **3,315,007** (1967).
(2) Ibid., **3,315,008** (1967).
(3) Ibid., **3,435,090** (1969).
(4) Berger, P. A., Roth, J. F., *J. Phys. Chem.* **1968,** 72, 3186.
(5) Bloch, H. S., U. S. Patent **3,391,218** (1968).
(6) Coughlin, R. W., *Ind. Eng. Chem., Prod. Res. Develop.* **1969,** 8, 12.
(7) Haensel, V., Hoekstra, J., U. S. Patent **3,293,319** (1966).
(8) Kearby, K. K., "Catalysis," P. H. Emmett, Ed., Vol. III, Reinhold, New York, 1955.
(9) Moore, R. N., Roth, J. F., U. S. Patent **3,274,287** (1966).
(10) Pines, H., Haag, W. O., *J. Am. Chem. Soc.* **1900,** 82, 2471.
(11) Roth, J. F., Abell, J. B., unpublished results, 1966.
(12) Roth, J. F., Abell, J. B., Schaefer, A. R., *Ind. Eng. Chem., Prod. Res. Develop.* **1968,** 7, 254.
(13) Roth, J. F., Schaefer, A. R., Belgian Patent **682,863** (1966).
(14) Roth, J. F., Schaefer, A. R., U. S. Patent **3,356,757** (1967).
(15) Schuikin, N. I., Timofeeva, E. A., Smirnov, V. S., "Proceedings of Third International Congress on Catalysis, Amsterdam," North-Holland Publishing Co., Amsterdam, 1964.
(16) Steiner, H., "Catalysis," P. H. Emmett, Ed., Vol. IV, Reinhold, New York, 1956.

RECEIVED January 12, 1970.

12

The Octafining Process for Isomerization of C_8 Aromatics Containing Ethylbenzene

H. F. UHLIG and W. C. PFEFFERLE

Engelhard Minerals and Chemicals Corp., Engelhard Industries Division, Newark, N. J. 07105

The growth in the use of individual xylene isomers, particularly p- and o-xylene, as chemical intermediates has led to the development of xylene isomerization and separation processes. The octafining process was developed to permit isomerization of C_8 aromatic mixtures without requiring removal of ethylbenzene. High conversions of ethylbenzene to xylenes are obtained. In conjunction with appropriate separation facilities any of the xylene isomers can be produced. Commercial applications to date have been for production of p- and/or o-xylene. Dependent on feedstock quality ultimate yields of xylene isomers greater than 85 wt % can be achieved. By-products are largely light and heavy aromatics. Catalyst is regenerable in situ, and catalyst lives greater than three years have been obtained.

C_8 aromatics, a mixture of the three xylene isomers plus ethylbenzene, have long been used as solvents. They have been obtained either from coal or petroleum refining. Although C_8 aromatics are still used as solvents, the growth in this area has been negligible. Since 1960 use as solvent has remained at about 50 million gallons per year (8). The substantial growth has been in the use of individual xylene isomers as chemical intermediates. As shown in Figure 1 each of the C_8 aromatics has essentially only one end use: ethylbenzene is used to make styrene, p-xylene is used to make terephthalic acid (or dimethyl terephthatlate), o-xylene is used to make phthalic anhydride, and m-xylene is used to make isophthalic acid. Ultimate end usage of each C_8 aromatic is also shown.

The growth in consumption of p- and o-xylene has been particularly spectacular, as can be seen from U.S. production figures shown in Figure

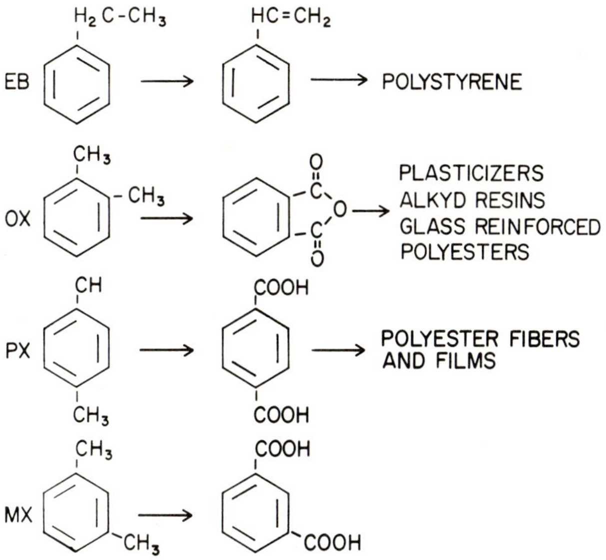

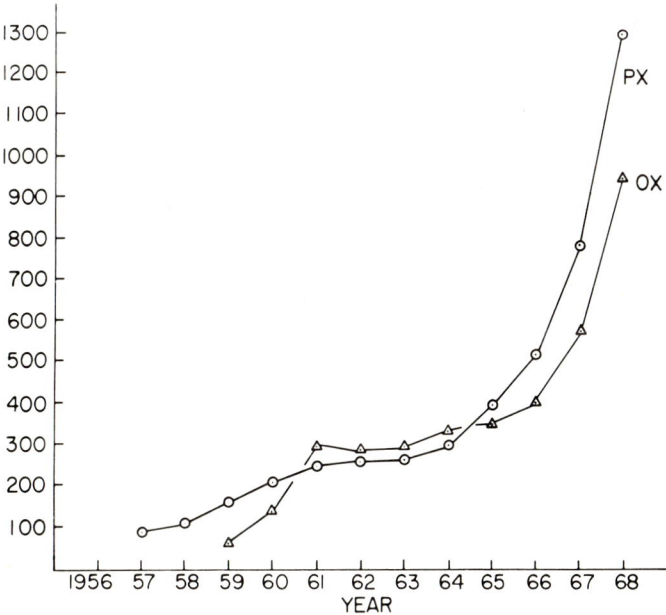

Figure 1. End usage of C_8 aromatics

Figure 2. U.S. production of p- and o-xylene in million pounds per year

2 (9). In 1968 U.S. production of p- and o-xylene was estimated at 1.3 billion and 0.97 billion pounds per year respectively. Production figures for m-xylene have never been published by the Tariff Commission. Its use has, however, remained quite small in relation to p- and o-xylene. It is expected that domestic demand for both p- and o-xylene will continue to increase. Terephthalic acid is the key component required for production of polyester film and fibers and is presently produced only from p-xylene. Phthalic anhydride is produced from both naphthalene and o-xylene. Although o-xylene is not expected to replace naphthalene entirely, its use for phthalic anhydride manufacture is expected to increase.

Both o-xylene and ethylbenzene can be removed almost completely from the mixed C_8 aromatics by fractionation. Low temperature crystallization is used to separate p-xylene. p-Xylene recovery, however, is limited by the formation of a eutectic mother liquor containing 7–12% p-xylene. If additional p-xylene is required, isomerization must be used to increase p-xylene production by converting m- and o-xylene and ethylbenzene in the mother liquor to p-xylene. The octafining process was developed to meet this demand and has become an important means of providing the p-xylene required by burgeoning development of the polyester fiber industry. With the growth in the o-xylene market the trend is towards use of octafining to increase both o- and p-xylene yields. Virtually all new plants are being designed for this dual purpose operation.

Although commercial use of the octafining process has been to isomerize C_8 aromatics, the process is equally suited to the isomerization of C_9 and C_{10} aromatics. A particularly promising application is the isomerization of mixed C_9 aromatics to produce pseudocumene.

Chemistry

The reactions of major importance in the octafining process are: isomerization of naphthenes and aromatics, hydrogenation of aromatics, dehydrogenation of naphthenes, disproportionation of aromatics, dealkylation of aromatics, and hydrocracking of saturates (Figure 3). The last three reactions, of course, result in loss of product xylenes. These reactions, like the desired isomerization reactions, are carbonium-ion catalyzed.

Catalysts which promote carbonium ion reactions have long been used for hydrocarbon isomerization (2, 6). Although such catalysts promote aromatics isomerization (3), in the case of ethylbenzene, the predominant reactions are disproportionation and dealkylation (5). Pitts, Connor, and Leum (7) demonstrated that hydrogenated intermediates were required to isomerize both ethylbenzene and cumene over platinum–

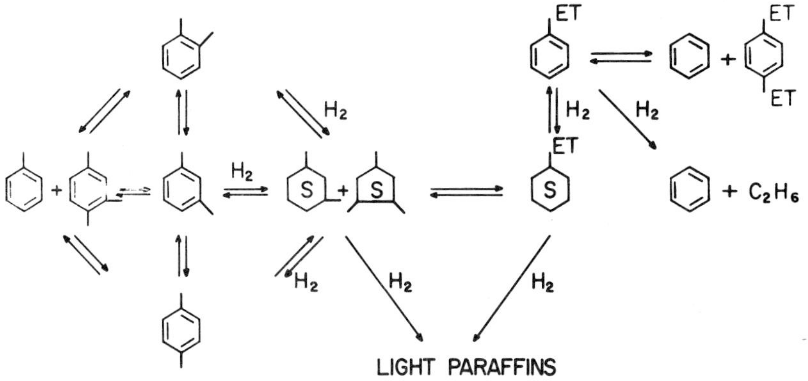

Figure 3. Reactions in octafining

Table I. Ethylbenzene Isomerization[a]

Yields, wt %[b]	Feed	Products
Saturates	—	6.4
p-Xylene	—	16.0
m-Xylene	—	35.1
o-Xylene	—	16.0
Ethylbenzene	99.9	16.3
Other aromatics	0.1	10.2
EB approach to equilibrium, %	—	88.0
Selectivity to xylenes, %	—	80.2

[a] Temperature = 850°F; H_2 partial pressure = 180 psia.
[b] 100% basis

silica–alumina catalysts. No isomerization was indicated for conditions where aromatics hydrogenation is forbidden.

Of course, the mere addition of hydrogenation activity to an acidic base is not sufficient for the selective isomerization of ethylbenzenes. If a catalyst is to be highly selective, the hydrogenation–dehydrogenation and the acidic functions must be balanced properly. Specifically, a high selectivity for the isomerization of ethylbenzenes (and of methylbenzenes as well) requires that the rate of naphthene cracking be low relative to that of the isomerization reactions (note reaction scheme, Figure 3). In addition, the catalyst acidity must be such that appreciable ethyl group dealkylation does not occur.

One of the important features of the octafining catalyst is its ability to convert ethylbenzene to xylenes. The data of Table I demonstrate the selectivity level at high conversions achievable with the present commercial octafining catalyst for the isomerization of ethylbenzene.

Although commercial designs are such that lower approaches to equilibrium are obtained, comparable selectivities are achieved.

In developing the present octafining catalyst, various acidic components were evaluated, including molecular sieves and halided aluminas. Silica–alumina has been found to offer a level of acidity which not only permits a high selectivity for the isomerization of ethylbenzene but high activity levels as well. In addition, the acidity level is stable and is not permanently affected by exposure to moisture levels sufficiently high to remove halogen from a halided alumina. Further, a silica–alumina catalyst, as prepared for the octafining process, is readily regenerable using the same procedures developed for catalytic reforming. For a description of the catalyst preparation *see* Ref. 4.

Octafining Process Description

The octafining process is used to restore to near equilibrium concentrations C_8 aromatic streams deficient in one or more of the xylene isomers. In conjunction with an appropriate separation process any of the xylene isomers can be produced by recycling to extinction the other less marketable xylene isomers and ethylbenzene.

The greatest application of octafining has been its use in conjunction with a crystallizer to produce *p*-xylene. The schematic in Figure 4 shows this application (Case A). The filtrate obtained from the crystallizer (generally containing 7–12 wt % *p*-xylene) is combined with recycle and make-up hydrogen and passed over octafining catalyst at 800°–900°F and 150–350 psig. Isomerization within the octafining reactor increases the *p*-xylene content so that it approaches its equilibrium value of about 24 wt % (xylene basis). Following the reactor the isomerizate is

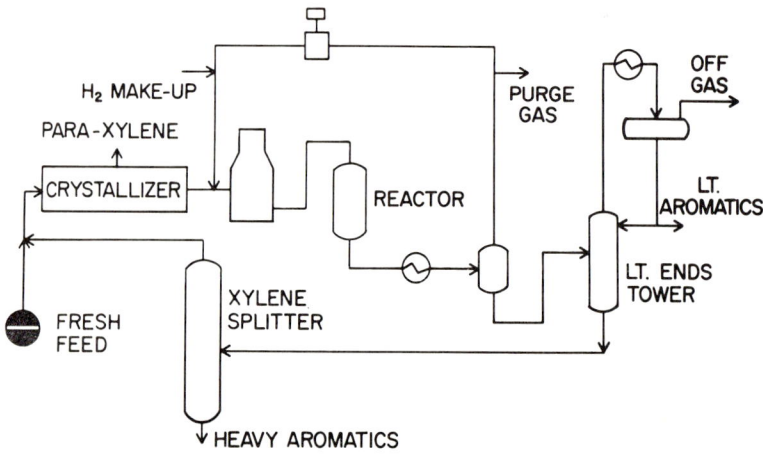

Figure 4. Octafining as used in production of p-xylene—*Case A*

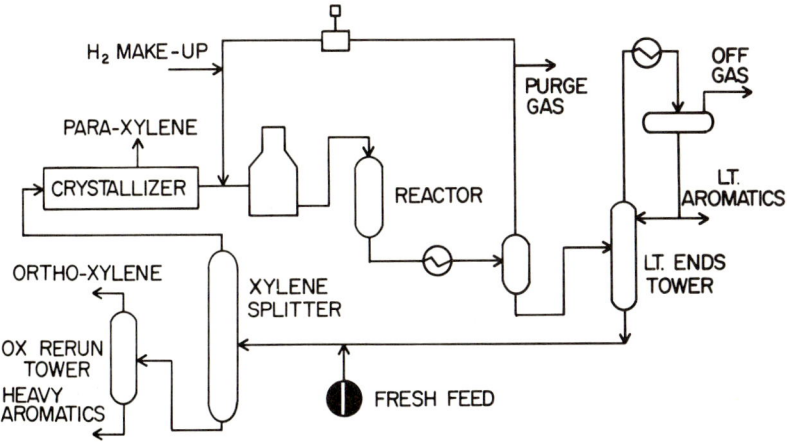

Figure 5. Octafining as used in production of p- and o-xylene—Case B

cooled, and a hydrogen-rich stream is separated for recycle. From the product separator the liquid effluent is fractionated successively in two towers to separate the C_8 aromatics from a C_7 and lighter fraction and a C_9 and C_{10} aromatics fraction. From the top of the xylene splitter the remaining C_8 aromatics mixture is recycled to join fresh feed before passing to the crystallizer, where the p-xylene product is removed.

Another application of octafining has been its use in conjunction with a crystallizer and appropriate fractionation to produce both p- and o-xylene, as shown in Figure 5 (Case B). In this processing scheme the fresh feed is combined with bottoms from the light ends tower as feed to the xylene splitter. Xylene splitter bottoms are fed to the o-xylene rerun tower where high purity o-xylene (96%+) is separated from the heavy aromatics. Xylene splitter overhead is routed to the crystallizer where the p-xylene product is removed. Crystallizer filtrate is combined with recycle and make-up hydrogen and passed over the octafining catalyst. Operating conditions are in the same range as those shown for Case A. Isomerization within the octafining reactor increases both the o- and p-xylene content to near equilibrium values (24 wt % for PX and 23.5 wt % for OX on xylene basis). Following the octafiner reaction section the liquid effluent is passed to the light ends tower where the C_7 and lighter by-products are removed overhead.

Octafining Material Balance

To illustrate typical commercial applications of octafining two material balances are shown in Table II. Case A corresponds to the schematic flow diagram in Figure 4, representing production of p-xylene only. Case B corresponds to the schematic flow diagram in Figure 5, representing production of both p- and o-xylene. Data shown are calculated values representing typical average commercial experience.

Table II. Octafining Material Balance

	Case A	Case B
Material Balance Summary		
Feed Streams, $\overline{M}/lbs/yr$		
Fresh feed	120.5	109.8
H$_2$ make-up	.9	.5
	121.4	110.3
Products, $\overline{M}/lbs/yr$	Calc.	Calc.
p-Xylene (100% basis)	100.0	41.7
o-Xylene (100% basis)	—	58.3
Gas make	6.4	3.5
Benzene	3.0	2.0
Toluene	1.9	.5
Naphthenes	5.6	2.3
C$_9$ and C$_{10}$	4.5	2.0
	121.4	110.3
Ultimate Xylene Yield		
p-Xylene, wt %	83.0	37.9
o-Xylene, wt %	—	53.2

Both cases are based upon a fresh feed having the following composition:

Ethylbenzene	20 wt %
p-xylene	20
m-xylene	40
o-xylene	20
	100

This composition represents an average C$_8$ aromatic mixture obtained by extraction from a catalytic reformate. Inclusion of aromatic extracts from pyrolysis gasolines and/or prefractionation to remove ethylbenzene can cause a wide variation in fresh feed composition.

Further assumptions used to prepare Cases A and B are as follows:

(1) Total xylene production is 100 million pounds per year. This level of production is fairly typical of the commercial units presently being designed.

(2) Crystallizer efficiency is 60% (for each pound of p-xylene in the crystallizer feed 0.6 lb is removed as p-xylene product). This efficiency is also fairly typical of commercial experience.

(3) For Case B the o-xylene removal in the xylene splitter has been set at 90%. (For each pound of o-xylene in the xylene splitter feed 0.9 lb is removed as product in the bottoms stream.)

Octafining is a highly selective process which can efficiently process C$_8$ aromatic feeds with high ethylbenzene contents. Prefractionation for ethylbenzene removal from the fresh feed is not required. With the feedstock shown, an ultimate yield of 83 wt % is obtained when producing only p-xylene. Producing both p- and o-xylene gives a combined ultimate yield of 91.1 wt %. Obviously, the relative yields of p- and o-xylene need not be fixed in the proportion shown. Based upon market requirements, the desired range in production rates of the individual isomers can be incorporated into the design of the octafiner and xylene separation facilities. The high selectivity of the octafining catalyst for converting ethylbenzene to xylenes is amply demonstrated in Case B. With a fresh feed containing 80% xylenes an ultimate xylene yield of 91.1% is obtained. An ultimate yield of this magnitude obviously requires low losses per pass for xylene isomerization as well as high selectivity for ethylbenzene conversion.

The reactions responsible for formation of individual by-products have already been discussed. Two gas make streams are obtained. The stream from the product separator has a high hydrogen content (generally 80–95 mole %), suitable for further downstream processing. The gas make stream from the light ends tower is relatively low in hydrogen content and is normally flared as fuel gas. Light aromatics obtained as a liquid overhead stream from the light ends tower consists of benzene, toluene, and C_8 naphthalenes. This stream can either be sent to the gasoline pool (typical clear octane no. = 97; 105.8 with 3 cc TEL) or returned to an aromatics extraction plant to separate benzene and toluene.

Heavy aromatics formed in the octafiner also have excellent octane number blending values and are generally returned to the refinery gasoline pool. The stream consists entirely of C_9 and C_{10} alkylbenzenes. Blending octane numbers (1) of these compounds range from 118 to 170. The C_9 fraction contains approximately 50 vol % pseudocumene and 20 vol % mesitylene.

Hydrogen make-up stream shown is the stoichiometric chemical consumption in the octafiner. Typical source of this stream is the off-gas produced from a catalytic reformer with a hydrogen content in the range of 75–95 mole %.

Investment and Operating Costs

Estimated investment costs for Cases A and B are shown in Table III. Investment costs are based upon erection of the plant on the U.S. Gulf Coast. Erected costs include the first catalyst charge but exclude associated off-site facilities, crystallizer, and royalty charges. Estimated operating costs are shown in Table IV. Utility costs in dollars per calendar day are based upon use of electric drivers for pumps and compressors, maximum use of air coolers, and fired furnaces for reboiling the light ends

Table III. Estimated Capital Investment

	Case A	Case B
Estimated Erected Cost, million dollars	2.4	1.8

Basis:
U.S. Gulf Coast
Includes first catalyst charges and fractionation.
Excludes off-site facilities, crystallizer, and royalties.

Table IV. Estimated Operating Costs

	Operating Costs, Dollars/cd	
	Case A	Case B
Utilities		
Fuel @ 20c/$\overline{M}$ BTU	336	187
Cooling water @ 1.5c/1000 gal	39	17
Power @ 1c/kwh	672	271
Labor		
2 operators/shift @ $4.40/hr	211	211
Supervision @ 25% of operating labor	53	53
Maintenance		
4% of erected cost (exclusive of catalyst)	211	162
Catalyst	271	180
Laboratory	35	35
Insurance and Taxes @ 2.5% of erected cost (exclusive of catalyst)	133	101
Total Direct Operating Cost (except depreciation)	1,961	1,217
Cents per lb Product	0.72	0.44

tower, xylene splitter and o-xylene rerun tower. Total direct operating costs include utilities, labor, maintenance, catalyst, and miscellaneous charges as shown but exclude any allowance for depreciation.

As can be seen from Table III and Table IV, appreciable savings in both capital investment and production cost can be realized in Case B where both p- and o-xylene are produced. As already noted, both Case A and Case B are based upon production of 100 million pounds per year of xylene products. For Case B capital investment is 1.8 million dollars as opposed to 2.4 million dollars for Case A and direct operating cost decreases from 0.72 cent per pound of product to 0.44.

The rapid increase in demand for the o-xylene since 1965 has resulted in virtually all new plants being designed to produce both p- and o-xylene. For constant p-xylene production, the high efficiency removal of o-xylene from the fresh feed results in appreciably smaller octafiner and crystallizer units and consequently in lower direct operating costs.

Commercial Experience

The first octafiner was placed on stream in 1960. Since then 10 additional plants producing approximately 1 billion pounds per year of p-xylene plus o-xylene have been placed on stream. Eight plants presently under design or construction will increase total xylene isomer production to almost 2 billion pounds per year.

Feedstocks processed commercially include C_8 aromatic extracts from catalytic reformates and pyrolysis liquids resulting from ethylene cracking plants. Composition of these stocks vary widely, the major differences being their ethylbenzene content. Fresh feeds from which ethylbenzenes have been removed typically contain 2–4 wt % ethylbenzene. Inclusion of C_8 aromatic extracts from pyrolysis liquids can increase ethylbenzene content to 30 wt % or higher. Fresh feeds with this range of ethylbenzene contents have been processed successfully over octafining catalyst.

Ultimate catalyst lives of greater than three years have been achieved commercially. The catalyst is regenerable *in situ* with no additional mechanical equipment required. Regeneration for coke removal is simple and straightforward. The recycle gas compressor is used to circulate nitrogen into which controlled quantities of plant air are added. Combustion of coke deposits is generally accomplished at 700°–900°F and 50–100 psig. A complete regeneration cycle from "feed-out" to "feed-in" can generally be completed in three days.

Commercial experience has led to optimization of catalyst performance characteristics and octafiner process design such that ultimate xylene yields have increased from the high 60's to greater than 90 wt % on feed.

Acknowledgments

The authors are indebted to the Atlantic Richfield Co., Engelhard Minerals & Chemicals Corp., and the many people who have contributed to the experimental work, design studies, and successful commercial application of the octafining process.

Literature Cited

(1) ASTM, *Spec. Tech. Publ. No. 225* (1958).
(2) Evering, B. L., *Advan. Catalysis* **1954**, 6, 197–239.
(3) Kilpatrick, M., Luborsky, F. E., *J. Am. Chem. Soc.* **1953**, 75, 577.

(4) Leum, Leonard N., Connors, James E., Jr., U. S. Patent **2,976,332** (March 21, 1961).
(5) Lien, A. P., "Abstracts of Papers," ACS, March-April 1954.
(6) Pines, H., *Advan. Catalysis* **1954**, 1, 215–272.
(7) Pitts, P. M., Connors, J. E., Jr., Leum, L. N., *Ind. Eng. Chem.* **1955**, 47, 770.
(8) Stanford Research Institute, "Chemical Economics Handbook," Research Information Center Economics Division, Menlo Park, Calif., 1967.
(9) U. S. Tariff Commission, "Synthetic Organic Chemicals—U. S. Production and Sales," 1968.

RECEIVED January 12, 1970.

13

Process Evaluation of Improved Solvents for Butadiene Recovery

G. D. DAVIS, E. C. MAKIN, JR., and C. H. MIDDLEBROOKS[1]

Monsanto Co., 800 North Lindbergh Blvd., St. Louis, Mo. 63166

Four solvents were evaluated for the recovery of 1,3-butadiene from crude C_4 fractions by extractive distillation. Furfural [5 wt % water], methyl Cellosolve [10% water], acetonitrile [10% water] and β-methoxypropionitrile [5% water] solvents were studied at comparable operating conditions in a 2 inch diameter, 140 tray column. Furfural and methyl Cellosolve solvents performed the desired separation between 1,3-butadiene and trans*-2-butene only at high solvent-to-C_4 feed ratios. Acetonitrile and β-methoxypropionitrile solvents were far superior to furfural and methyl Cellosolve at equivalent solvent-to-C_4 feed ratios. The superior solvents could perform the desired separation at half the solvent ratio. When related to a plant scale operation, β-methoxypropionitrile solvent could double the capacity of an existing butadiene plant using furfural.*

Butadiene is one of the most valuable chemicals produced in the modern chemical refining complex. The pure monomer is recovered from crude C_4 streams by liquid-liquid extraction or extractive distillation with a selective solvent. A simplified diagram of an extractive distillation process is shown in Figure 1. The C_4 feed is separated into two product streams in the extractive distillation column—a butane–butene stream which goes overhead from the column and a butadiene concentrate which is carried out the bottom of the column dissolved in the solvent. The butadiene concentrate contains *cis*- and *trans*-2-butene, and after being stripped from the solvent, it is subjected to normal distillation. Butadiene is distilled overhead at 99+% purity.

[1] Present address: Monsanto Co., Box 711, Alvin, Tex. 77511.

In extractive distillation systems, good performance in economy and efficiency of operation can be achieved only with solvents having high selectivity for the desired component. To extract one component having a higher volatility from other components having lower volatilities requires a high selectivity because the relative volatilities must be reversed. For a given solvent, selectivity determines the number of stages required to achieve the desired separation. For a given number of stages and non-solvent reflux ratio, selectivity establishes the solvent-to-feed ratios which are permissible with that particular solvent. In an existing purification plant, solvents with higher selectivities allow efficient operation at lower solvent to feed ratios [*e.g.*, 6 or 8 to 1]. Plant operation at low solvent ratios is obviously preferred to attain higher production rates at lower energy requirements.

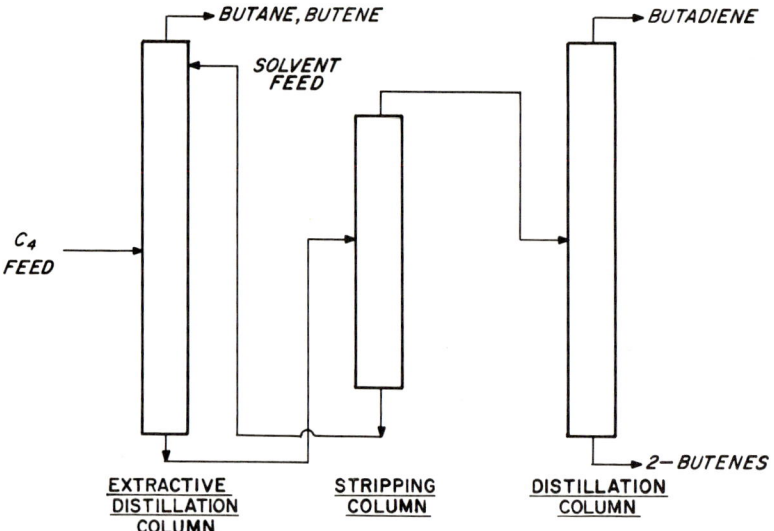

Figure 1. Schematic of a butadiene extractive distillation and purification system

Other important solvent properties which must be considered are solute carrying capacity, ease of solute–solvent separation, thermal and chemical stability, and corrosivity. Properties of secondary importance include cost, density, viscosity, specific heat, and heat of vaporization.

A number of solvents were studied as candidates for replacing furfural in existing butadiene extraction plants. The significant data from that study have been reported (3), and the most promising solvents are compared in Table I.

Table I. Comparison of Butadiene Solvents at Optimum Water Concentration

	Relative Volatility of trans-2-Butene at 70 psig	
Solvent	At Equivalent[a] Solvent Load	At Equivalent[b] Temperature
Furfural (5% water)	1.292	1.269
Methyl Cellosolve (10% water)	1.302	1.303
Acetonitrile (10% water)	1.349	1.350
β-Methoxypropionitrile (5% water)	1.442	1.408
Normal fractionation, no solvent	—	0.848

[a] This represents 20 mole % C_4's in the solvent on a water-free basis.
[b] Temperature of solvent systems equivalent at 140°F.

The solvent selectivities for butadiene over *trans*-2-butene (key pair) are compared at optimum water concentration and at equivalent temperature and solvent loading. Solvent selectivity is expressed in terms of the relative volatility (α) for *trans*-2-butene over butadiene. Relative volatility of each C_4 component was calculated from its equilibrium ratio (K value). Equilibrium ratios were calculated from the analyzed mole fraction of the components in the system as follows:

$$K_i = y_i/x_i$$

where K_i is the equilibrium constant, y_i is the vapor-phase mole fraction of the *i*th component, and x_i is its liquid-phase mole fraction. Since the equilibrium ratio is a direct measure of component volatility, relative volatility (α) may be expressed as:

$$\alpha_{ij} = \frac{K_i}{K_j} = \frac{y_i/x_i}{y_j/x_j} = \frac{y_i x_j}{y_j x_i}$$

where *j* is butadiene and *i* represents all other components in the system taken individually.

The specific objective of this work was to confirm the predicted improved performance of the solvents in an extractive distillation column. Another objective was to relate the experimental data to plant scale operations.

Experimental

Materials. Furfural was obtained from Quaker Oats Co. Methyl Cellosolve was obtained from Union Carbide Corp. Sohio Chemical Co. supplied the acetonitrile, and β-methoxypropionitrile was an Eastman Kodak Co. product.

Table II. Physical Properties of Some Selective Solvents

Property	Furfural	Methyl Cellosolve	Acetonitrile	β-Methoxy-propionitrile
Cost, cents/lb.	14–16	18–22	24	20–25
Molecular weight	96.08	76.10	41.05	85.10
Boiling point, °F at 1 atm	323	255.7	178.9	320
Density, lb/gal at 68°F	9.6	8.03	6.7	7.8
Viscosity (cp) at 100°F	1.35	1.23	0.30	0.99
Specific heat, BTU/lb/°F at (°C)	0.416 (20°–100°)	0.534 (124.5°)	0.541 (21°–76°)	0.42 (60°)
Heat of vaporization, BTU/lb at 1 atm	193.4	223	313	243.4

All solvents were distilled, and a middle fraction was collected and analyzed for purity. Furfural, 5% water; methyl Cellosolve, 10% water; acetonitrile, 10% water and β-methoxypropionitrile, 5% water were the solvent systems selected for study. Stated water concentrations are weight percent.

The physical properties of the pure solvents are shown in Table II. The solvents range in price from 14 to 25 cents/lb. High priced solvents (40 to 60 cents/lb) were not studied because solvent replacement becomes a considerable operating expense on a plant scale. The solvent boiling point should not be low enough to contaminate the products. The boiling point of acetonitrile is low enough to present a problem in this respect. Density is another important consideration. Since most extraction plants operate at maximum volumetric flow capacity, it is advantageous to have as many pounds of solvent circulating as possible. Relatively low viscosity, specific heat, and heat of vaporization are favored solvent properties.

All experimental runs were made on a C_4 feed of known composition. The feed composition ran from 37 to 40 wt % 1,3-butadiene. A typical analysis of this feed is shown in Table III. This composition is similar to C_4 streams obtained by thermal cracking of petroleum fractions.

Equipment. The equipment used is shown diagramatically in Figure 2. The extractive distillation column was made from 2-inch diameter (id) 316 stainless steel seamless tubing (0.218-inch wall). It was made in 5, four-foot sections, which were polished inside to remove surface imperfections. The ends were beveled to give a metal-to-metal, gas-tight seal. A slip-ring and threaded nut arrangement was used to join the column sections. Each section contained 28 sieve trays with segmental

Table III. Typical C_4 Feed Composition and Component Vapor Pressures

Component	Wt %	Vapor Pressure, psia at 70°F	Vapor Pressure, psia at 120°F	Relative Volatility, $V_p/V_p\,C_4H_6$ at 120°F
Isobutane	2.7	44.9	95.6	1.202
Isobutene	43.0	39.0	84.9	1.068
1-Butene		37.4	82.9	1.043
1,3-Butadiene[a]	37.0	35.8	79.5	1.000
n-Butane	3.2	30.9	69.0	0.868
trans-2-Butene[b]	6.5	29.8	67.9	0.854
cis-2-Butene	7.6	27.1	62.4	0.785

Average column head pressure at 120°F averaged 68 psig

[a] Heavy key component in extractive distillation.
[b] Light key component in extractive distillation.

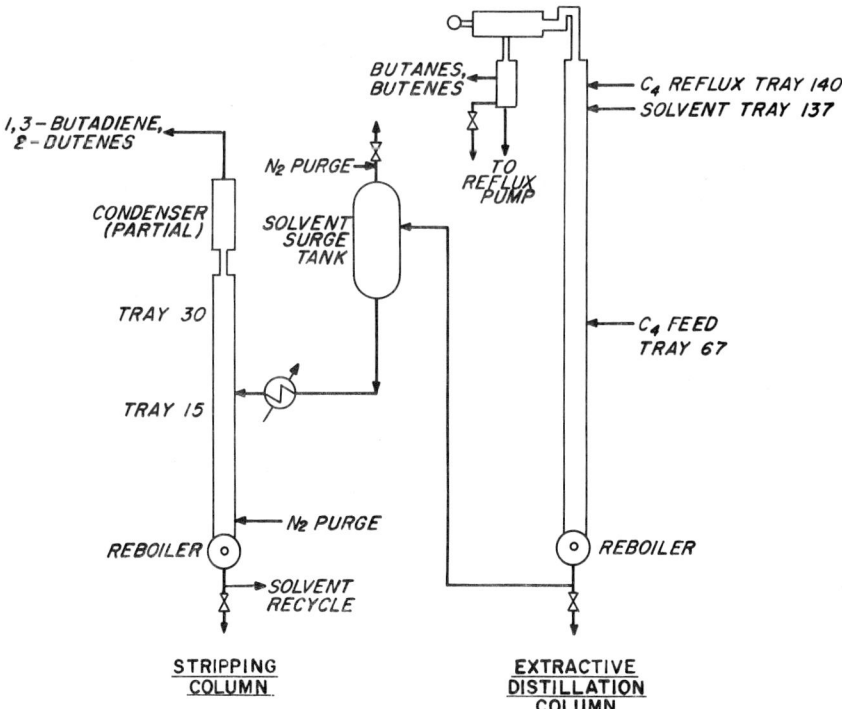

Figure 2. Extractive distillation unit

downcomers. The tray spacing was 1.5 inches, except for the feed tray, where the spacing was 3.0 inches. Each tray contained 80 holes, 0.043 inch in diameter, providing a hole area 4.3% of the total tray area.

Table IV. Comparison of Furfural and Furfural, 5% Water

Solvent-to-C$_4$ feed ratio	13:1			12:1		
Solvent feed, grams/hr	3308			3060		
C$_4$ feed, grams/hr	225			250		
C$_4$ reflux, grams/hr	425			475		
Reboiler temp., °F	306			308		
C$_4$'s in reboiler liquid, wt %	3.11			3.42		

Component, wt %	Dist.	Btms.	% Extd.	Dist.	Btms.	% Extd.
Isobutane	4.7			4.8		
n-Butane	5.6			5.7		
1-Butene, isobutene	74.7			76.0		
trans-2-Butene	10.8	0.7	3.6	10.8	1.0	6.2
cis-2-Butene	4.0	12.5	69.8	2.2	14.5	82.9
1,2-Butadiene	0.2	86.8	99.7	0.5	84.5	99.2

Downcomer area was 6.88% of the total tray area, while the weir height was 0.1875 inch. Pipe couplings were welded at various locations for thermocouple and feed inlet ports. The solvent feed tray was 137 trays from the bottom. C$_4$ reflux was returned to the 140th tray. The fully assembled extraction column contained 140 sieve trays and stood 21 feet high.

The column could handle up to 4500 grams/hour of solvent at operating conditions. The C$_4$ feed rate varied from 200 to 500 grams/hour. Butene reflux rates of 400 to 700 grams/hour were sufficient for the range of C$_4$ feed and solvent flows studied.

The stripping column construction was identical to that of the extractive distillation column. It contained 30 trays with the feed point on the 15th tray.

Operation. Typical operating conditions for butadiene recovery plants have been described by Buell and Boatright (1). These data were used as a guide for setting the conditions used in this study. At the start of a run, the column was preheated electrically to the desired column temperature profile. The C$_4$ feed was pumped into the column until the pressure reached 30–35 psig. Solvent (preheated to 130°F) was then circulated through the column, and the reboiler temperature increased to about 275°F. The hydrocarbon feed was pumped in at a rate which gave the desired solvent-to-hydrocarbon feed ratio. A portion of the butene stream was returned to the column to provide reflux. Temperature of the solvent feed tray was used to control the amount of reflux. Only that portion of C$_4$'s which dissolved in the solvent could be carried down the column. The excess C$_4$'s were revaporized to the hydrocarbon trays.

β-MOPN Solvents at Limiting Solvent Ratios

β-MOPN, 5% Water

	6:1				5:1	
	3205				2670	
	530				510	
	600				600	
	304				254	
	6.33				8.35	
Dist.	Btms.	% Extd.	Dist.	Btms.	% Extd.	
4.7			4.8			
5.6			5.7			
74.9			76.4			
11.0	0.5	3.0	10.6	1.2	7.7	
3.7	12.9	72.3	2.1	14.6	84.1	
0.1	86.6	99.9	0.4	84.2	99.5	

After the column was lined out, liquid samples were taken from the butenes condenser and from the reboiler. The composition of these samples served as a guide to indicate which adjustments in operating conditions were necessary to obtain the desired column operation. If the reboiler liquid contained 1-butene or isobutene, its temperature was increased, provided there was no butadiene in the overhead. If butadiene appeared in the overhead, the internal hydrocarbon reflux was increased by increasing the pressure or decreasing the solvent feed tray temperature. Solvent-to-hydrocarbon feed rate adjustments were also effective in giving desired changes in bottoms and overhead compositions.

After equilibrium in the extractive distillation column was reached, the overhead pressure stabilized and the column temperature gradient increased smoothly from tray to tray down the column. Samples were taken, and column conditions were recorded at two-hour intervals. When the desired data had been obtained at one solvent-to-C_4 feed ratio, solvent and feed flows were adjusted to a new ratio, and additional data were obtained. Table IV and Figures 3 and 4 show typical data obtained for each solvent.

Precision and Accuracy. Analytical precision was believed to be accurate to ±0.1% to component concentrations as low as 0.5%. Temperature measurement was accurate to ±0.5°F, and the pressure measurement was accurate to ±1 psig.

Analytical Methods. The distillate and bottoms C_4 compositions from the extractive distillation column were determined by VPC (F and M Model 810) using a 200-foot column coated with 1-dodecyne. Resolu-

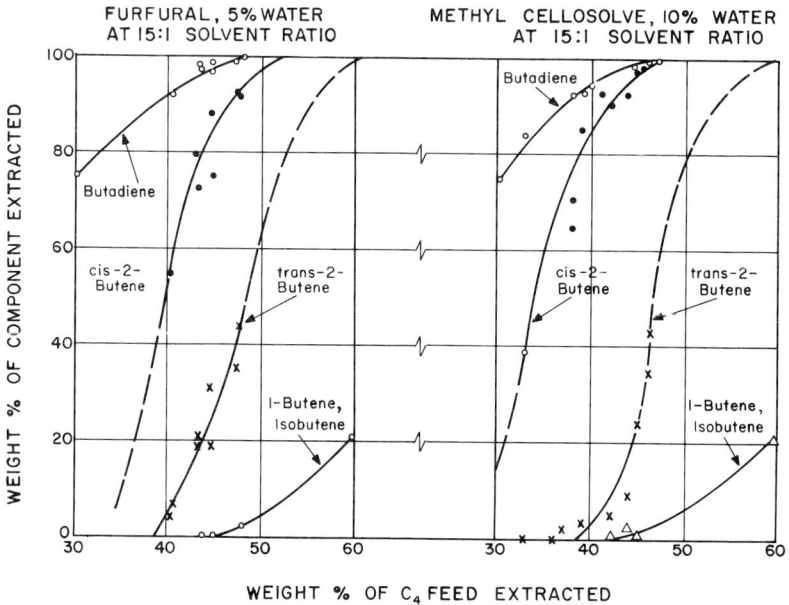

Figure 3. Extraction data for furfural and methyl Cellosolve solvents

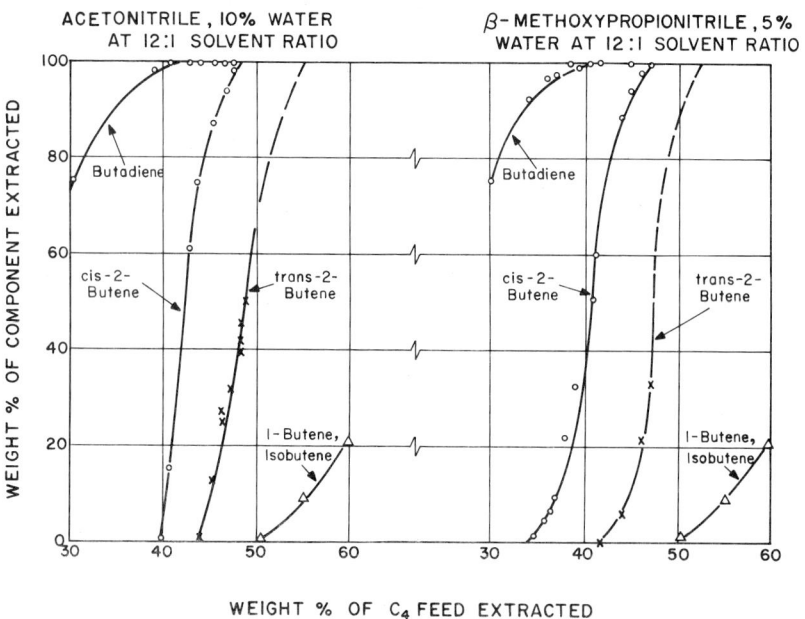

Figure 4. Extraction data for acetonitrile and β-methoxypropionitrile

tion of the key components was good at 0°C, and the analyses were fast. Only 20 minutes were required for scanning both samples. The bottoms samples were analyzed by VPC (F and M Model 720) to determine the percent total C_4's. A 12-foot by 1/2-inch column packed with SE 52 gum rubber on Chromosorb W programmed to 300°C was used for this analysis. This column did not resolve the individual C_4 components. The water concentration in acetonitrile solvent was monitored by VPC (Fisher-Gulf Partitioner) using a 12-foot by 1/4-inch column packed with 5% polyethylene glycol on powdered Teflon. Resolution was good, and only 5 minutes were required for elution at 45°C. Correction factors for area response were: 5% water = 0.777, 10% water = 0.822, 15% water = 0.864, 20% water = 0.910 and 25% water = 0.954. Corrected areas for water peaks were obtained by multiplying the response factor times the measured area. Corrected reproducible analyses for water were obtained.

Water content of furfural, methyl Cellosolve, and β-methoxypropionitrile solvents was determined gravimetrically by azeotropic distillation with benzene. One hundred grams of solvent and 50 grams of benzene were refluxed until no more water could be extracted. The extracted water collected in a separate phase in a sidearm receiver was measured and related to the original solvent fraction to determine the water concentration.

Discussion of Results

The performance of butadiene recovery plants improves as solvent selectivity increases. The greater the solvent's selectivity, the better the separation of *trans*-2-butene (light key) and butadiene (heavy key) in the extractive distillation column. Relative volatility of the key pair in β-MOPN was 1.337 at 70 psig and 130°F. This results from inversion of the volatilities. In normal distillation *trans*-2-butene is the heavy key ($\alpha = 0.848$). Lower *trans*-2-butene levels in the butadiene concentrate require less reflux and enables handling of increased loads in the final purification column.

Figures 3 and 4 show that furfural and methyl Cellosolve solvents cannot produce butadiene concentrates free of *trans*-2-butene at solvent-to-feed ratios where the more selective solvents acetonitrile and β-methoxypropionitrile (β-MOPN) are able to reject all of this isomer from the concentrate. In fact, with furfural solvent, considerable loss of butadiene must be accepted to approach a low level of *trans*-2-butene in the butadiene concentrate. On the other hand, it is possible to recover 100% of the butadiene in the feed while no *trans*-2-butene is extracted with the β-methoxypropionitrile or acetonitrile solvents. As much as 40% of the higher boiling *cis*-2-butene isomer is rejected at 100% butadiene recovery at solvent ratios as low as 12 to 1. These data indicate much lower solvent ratios could be achieved with these solvents.

The extractive distillation data show that in general, for satisfactory operation, furfural is limited to high solvent-to-C_4 feed ratios. The data in Table IV indicate the limiting solvent ratio to be *ca.* 12 to 1. In addition to its poor selectivity compared with other solvents, furfural reacts with itself to form a polymer (4) and with butadiene to form codimers which have been characterized in detail by Hillyer (5). Solvent losses to polymer and in the associated solvent purification system amount to a considerable yearly operating expense.

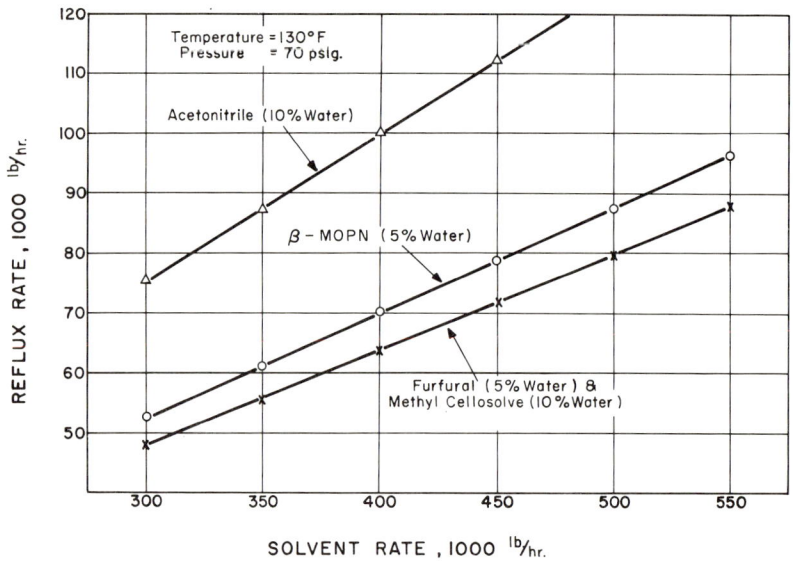

Figure 5. Minimum reflux required for solvent saturation

Methyl Cellosolve was rated in our studies to be equivalent to furfural. It had a similar selectivity for butadiene and therefore could offer no additional plant capacity and little economic advantage. It does, however, offer an advantage in thermal and chemical stability, which would reduce solvent replacement expense. Its lower density is a disadvantage in that at equivalent volumetric flow rates the plant capacity using methyl Cellosolve would be lower than that of furfural.

Acetonitrile was shown to have excellent selectivity for butadiene and could produce the desired separation of the key components at low solvent ratio. However, acetonitrile has several serious disadvantages. It is thermally unstable above 290°F. At this temperature and above, highly corrosive acetic acid rapidly forms in the presence of water. The acid can be neutralized easily, but ammonia (a by-product of the decomposition) contaminates both the butene and butadiene products. This necessitates additional purification of products by water scrubbing. Another disadvantage of acetonitrile is the higher reflux rate required. Minimum reflux requirements for the extractive distillation column operation are set conveniently by the butene solubility in the solvent at top tray pressure and temperature. For example, at 130°F and 70 psig solubility in acetonitrile is 25 wt %, while furfural is saturated with 17 wt % C_4's at the same conditions. Minimum reflux requirements in pounds per hour for all the solvents are shown in Figure 5. These rates are based on the minimum saturation requirements for the respective solvents at comparable tray conditions.

The superior selectivity of β-methoxypropionitrile (β-MOPN) was demonstrated in the experimental operation at low solvent-to-C_4 feed ratios. Hot β-MOPN is non-corrosive to mild steel and thermally stable up to 338°F. The solvent does not react with butadiene or with itself. The advantage of using β-MOPN to increase extractive distillation column capacity may be seen in Figure 6. Calculated butadiene capacity in millions of pounds per year is shown as a function of solvent-to-C_4 feed ratio. Maximum butadiene capacity with furfural at 12 to 1 solvent ratio is limited to about 100 million lbs/year. With β-MOPN, butadiene extraction capacity approaches 160 million lbs/year at 6 to 1 solvent ratio. Lower solvent ratios were achieved with this solvent in experimental operation as shown in Table IV. The curve in Figure 6 shows that at

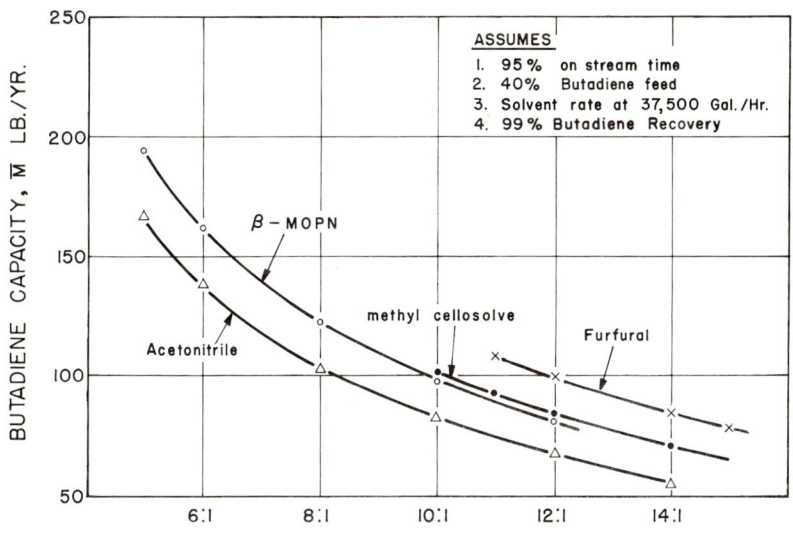

Figure 6. Calculated butadiene production rates with various solvents

5 to 1 solvent ratio, the butadiene capacity would be approximately 200 million lbs/year. This represents 100 million pounds more production capacity than is possible with furfural in the same extractive distillation column.

The economic advantage of using β-methoxypropionitrile is shown in Table V. The solvents are compared at a constant volumetric flow rate which may be considered as a limiting factor in plant capacity. The solvent flow in pounds per hour becomes a function of solvent density.

Table V. Calculated Performance of Butadiene Plant With Various Solvents (Solvent Flow at 37,500 gal/hr)

Property	Furfural	Methyl Cellosolve	Acetonitrile	β-Methoxy-propionitrile
Solvent density, lb/gal	9.6	8.03	6.7	7.8
Solvent flow, lb/hr	360,000	301,225	251,250	292,500
Minimum solvent-to-C_4 feed ratio	12:1	12:1	6:1	6:1
C_4 feed rate, lb/hr	30,000	25,094	41,875	48,750
Butadiene rate, lb/hr (at 40% in feed)	12,000	10,037	16,750	19,500
Butadiene rate $\overline{M}^a$ lb/yr (at 8320 hrs/yr)	100	83	139	162
Sales value, $\overline{M}$ $/yr (BD at 9.5 cents/lb)	9.5	7.9	03.2	15.4

^{a} $\overline{M}$ = million.

The minimum solvent-to-C_4 feed ratio is the only experimental variable needed to calculate solvent capabilities. This variable is defined as the lowest solvent ratio which will give the desired separation of the key pair. The balance of the calculation is based on physical property data and on variables which are independent of solvent character.

At a solvent flow rate of 37,500 gal/hour, furfural is capable of processing only 30,000 lbs of C_4 feed/hour, while β-MOPN can process 48,750 lbs/hour. Assuming the C_4 feed contained 40% butadiene, these rates of processing would produce 100 million lbs and 162 million lbs of butadiene, respectively, per year (8320 hours). The sales value of the increased butadiene production represents 5.9 million dollars. This is a conservative economic estimate of the advantage of converting to the more selective solvent. Since the physical properties of β-methoxypropionitrile and furfural are so similar, only minor revision of a plant now using furfural is needed.

The availability and cost of β-methoxypropionitrile should be considered since it has been produced only on a commercial scale as a chemical intermediate. Acrylonitrile reacts readily with aliphatic alcohols in the presence of strong alkaline catalysts to form a broad spectrum of compounds known as alkoxypropionitriles. These reactions, commonly known as cyanoethylation reactions, have long been of interest to organic chemists because of the ease with which the addition takes place (6). The availability of low cost acrylonitrile and methanol has now given β-methoxypropionitrile an attractive price tag. Cost estimates for producing this solvent are in the range of 20 to 25 cents/lb.

Conclusions

The over-all performance of β-methoxypropionitrile solvent in the pilot plant tests qualify it as a superior replacement solvent for furfural in butadiene extractive distillation plants. It offers distinct economic and operational advantages. Operation at lower solvent-to-C_4 feed ratios greatly increases existing extractor capacity. In addition, the improved separation of *trans*-2-butene and butadiene in the extractive distillation column reduces the load on the final butadiene purification column. Operation at lower solvent-to-C_4 feed ratio and lower reboiler temperature provides substantial utility savings. The lower reboiler temperature also reduces the rate of butadiene dimer formation.

Perhaps the most outstanding advantage of β-MOPN over other solvents is that its physical properties are close enough to those of furfural that little or no changes in solvent-related plant equipment are required.

Literature Cited

(1) Buell, C. K., Boatright, R. G., *Ind. Eng. Chem.* **1947**, 39, 695.
(2) Davis, G. D., Makin, E. C., U. S. Patent **3,372,109** (March 5, 1968).
(3) Davis, G. D., Makin, E. C., *Ind. Eng. Chem. Process Design Develop.* **1969**, 8, 588.
(4) Dunlop, A. P., Stout, P. R., Swadesh, S., *J. Chem. Phys.* **1939**, 7, 725.
(5) Hillyer, J. C., *Ind. Eng. Chem.* **1948**, 40, 2216.
(6) MacGregor, J. H., Pugh, C., *J. Chem. Soc.* **1945**, 535.

RECEIVED January 12, 1970.

14

Tetraethylene Glycol—A Superior Solvent for Aromatics Extraction

G. S. SOMEKH and B. O. FRIEDLANDER

Union Carbide Corp., Chemical and Plastics, P. O. Box 65, Tarrytown, N. Y. 10591

> *The separation of benzene, toluene, and C_8 aromatics (BTX) from gasoline fractions is reviewed from a fundamental and historical point of view, and the disadvantages of the old distillation processes are presented. The Udex process, which has been the dominant liquid extraction and extractive distillation process for effecting this separation, is being used more and more. Phase equilibrium data show the differences between the various glycol solvents that can be used in this process. These data have shown that tetraethylene glycol is superior to other solvents. There is a significant trend in the industry to employ tetraethylene glycol.*

The benzene–toluene–C_8 aromatics fraction is presently the principal raw material used to manufacture petrochemicals. This fraction (usually referred to as BTX) is even more important than ethylene as a raw material. Reformed gasolines contain these aromatics in concentions ranging from about 30 to 60%. BTX is also available from hydrogenated coke oven light oils, cracked gasolines, or hydrogenated dripolenes in which the concentrations can vary from about 70 to 97%.

The earliest large scale process used to separate BTX from aliphatics was straight distillation. However, it is impossible to obtain high-purity benzene by straight distillation, as can be seen from Figure 1. This figure shows the large number of homogeneous, binary azeotropes that exist between aliphatics and benzene. The boiling point of the pure aliphatic is shown on the left ordinate, and a line is drawn to connect each aliphatic to the composition and boiling point of its binary azeotrope with benzene. Thus, for each binary (aliphatic–benzene pair) there are three points on the vapor–liquid equilibrium curve: the boiling point of each of the two pure components and the boiling point of the azeotrope.

What happens if benzene is recovered by batch fractional distillation in the presence of aliphatics? First, pure light aliphatics—branched hexanes and all the C_5's—would be removed overhead since they do not form azeotropes with benzene. Then, aliphatics in the 70°–80°C boiling range would come off and would contain roughly 25% benzene. Thus, it is immediately apparent that one disadvantage of this process is that a good deal of benzene is lost in the light ends. The benzene that is distilled next is only fairly pure and comes off with the aliphatics that boil in the 80°–95°C range. With luck, the gasoline fraction might not have too many of these components. However, a substantial amount of high quality benzene cannot be obtained by this process because aliphatics that have boiling points even as high as 100°C are very difficult (if not impossible) to separate from benzene by fractional distillation. Furthermore, the whole benzene fraction (aliphatics as well as benzene) had to be distilled to get part of the benzene at only a good purity.

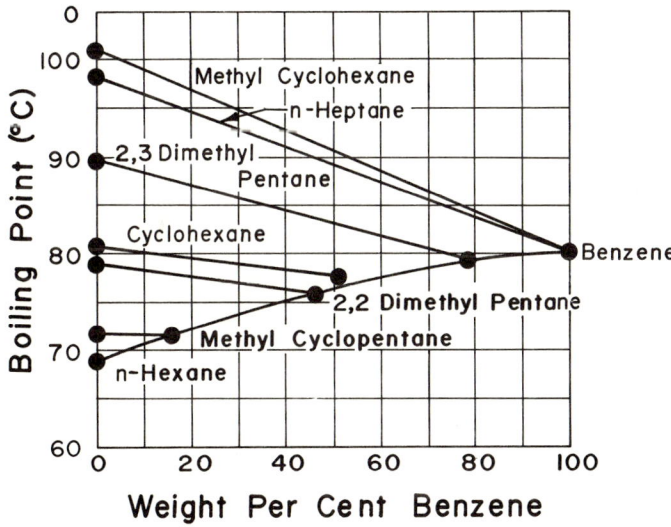

Figure 1. Azeotropy in nonaromatic hydrocarbon–benzene binary systems

Attempts to improve benzene purity by changing the pressure in hopes of altering the relative volatilities will not be very successful, as shown in Figure 2. The slopes of the vapor pressure curves of benzene and aliphatics are essentially parallel. Thus, the relative volatilities are almost unaffected, and the compositions of the binary azeotropes will not change much with change in pressure.

In the 1940's extractive distillation was used to increase aliphatic–benzene relative volatilities, thereby increasing benzene recovery and

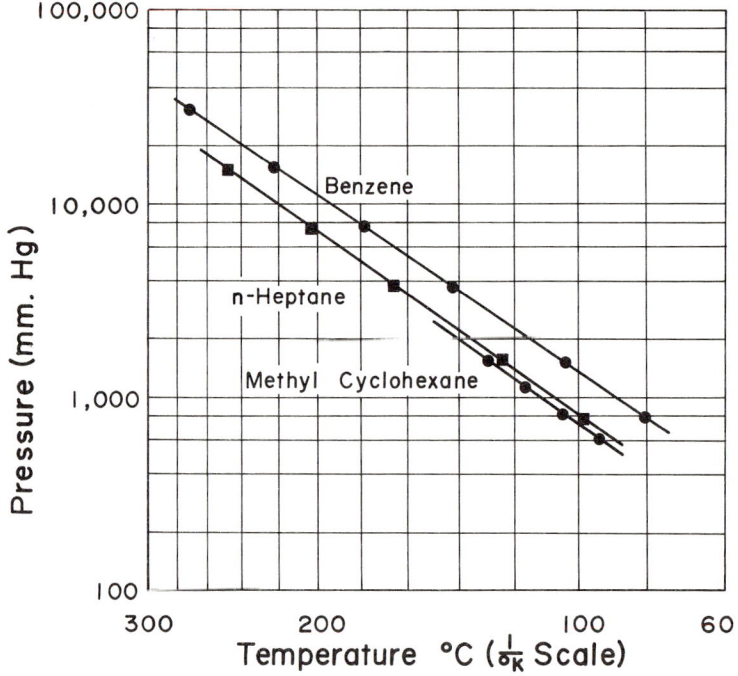

Figure 2. Vapor pressure vs. temperature benzene–n–heptane–methylcyclohexane

purity. Phenol was a favorite solvent for this purpose. However, the whole benzene fraction still had to be distilled. Furthermore, for these distillation procedures to be practical, a separate facility was needed to treat each aromatic-containing fraction—C_6, C_7, and C_8.

The Udex Process

In the early 1950's the Udex extraction process came into commercial existence and began to take over this separation. Liquid extraction can be used to separate hydrocarbons by type. Thus, benzene as well as toluene and C_8 aromatics could be recovered from aliphatics in one process. By extracting the aromatics selectively and then distilling them from the solvent, distillation of the aliphatics is avoided so that operating costs are lower than in the fractional and extractive distillation processes. High purity aromatics can be obtained because the aromatics are not only purified in the extraction step but in the distillation step as well. The Udex process was, therefore, a significant breakthrough, and it has been used widely ever since its introduction.

Diethylene glycol (DEG) was the first solvent employed in the Udex process. The structural formula of DEG, as well as those of the other presently used glycols, are shown in Table I. Diethylene glycol is ethylene glycol with an extra ethylene oxide group. Triethylene and tetraethylene glycols merely have additional ethylene oxide groups.

A description of the liquid–liquid equilibria in the benzene–heptane system with glycols is given now using DEG as the model glycol. Figure

Table I. Glycols Employed in the Udex Process

HO [CH$_2$CH$_2$O]$_1$CH$_2$CH$_2$OH	Diethylene glycol	DEG
HO [CH$_2$CH$_2$O]$_2$CH$_2$CH$_2$OH	Triethylene glycol	TEG
HO [CH$_2$CH$_2$O]$_3$CH$_2$CH$_2$CH	Tetraethylene glycol	TETRA

$$\begin{array}{l} \text{H} \\ \text{HOC} \leftarrow \text{CH}_3 \\ \quad | \qquad \quad \text{H} \\ \text{CH}_2\text{OCH}_2\text{C} \leftarrow \text{OH} \\ \qquad \quad \text{CH}_3 \end{array}$$
Dipropylene glycol   DPG

$$\text{HO} \begin{bmatrix} \text{CH}_3 \\ | \\ \text{C CH}_2\text{O} \\ | \\ \text{H} \end{bmatrix}_1 \begin{matrix} \text{CH}_3 \\ | \\ \text{CH}_2\text{C OH} \\ | \\ \text{H} \end{matrix}$$

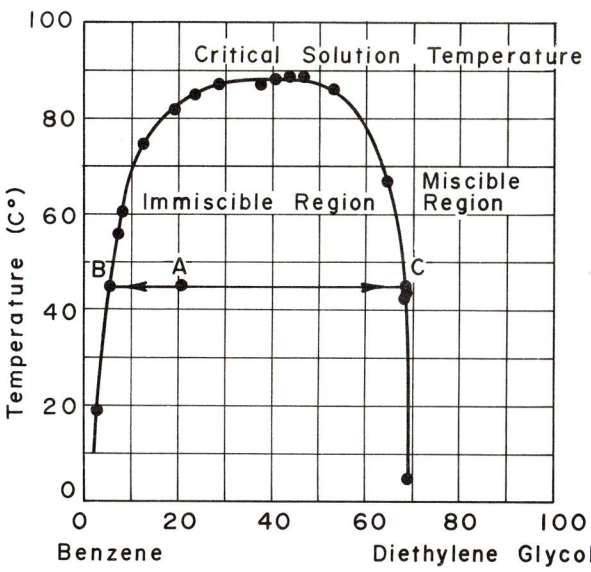

Figure 3. Temperature–composition data for the system: benzene–diethylene glycol

3 is a temperature–composition diagram; benzene is the left ordinate, DEG is the right ordinate. The curve is the miscibility or solubility curve. The area inside the curve represents the compositions at which two liquids exist in equilibrium. The area outside the curve represents the compositions of the two components which form one liquid. The

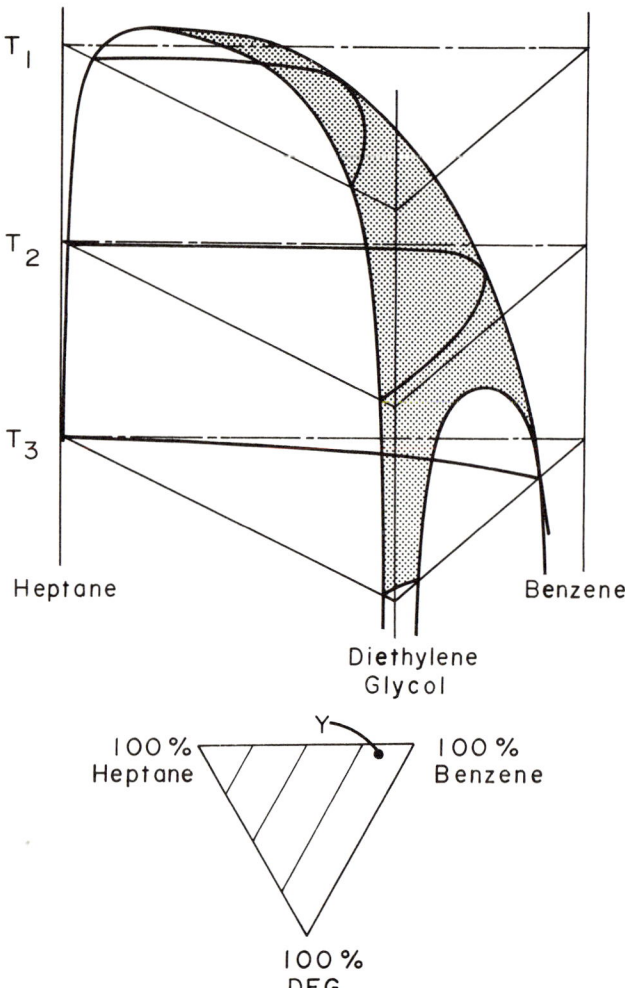

Figure 4. Ternary temperature–composition diagram

critical solution temperature (CST) is the temperature above which only one liquid phase is formed, no matter what the composition. This sort of curve is typical for all aromatics and actually with all solvents. Below the CST the solubility of the aromatic in the solvent tends to be greater than the solubility of the solvent in the aromatic. At 45°C, over-all composition A (containing about 20 wt % DEG), separates into upper phase B (containing about 95 wt % benzene) and lower phase C (containing about 68 wt % DEG).

With aliphatics the curve is much higher, and the solubility of the aliphatic in the solvent is much less than that of the aromatic. This is illustrated better in Figure 4.

This is a ternary temperature–composition diagram—a triangular prism. It is easy to read. Each corner represents 100% of one component, and the side opposite 0%. Lines parallel to the base represent the various percentages of that component. The 25, 50, and 75% lines are shown for heptane. Point Y has the composition of about 80% benzene, 15% heptane, and 5% DEG. A feel for ternary diagrams can be obtained by realizing that the closer the composition is to a particular corner, the more of that component there is in the solution or mixture. Thus, Y is near the benzene corner and far away from the heptane and DEG corners.

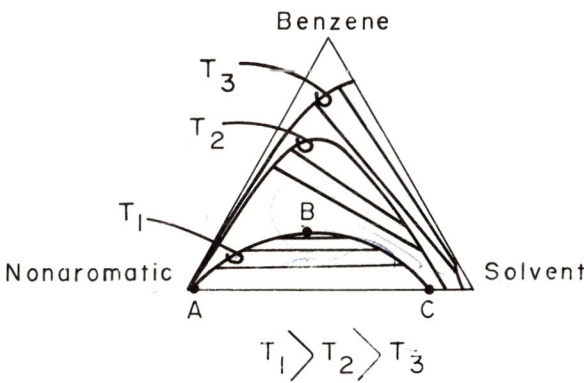

Figure 5. *Liquid–liquid equilibria in the benzene-nonaromatic solvent system at various temperatures*

Referring back to the prism, the benzene–DEG binary is that presented in the previous figure. The heptane–DEG binary has a similar curve, except that it is shifted much higher. Thus, at any temperature below the critical solution temperatures, the mutual solubilities of benzene and DEG are much higher than those between heptane and DEG. The equilibria at any particular temperature can be determined by taking a horizontal slice, which can be either one solubility curve (slice near the top) or two separate solubility curves (slice near the bottom).

Figure 5 shows some possible configurations. T_1 is at a higher temperature than T_2, and T_2 is at a higher temperature than T_3. Referring

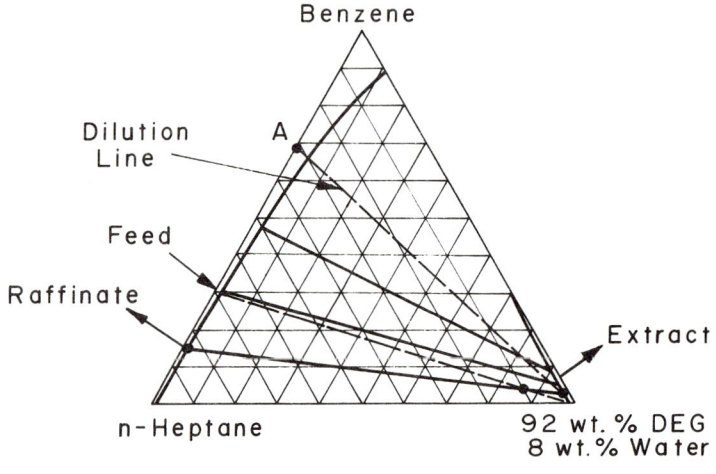

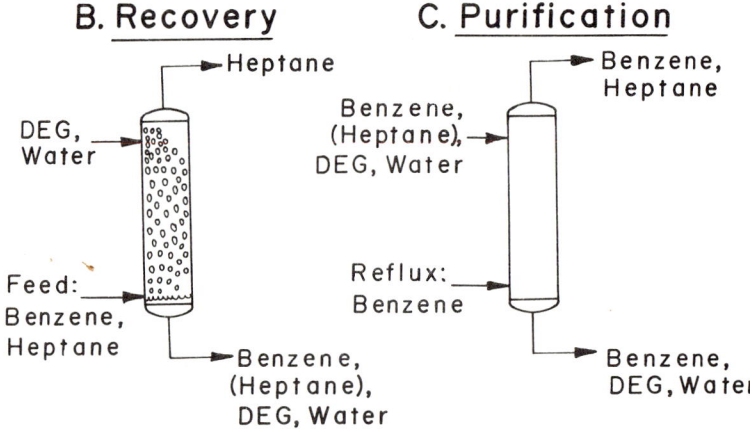

Figure 6. Liquid-liquid equilibria in the system: benzene-heptane-92 wt % DEG-8 wt % water at 125°C volume %

to the curve at temperature T_1, the three components are completely miscible in the area above curve ABC. Inside the miscibility curve two liquid phases are formed, and the compositions of these phases lie at the ends of tie lines. Curve AB represents compositions of all possible hydrocarbon raffinate phases, and curve CB represents compositions of the solvent extract phases. At point B (the Plait point) both phases have the same composition and density. As the temperature is lowered below the benzene–DEG critical solution temperature, two separate solubility curves are formed, as is illustrated at the low temperature (T_3).

Figure 6 shows an actual extraction system—benzene, heptane, 92% DEG–8% water at 125°C under pressure. Water is included to reduce the boiling point of the DEG solvent to about 140°C as well as to increase selectivity. If water had been excluded, one continuous solubility curve would have been obtained at 125°C. However, in this system there are two separate curves. The tie lines are shown in solid lines. A single-stage extraction is depicted on the ternary diagram. The feed contains 30% benzene. Extracting with fresh solvent at a solvent/feed ratio of roughly 8/1 produces a hydrocarbon raffinate containing only 15% benzene in just one extraction stage. The benzene content of the raffinate can be reduced to 0% by countercurrent, multistage extraction, depicted in flow diagram B. Referring back to the ternary diagram, if the extract of the one-stage extraction were distilled, the hydrocarbon distillate obtained would be almost 70% benzene. This composition (point A) is indicated at the end of the dilution line that goes through the extract composition. The dilution line represents all possible compositions of the extracted hydrocarbons and glycol.

The ternary diagram also indicates that pure benzene can be obtained in the extract by treating it countercurrently with additional pure benzene—*i.e.*, reflux. If the benzene content of the extract were built up to about 30%, there would be no heptane in the extract. This multistage, countercurrent purification is depicted in flow diagram C.

Column B can be put on top of column C, as depicted in Figure 7. The benzene–heptane raffinate of the purification section is merely additional feed for the recovery section.

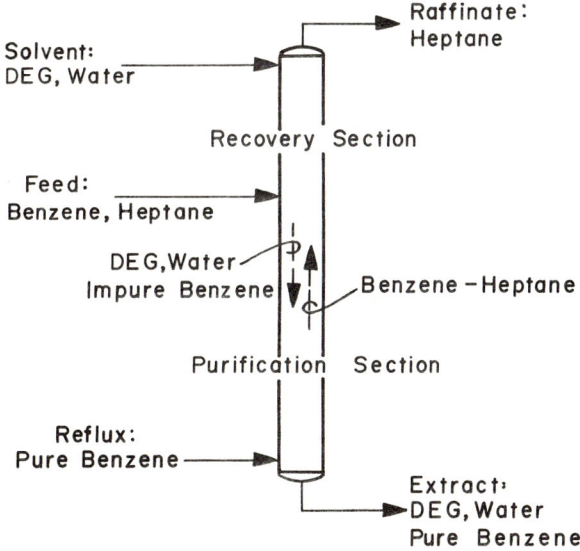

Figure 7. Schematic of extractor with recovery and purification sections

Figure 8 is the actual flow diagram of the Udex system (2, 4). The extract is transferred to the top of the distillation column. The distillate obtained overhead is condensed and decanted. The bottom layer, containing water and a small proportion of glycol, is recycled to the bottom of the distillation column. The top layer consists of benzene along with some light aliphatics (C_5's and C_6's) and is recycled to the bottom of the extractor as reflux. The aromatics (along with some water) come off as a vapor side stream. These vapors are condensed and decanted. BTX is removed, and the water phase is recycled. As mentioned earlier, water is included in solvent primarily to reduce the boiling point at the bottom of the distillation column.

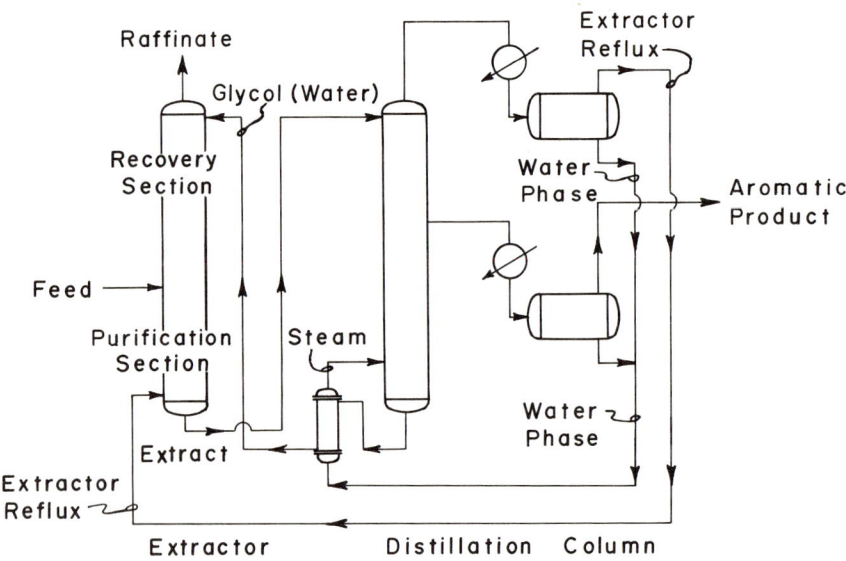

Figure 8. Schematic flow diagram of Udex process

Both vapor distillates also contain some glycol. In the decantation steps most of the glycol goes into the water phases, and hence a separate distillation column is not needed and BTX reflux is not needed. The raffinate and BTX are washed further with small proportions of water to remove the traces of glycol.

Increasing the Aromatics Capacity

With the growing demand for petrochemicals and high octane gasoline, more and more extraction capacity has been and will be needed throughout the years. It has been the trend in the industry to employ solvents of increasing solvency to achieve this capacity increase. The

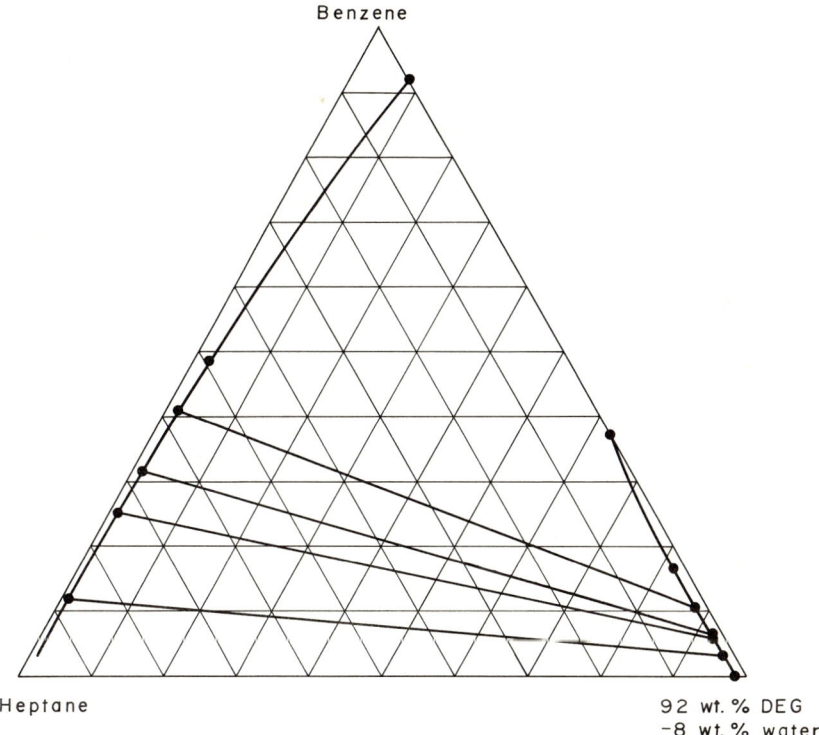

Figure 9. Liquid–liquid equilibria in the system: benzene–heptane–92 wt % DEG–8 wt % water at 125°C volume %

industry started with diethylene glycol. The liquid–liquid equilibria in the benzene–heptane system with DEG were presented above and are shown again in Figure 9. The data are at use conditions: 125°C with the solvent containing 8% water. At this water concentration the solvent has a boiling point of about 140°C. Diethylene glycol solvent is very selective in that it does not dissolve much aliphatics. However, it does not dissolve much benzene either. The limited solubility with benzene reduces extraction capacity, so that benzene distributes in only about 1 to 4 ratio betwen extract and raffinate. The distribution coefficients with toluene and C_8 aromatics (which determine the effective capacity of the solvent) are substantially lower.

With triethylene glycol at 121.5°C (Figure 10) the results are better. In this case the solvent contains 5% water, and its boiling point is about 140°C. This solvent is also very selective, and the extracts have low heptane solubilities. Benzene distributes in a ratio just under 1 to 2 between extract and raffinate. Triethylene glycol has nearly twice the extraction capacity of diethylene glycol. It is also important to realize that even highly aromatic feeds can be treated.

However, tetraethylene glycol is the best of the glycols studied (Figure 11). The solvent contains 3.9% water so that it also has a boiling point of about 140°C. Benzene distributes quite favorably in tetraethylene glycol. The tie lines are rather flat. The data are at 100°C only. Highly aromatic feeds can be treated also.

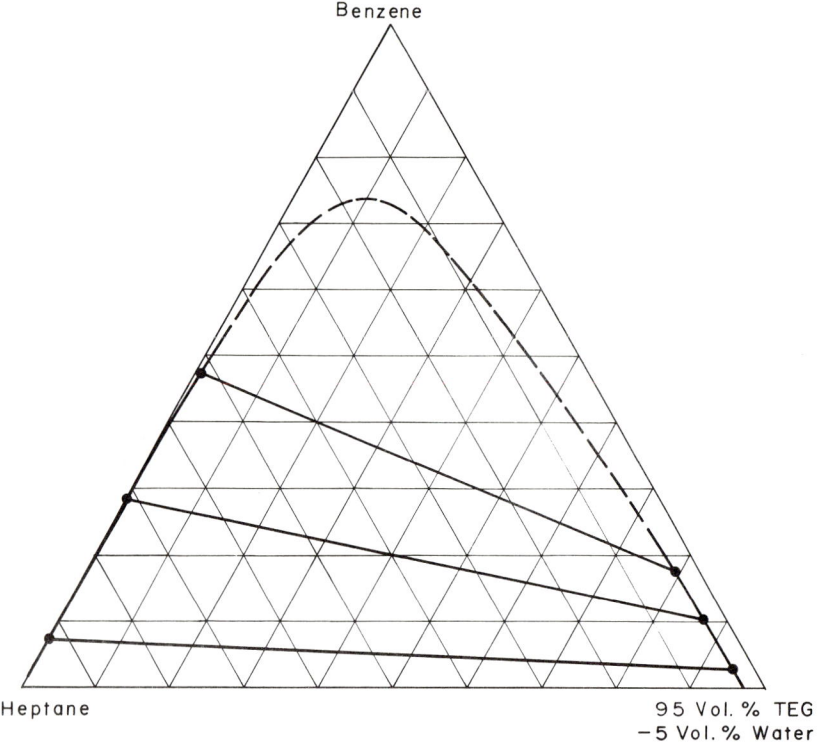

Figure 10. Liquid–liquid equilibria in the system: benzene–heptane–95 vol % TEG–5 vol % water at 121.5°C volume %

The data are summarized in Figure 12, which depicts the distribution coefficient as a function of benzene concentration in the raffinate. The average distribution coefficient with diethylene glycol is about 0.25 and that with triethylene glycol is nearly 60% higher at about 0.4. However, toluene and C_8 aromatics are so much more easily extracted with triethylene glycol that about half as much solvent is needed compared with DEG. Some refiners have increased capacity by adding dipropylene glycol to their diethylene glycol solvent. At about 25% DPG, the solvent is similar to triethylene glycol. However, individual solvents are pre-

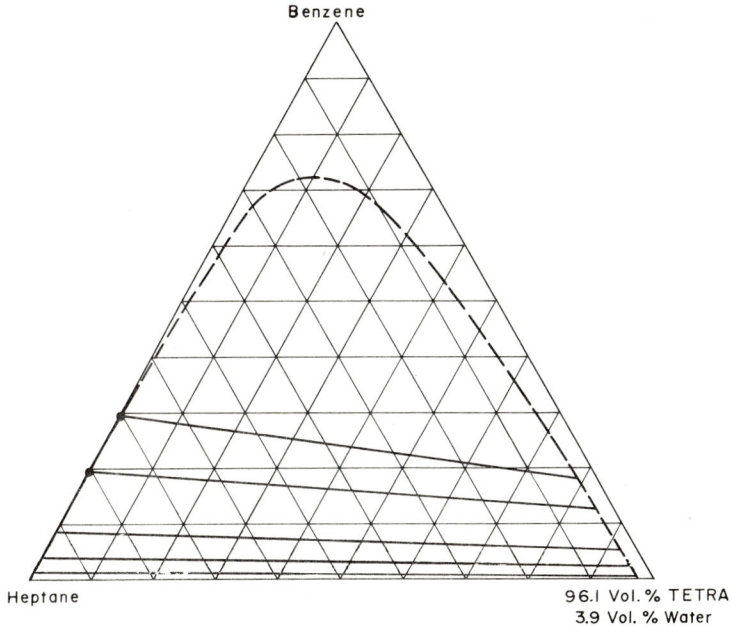

Figure 11. Liquid–liquid equilibria in the system: benzene–heptane–96.1 vol % TETRA–3.9 vol % water at 100°C volume %

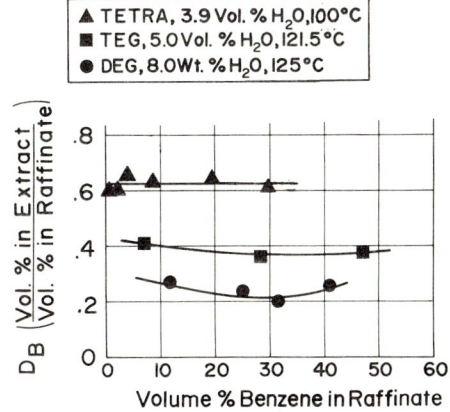

Figure 12. Distribution of benzene between extract and raffinate for various glycols

ferred over mixtures to avoid having to control solvent composition. The distribution coefficient with TETRA is about 0.6 compared with the 0.4 with triethylene glycol. Here again the comparison is that the solvent/feed ratio with TETRA would be nearly half that with triethylene glycol. All of these results have been confirmed in our pilot plant (Figure 13).

This pilot plant is a fully integrated, continuous system. It can be used to simulate any extraction and distillation process using one or more extractions and one or more distillations.

Figure 13.   Pilot plant

The pilot plant was operated for long periods on various feed stocks, both synthetic and real, and the results confirmed all of the laboratory bench scale findings.

The superiority of tetraethylene glycol has been shown by low S/F ratios, low R/F ratios, high recoveries of all aromatics, high purity of benzene and other aromatics and low operating costs.

Processes using the more recent solvents, sulfolane (*1*) and N-methylpyrrolidone (*3*), have been studied. A simplified flow diagram of the processes using these solvents is shown in Figure 14. The distillations are done in two columns, not one, and distillation reflux is usually employed. The extractor reflux is distilled in the first distillation column,

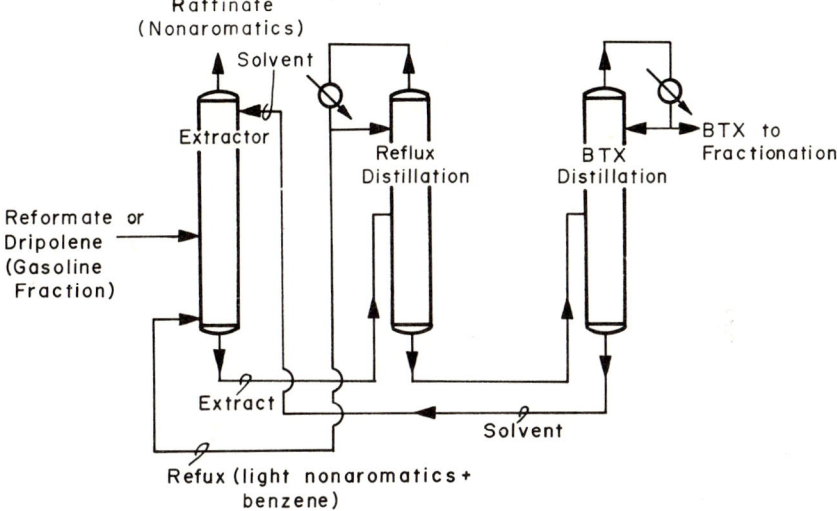

Figure 14. Generalized schematic flow diagram of "conventional" process to separate light aromatics from gasoline fractions

and aromatics are distilled from the solvent in the second distillation column. Though these solvents are beneficial in the extractor (high capacity for aromatics), there are significant disadvantages in the distillations (difficult to purify aromatics and difficult to separate aromatics from solvent). On the other hand, the authors have found that the tetraethylene glycol Udex exhibits superior economics.

Conclusion

A significant trend has developed recently in the industry toward using tetraethylene glycol. Plant capacities have been increased considerably. Recoveries of C_8 aromatics are very high. Aromatic purities are even better than they were with the lower glycols. Furthermore, operating costs have been cut by up to half. The commercial performance of tetraethylene glycol has indeed been excellent.

Literature Cited

(1) Deal, G. H. *et al., Petrol. Refiner* **Sept. 1969,** 185–192.
(2) Grote, H. W., *Chem. Eng. Progr.* **Aug. 1958,** 43–48.
(3) *Hydrocarbon Processing* **Nov. 1965,** 181.
(4) Read, Davis, API Meeting, San Francisco, Calif., May 14, 1952.

RECEIVED January 12, 1970.

15

A New Solvent for Aromatics Separation

JULIAN FELDMAN

Research and Development Dept., U. S. Industrial Chemicals Co., Cincinnati, Ohio

> *1,3-dicyanobutane derived from acrylonitrile shows a higher capacity for light aromatic hydrocarbons in liquid–liquid extraction than does sulfolane. As a solvent in extractive distillation its low volatility, high thermal stability, and low freezing point, in combination with its high selectivity, make it unusually attractive. Data are presented in comparison with identical operations with sulfolane. The experiments include equilibrium measurement on synthetic mixtures and multistage extractions of a typical light catalytic reformate. The effectiveness of using a light counter-solvent is demonstrated especially when followed by extractive distillation. A combination of methods gives high purity aromatics or improved octane gasoline. A flow diagram illustrates a typical application in a continuous cyclic operation whereby a reformate is separated and solvent is recovered.*

The separation of aromatic components from a hydrocarbon stream is such a well established part of petrochemical processing that there is little need here for reviewing the considerable art in this field. It is sufficient to state that solvent extraction has been the method of choice and that the number of solvents which have attained successful commercial application is small. For many years, glycols have dominated the field, and practically every gallon of petroleum-derived benzene, toluene, and xylene produced in the United States went through the Udex (10) process. Recently other solvents have been demonstrated to be more efficient so that the technology is now displacing glycols with sulfolane (4). Other solvents lately promoted by developers of competitive processes are N-methylpyrrolidone (Arosolvan) (20), dimethyl sulfoxide (21), morpholine (1), substituted morpholine (12), methyl carbonate (Carmex) (3), and dimethylformamide (11).

The proponents of the various extraction methods as well as other serious investigators have described the properties required of the ideal solvent and the practical solvent. In an extraction process the three most significant properties which determine the suitability of a solvent are the group selectivity, the light–heavy selectivity and the solvent power. Voetter and Kosters (22) have described these and their significance. While the solvents currently in use are generally efficient, there is room for improvement with subsequent cost savings. This presentation describes a new solvent—1,3-dicyanobutane—and compares it with the widely accepted sulfolane.

Results and Discussion

Solvent Manufacture and Preparation. 1,3-Dicyanobutane is a new material which can be produced at low cost by a process developed in the laboratories of U. S. Industrial Chemicals Co. (6, 7, 14). The catalytic dimerization of acrylonitrile produces methyleneglutaronitrile which is readily hydrogenated to give methylglutaronitrile, also called 1,3-dicyanobutane. Thus,

$$N\equiv C-CH=CH_2 + \underset{CH_2}{\overset{CH_2}{\underset{\|}{CH}}}-CN \xrightarrow{cat.} NC-CH_2-CH_2-\underset{\|}{\overset{CH_2}{C}}-CN$$

$$NC-CH_2-CH_2-\underset{CH_3}{\overset{|}{CH}}-CN \xleftarrow{\underset{Pd}{H_2}}$$

The price of producing the dimer in the pure state depends largely on the scale of operation, but for a moderate size plant it would be about a 10-cent surcharge on the price of the acrylonitrile. Hydrogenation adds a few more cents to bring the selling price of the saturated dinitrile in the low 30's.

Properties of 1,3-Dicyanobutane. PHYSICAL. The physical properties of 1,3-dicyanobutane are:

Molecular weight	107.14
Boiling point	139°C at 15 mm Hg
Density, D_4^{25}	0.9501
Refractive index, n_d^{25}	1.3701
Freezing point	−38°C
Viscosity at 25°C, cp	5.687
Specific heat	0.4899 cal/gram/degree
Flash point, closed cup	86.5°C

Vapor pressure:
$$P = 760 \text{ antilog } (66.4219 - \frac{7133 \cdot 6889}{T} - 19.4926 \log T)$$

where $P = $ mm Hg and $T = $ °K

Heat of vaporization, cal/gram
140 at 135°C
125 at 205°C

SOLUBILITY AT 25°C BY WEIGHT %. For the system 1,3-dicyanobutane (DCB) and compound S, the following values were found:

DCB in S	S	S in DCB
11.9	water	4.74
0.1	n-pentane	14.84
1.2	n-heptane	2.53
0.1	n-decane	2.68
0.5	cyclohexane	8.42
0.1	Decalin	1.88
9.0	isoprene	4.66
∞	benzene	∞

CHEMICAL PROPERTIES. *Reduction.* Catalytic hydrogenation over cobalt in the presence of ammonia gives 2-methylpentamethylene-1,5-diamine along with minor amounts of 3-pipecoline (8). Under special conditions of limiting hydrogenation, the major product is 2-methyl-5-aminovaleronitrile (9).

Hydrolysis. Acid or alkaline hydrolysis yields 2-methylglutaric acid.

Thermal Stability. Prolonged heating in the absence of air or water has no effect. Thus, a sample refluxed at 271°C under nitrogen for 500 hours showed less than 5% loss.

Explosibility. Tests by an outside consultant demonstrated that dicyanobutane is not detonatable.

Applications of Solvent to Aromatics Separation. There are two distinct techniques for using solvents in separation processes (2): extractive distillation and liquid–liquid extraction. Frequently these techniques or special variants can be combined in specified sequences to give a separation unobtainable by either method alone. Such has been the case for many of the solvents used or suggested in aromatics separation. The remainder of this discussion demonstrates the applicability and advantage of using dicyanobutane as a solvent for aromatics separation in petroleum refining.

EXTRACTIVE DISTILLATION. A solvent used in extractive distillation must have an appreciably lower volatility than the components being separated and must have a more decided effect on changing the activity coefficient of one component than it has over the other.

Vapor–liquid equilibrium data were obtained for the system benzene–cyclohexane–1,3-dicyanobutane (Table I). These results demonstrate the great effect that the solvent has even at relatively low solvent loadings.

Table I. Vapor-Liquid Equilibrium

	Pot Charge, Wt %		Temp., °C	Cyclohexane Wt %[a]		
DCB	Benzene	Cyclohexane		Liquid	Vapor	K^b
0	55.7	44.3	77.6	55.1	44.9	1.01
15.1	58.0	26.9	76.5	31.5	39.8	1.26
26.0	51.5	23.5	77.0	31.2	42.2	1.35
41.0	40.2	18.7	—	27.6	52.5	1.90
88.0	7.8	4.0	83.5	29.0	61.0	2.10

[a] Solvent-free basis.
[b] Relative volatility.

It is well established that gas–liquid chromatographic columns are essentially multiplate extractive distillation columns in which the immobile liquid phase functions comparably with an extractive distillation solvent (*15, 16, 17, 19*). The retention time, R_t, of substances miscible with the immobile phase, upon passage through the column containing the solvent are inversely proportional to the volatilities which these compounds would exert in an extractive distillation with high solvent loading. Consequently, compounds having different retention times should be separable by extractive distillation. In Table II, retention times are given for a number of hydrocarbons injected through a hot vaporizer onto a column consisting of 8 feet of 1/4 inch copper tubing packed with 19 grams of a mixture made from 75% of an inert support of 60/80 mesh silanized Chromosorb W and 25% 1,3-dicyanobutane. This column was used in an Aerograph 700 GLC apparatus. Helium at 30 psi was the carrier gas.

The data indicate that dicyanobutane discriminates on the basis of degree of unsaturation and relatively little on other structural effects in hydrocarbons. The volatility of hydrocarbons is reduced in the order of the number of double bonds and cyclic structure. Thus, it would be possible to separate monoolefins from paraffins, diolefins from monoolefins, naphthenes from paraffins, and aromatics from olefins, naphthenes, and saturates by using dicyanobutane as an extractive solvent with mixtures having selected boiling ranges.

While the use of nitriles and dinitriles as selective solvents for hydrocarbon separation by extractive distillation has been described (*13*), no

Table II. GLC on a Dicyanobutane Column

Hydrocarbon	Boiling Point, °C	Column Temp., °C	Gas Flow ml/min	Retention Time, min
1-Butene	−6.5	25	60	0.8
trans-2-Butene	0.3	25	60	1.2
cis-2-Butene	3.7	25	60	1.3
1,3-Butadiene	−4.5	25	60	2.1
n-Hexane	69.0	25	55	2.1
1-Hexene	64.1	25	55	4.2
2-Hexene	68.0	25	55	4.5
Cyclohexene	83.0	25	55	8.5
Cyclohexene		50	55	5.7
Cyclohexane	81.4	50	55	3.1
Benzene	80.1	50	55	21.7
Benzene		72	100	6.3
n-Heptane	98.4	72	100	0.8
n-Heptane		25	55	5.5
2-Heptene	98.2	25	55	10.5
Methylcyclohexane	100.9	72	100	1.5
2,2,4-Trimethylpentane	99.3	72	100	0.8
2,4,4-Trimethylpentene	101.0	72	100	1.5
Toluene	110.6	72	100	12.6
n-Octane	125.7	72	100	1.7
1-Octene	121.2	72	100	2.6
2-Octene	125.2	72	100	3.9
Ethylcyclohexane	131.8	72	100	3.2
4-Vinylcyclohexene	129.3	72	100	7.7
cis-1,2-Dimethylcyclohexane	129.0	72	100	3.4
2,6-Octadiene	122.0	72	100	4.6
1,3,7-Octatriene	124.0	72	100	9.4
1,3,6-Octatriene	130.0	72	100	11.2
p-Xylene	138.4	72	100	24.4
m-Xylene	139.1	72	100	24.2
o-Xylene	144.4	72	100	32.3
Cyclooctane	151.0	72	100	7.6
Cyclooctene	144.0	72	100	9.0
1,3-Cyclooctadiene	141.0	72	100	12.8
1,5-Cyclooctadiene	150.9	72	100	23.0

commercially available dinitriles have been specifically proposed for this application.

LIQUID–LIQUID EXTRACTION. Dinitriles also have been suggested as solvents in liquid–liquid extraction (5, 13, 18). Therefore, it was not unexpected to find dicyanobutane a possible candidate as a selective solvent in liquid–liquid separation of hydrocarbon mixtures. To determine just how good it was, we compared its properties with respect to the newly accepted commercial standard for hydrocarbon separation—sulfolane.

Phase equilibrium studies were made on relatively simple systems. Thus, at room temperature when one volume of hexane and one volume of toluene were mixed with two volumes of dicyanobutane, the mixture separated rapidly into two phases on cessation of mixing. The upper phase showed a 40% volume decrease with a corresponding

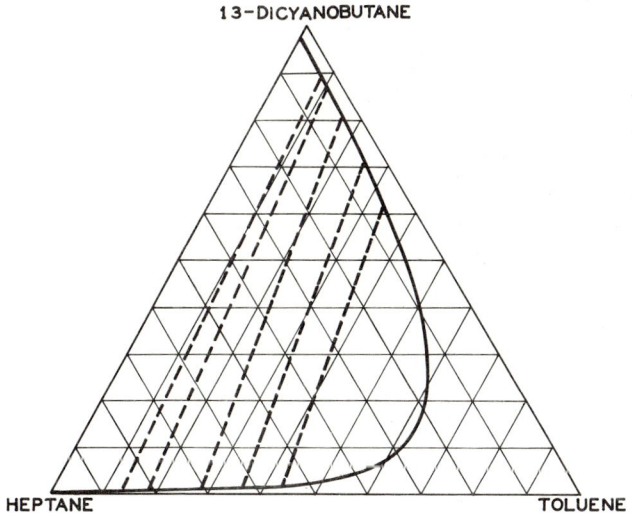

Figure 1. Phase diagram of heptane, toluene, and 1,3-dicyanobutane at 25°C

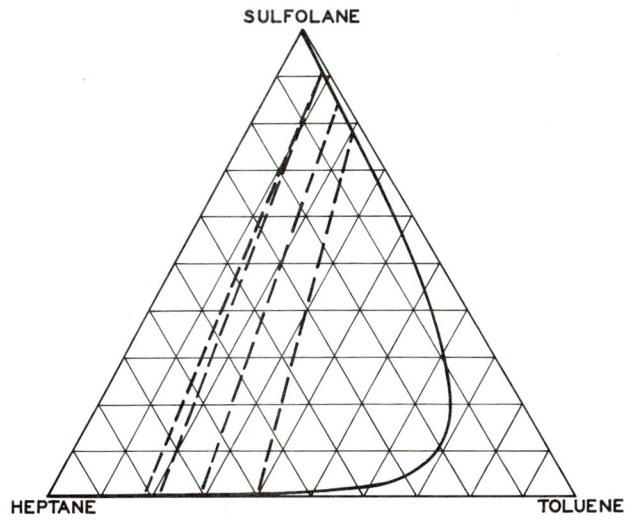

Figure 2. Phase diagram of heptane, toluene, and sulfolane at 25°C

increase in the lower phase. Solubility of the solvent in the remaining hydrocarbon solvent was 2%, whereas solubility of the hydrocarbon mixture in the dicyanobutane was 32%. The equilibrium relative distributions of hexane to toluene were 63 to 37 in the upper phase and 23 to 77 in the lower phase.

The system toluene–heptane–dicyanobutane was studied at 25°C and compared with the system toluene–heptane–sulfolane at 25°C. The ternary phase diagrams (Figures 1 and 2) resemble each other closely, with complete miscibility of phases occurring at somewhat lower aromatics content for the dicyanobutane than for sulfolane. This higher capacity for hydrocarbons by the dicyanobutane compensates for its slightly lower selectivity as shown in Figure 3.

Mixtures of pure hydrocarbons were made up and analyzed by GLC. The mixtures were shaken with equal volumes of polar solvent (dicyanobutane or sulfolane), allowed to separate, and the volumes were measured. Samples of each phase were analyzed by GLC for the hydrocarbons present. The approximate concentrations of hydrocarbons in the polar phase, C_L, were determined by relationships existing between sample size and integrated areas of the hydrocarbon peaks. These values

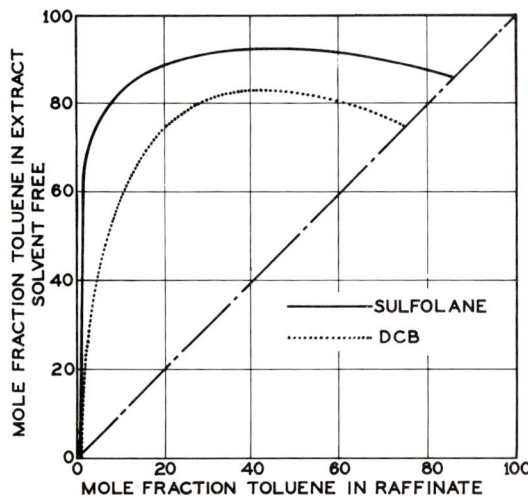

Figure 3. Selectivity curves of toluene–heptane at 25°C in solvent

were divided into the corresponding concentrations found by GLC in the hydrocarbon phase, C_U, to give the distribution constant $K = C_U/C_L$. The data for various systems studies are given in Table III. Where pairs of values appear in neighboring columns, they were obtained by adding to the equilibrated mixture a given volume of a specific component—

usually toluene—remixing, and making additional analyses. While the data obtained in this way are approximate, the results bear a reliably relative ratio to each other and may be used for comparisons.

The data indicate a marked selectivity of the dicyanobutane for aromatics as compared with other hydrocarbons. This selectivity is greatest over paraffins, somewhat less over naphthenes, and appreciably less over unsaturated naphthenes. Selectivity for aromatics increases with increasing molecular weight of paraffins and naphthenes and with decreasing molecular weight of the aromatic. In comparison, sulfolane shows a lower solvency for all the hydrocarbons. Thus, dicyanobutane on an equal volume basis has a higher capacity for aromatic hydrocarbons, while the selectivity relationships are comparable.

The effect of temperature on the relative distribution of mixtures of normal paraffins and toluene is shown in Table IV. There appears to be a small decrease in selectivity with increasing temperature.

The remainder of the studies were done on a light catalytic reformate supplied by one of the oil companies which had the following composition according to their analysis:

PONA Analysis	%	GLC	%
Paraffins	54.0	Paraffins	58.7
Olefins	1.4	Benzene	4.9
Naphthenes	7.0	Toluene	13.6
Aromatics	37.6	Xylenes	12.0
		Heavies	10.8

Distillation analysis gave the distribution of products shown in Figure 4.

This reformate was extracted with successive portions of fresh solvent with the results shown in Tables V and VI. These data show that dicyanobutane has a greater capacity for aromatics than does sulfolane and also a higher solubility for nonaromatics. Thus, only six extractions with DCB were required to reduce the aromatics to the same level as eight extractions with sulfolane.

Studies were made at room temperature (27°C) of the continuous countercurrent extraction of catalytic reformate with sulfolane and with 1,3-dicyanobutane. In this study a 22-stage, 1 inch diameter York-Scheibel extraction column was used. The hydrocarbon mixture was introduced into the bottom of the column, and the polar solvent was introduced at the top of the column. The effluent streams emerged just beyond the feed points. The streams were metered into the column under a hydraulic head plus 2 psi of air from reservoirs through needle valves and calibrated rotameters. The stream rates were well below flooding

Table III. Equilibrium Studies

Polar Phase:	Dicyanobutane							
Hydrocarbon:	C_U	K	U/L_i	U/L_f	C_U	K	U/L_i	U/L_f
Pentane	25	16			19	9		
Hexane	24	18			18	15		
Heptane	25	39	1.00	.92	20	17	1.40	.90
Octane	26	53			20	25		
Toluene	0	—			23	1.5		
Heptane	0	—			39	14		
Cyclohexene	57	1.8	1.00	.25	41	5	1.40	.85
Toluene	43	1.5			20	1.5		
Octane	39	32			26	26		
Nonane	39	75	1.00	.82	57	47	1.40	1.26
Cymene	23	3.5			16	3.3		
Ethylcyclohexane	70	17	1.00	.67				
Cymene	30	5						
Cyclohexane	74	10	1.00	.39				
Toluene	26	1.7						
Methylcyclopentane	24	16			18	11		
Cyclohexane	35	13			28	9		
Methylcyclohexane	16	16	1.00	.89	12	10	1.40	.94
Ethylcyclohexane	25	30			19	19		
Toluene	0	—			23	1.8		

Table IV. Equilibration Studies

Polar Phase:		Dicyanobutane				Sulfolane			
Hydrocarbon:	Temp., °C	C_U	K	U/L_i	U/L_f	C_U	K	U/L_i	U/L_f
n-Heptane	27	26	15		.67	28	—		.78
	51	25	16	1.0	.65	26	33	1.0	.74
	100	26	12		.74	25	16		.72
n-Octane	47	27	21		.67	28	—		.78
	51	28	35	1.0	.65	26	—	1.0	.74
	100	26	17		.74	28	17		.72
n-Nonane	27	30	50		.67	48	—		.78
	51	29	36	1.0	.65	27	—	1.0	.74
	100	29	19		.74	29	26		.72
Toluene	27	17	1.3		.67	20	1.5		.78
	51	18	1.6	1.0	.65	22	1.7	1.0	.74
	100	19	2.0		.74	19	1.9		.72

at Room Temperature[a]

			Sulfolane				
C_U	K	U/L_i	U/L_f	C_U	K	U/L_i	U/L_f
25	38			21	16		
23	42			18	36		
25	150	1.00	1.00	20	40	1.40	.94
26	—			23	63		
0	—			23	1.7		
0	—			36	16		
53	2.4	1.00	.39	45	7	1.40	.94
47	2.0			19	1.5		

[a]Symbols:
   U/L_i = Volume ratio of upper hydrocarbon phase to polar phase before mixing.
   U/L_f = Volume ratio of upper phase to lower phase after equilibration.
   C_U = Approximate weight concentration of component in hydrocarbon phase (polar component-free basis).
   K = Ratio of concentration of component in hydrocarbon phase to its concentration in polar phase.

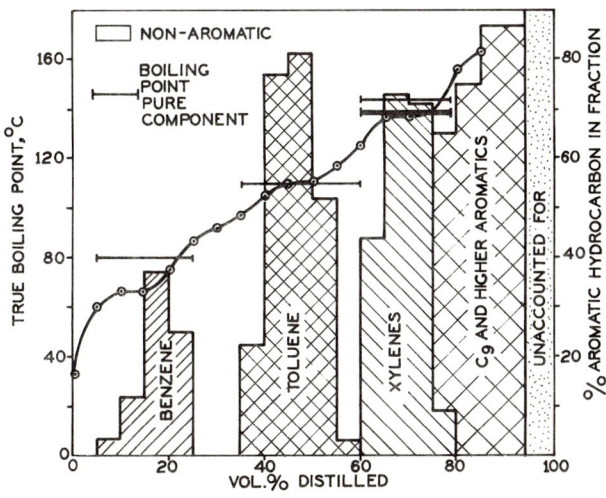

Figure 4. Composition of catalytic reformate

Table V. 500-ml Reformate Extracted with

No. of extractions	0	1	2
Vol. of hydrocarbon phase, ml	500	450	400
Vol. of polar phase, ml	0	250	240
Composition of hydrocarbon phase, %			
Nonaromatics	65	70.5	73.5
Benzene	5	4.7	3.6
Toluene	14.6	12.1	10.0
Xylenes	11.8	10.0	9.5
C_9 aromatics	3.8	2.7	3.4
Composition of polar phase, %			
Nonaromatics		5	4.9
Benzene		3	2.3
Toluene		7.4	5.8
Xylenes		5.4	4.6
C_9 aromatics		1.1	1.0

Table VI. 500-ml Reformate Extracted with

No. of extractions	0	1	2
Vol. of hydrocarbon phase, ml	500	465	425
Vol. of polar phase, ml	0	230	230
Composition of hydrocarbon phase, %			
Nonaromatics	65	71.5	72.5
Benzene	5	5.1	3.5
Toluene	14.6	10.4	10
Xylenes	11.8	11.4	10.2
C_9 aromatics	3.8	1.8	3.5
Composition of polar phase, %			
Nonaromatics		2.6	2.4
Benzene		3.2	2.0
Toluene		7.2	5.2
Xylenes		5.3	4.6
C_9 aromatics		.9	.8

rates and were maintained constant by surveillance of the operator. The interface was held approximately at the 3/4-column height level. The column was assumed to approach steady-state operation after at least three inventories passed through at which time samples were collected and effluent rates were measured by collection in graduated cylinders. Solvent ratios were then changed, usually by changing the rate of the polar solvent. Samples were analyzed by GLC. Although benzene and C_9 aromatic hydrocarbons were present, uncertainties in analysis because of their smaller concentrations excluded them from the graphical interpretation presented below. However, their behavior was similar qualitatively to those of toluene and the xylenes. The sulfolane contained 1.5% water.

Successive 200-ml Portions of Dicyanobutane

3	4	5	6	7	8
360	325	300	280	260	245
230	220	220	215	210	205
78.5	80	89	92.5	94.5	96.5
3.1	2.4				
8.2	6.2	4.8	3.5	2.6	1.3
8.2	5.5	5.0	3.3	2.1	1.8
1.8	1.2	1.5	.7	.6	.7
4.2	3.9	4.3	4.1	4.1	4.3
1.4	.8	1.0	.4	.2	
5.3	2.9	2.4	1.8	1.2	1.0
3.2	2.2	2.5	1.5	1.0	.9
.8	.5	.4	.2		.2

Successive 200-ml Portions of Sulfolane

3	4	5	6	7	8
393	365	346	325	310	300
222	220	212	210	206	205
81	82.5	87.7	88.5	91	92.5
2					
8.4	6.4	4.8	4.6	3.4	2.5
6.2	8.0	5.7	5.0	4.5	3.3
2.4	2.8	2.3	2.0	1.3	1.5
2.5	2.2	2.2	2.0	2.0	2.4
1.9	1.1	.8	.5	.2	.1
4.8	3.6	2.7	1.9	1.5	.9
4.4	3.1	2.2	1.6	1.5	1.0
.5	.4	.3	.3	.2	.1

The results of these studies are presented in the following figures. Figure 5, where the concentrations of toluene are displayed, shows that it takes about 75% as much dicyanobutane as sulfolane to remove most of the toluene; Figure 6 (xylene extraction) shows it takes about 63% as much dicyanobutane as sulfolane. On a weight basis, 47–54% as much DCB as sulfolane is required in these two extractions. The reasons for this are demonstrated adequately in Figure 7, which shows larger distribution coefficients for both toluene and xylene in comparing the ratio of steady-state concentrations in the polar solvent with the hydrocarbon phase.

In Figure 8, the solubility of the nonaromatic portion in the polar phase also is significantly greater in dicyanobutane than in the sulfolane.

This solubility is increased in dicyanobutane by the presence of aromatic components, whereas these seem to have little influence on solubility in sulfolane. It is expected that these nonaromatic hydrocarbons will be removed in the subsequent extractive distillation step.

The effect of temperature on the countercurrent extraction shown in Table VII is negligible on the relative capacities and selectivities of the two solvents.

In the countercurrent extraction of aromatic hydrocarbons from a catalytic reformate by dicyanobutane, we have found that a solvent-to-

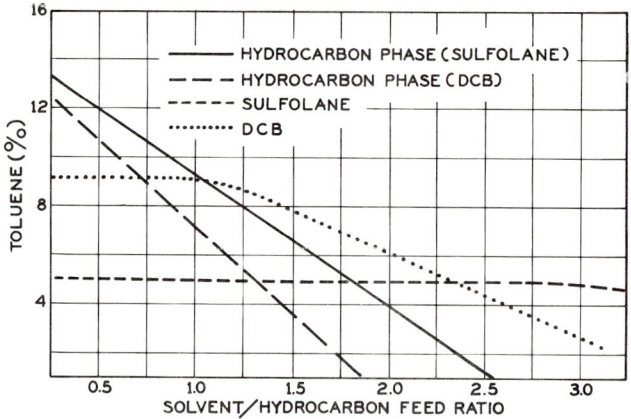

Figure 5. Concentration of toluene after countercurrent extraction of catalytic reformate at 27°C

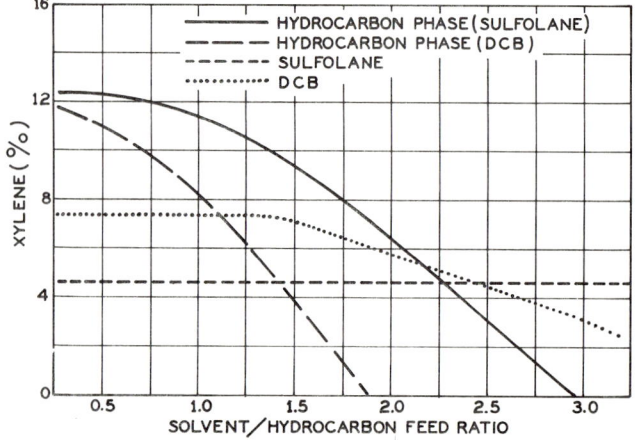

Figure 6. Concentration of xylene after countercurrent extraction of catalytic reformate at 27°C

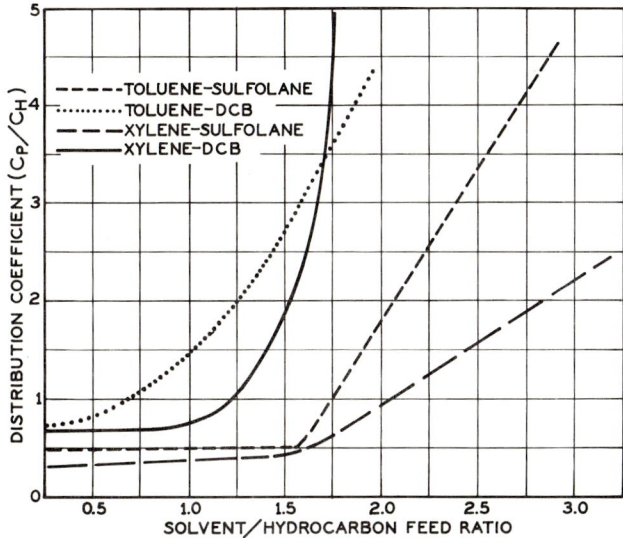

Figure 7. Distribution coefficients in countercurrent extraction of catalytic reformate at 27°C

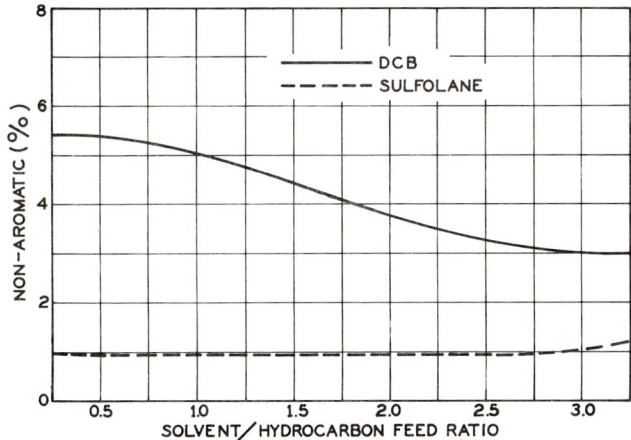

Figure 8. Solubility of nonaromatics in solvent after countercurrent extraction of catalytic reformate at 27°C

feed ratio of 3:1 by volume was sufficient to remove all the aromatics from the hydrocarbon phase. In contrast, solvent-to-feed ratio for sulfolane is 7:1 and for Udex 20:1.

As already mentioned the increased capacity of DCB for aromatics has the corollary effect of a somewhat lower selectivity. However, this

Table VII. Countercurrent Extraction of Reformate at 70°C [a]

Solvent/Hydrocarbon	Hydrocarbon Phase					Dicyanobutane Phase				
	A	B	T	X	C_9	A	B	T	X	C_9
2.1	89		3.4	3.4	1.4	5.3	2.7	9.4	6.9	3.3
3.3	94		2.5	2.8	1.5	5.0	2.2	7.8	6.7	3.3
6.1	100	0.0	0.0	0.0	0.0	5.7	2.1	6.0	5.0	2.5
10.3	100	0.0	0.0	0.0	0.0	5.2	1.3	3.8	2.8	1.7
11.3	100	0.0	0.0	0.0	0.0	4.4	1.4	3.8	3.9	1.7
						Sulfolane Phase				
.77	76		10.2	10.5	3.0	2.4	2.6	6.5	4.9	1.3
1.65	83.5		5.8	6.5		2.8	2.3	6.4	3.9	1.5
1.74	82		6.5	7.7	1.5	2.8	3.1	5.6	4.8	1.5
5.0	98		0.0	.9	1.2	3.2	1.2	4.1	3.8	
6.3	100		0.0	0.0	0.0	1.5	1.0	2.1	2.2	

[a] Symbols: A = Nonaromatics
B = Benzene
T = Toluene
X = Xylene
C_9 = Nine carbon aromatics

Table VIII. Countercurrent Extraction of

Countersolvent	Feed Rates, ml/min		
	HC	DCB	C.S.
None	9	36	0
7% Benzene in n-Heptane	9	36	3
7% Benzene in n-Heptane	12	36	3
Cyclohexane	12	36	3.6
Isoheptane	12	36	3
n-Hexane	12	36	2

Table IX. Extractive

Countersolvent in Extraction	Feed Rates, ml/min		Reflux Ratio
	Extract	DCB	
0	26	5	5
7% Benzene in n-Heptane	25	6.7	10
7% Benzene in n-Heptane	33	8.5	25
Cyclohexane	33	8.5	16
Isoheptane	33	8.5	16
n-Hexane	33	8.5	30

is relatively unimportant because of the marked effect of molecular weight on hydrocarbon solubility. Thus, when a light aliphatic hydrocarbon (C_6 cut) is used as a countersolvent in the extraction portion, it is sufficient to displace all the heavier aliphatics which would contaminate the aromatic products if not removed. In our work this was accomplished by using a feed-to-countersolvent volume ratio of 6:1. The limitations of these ratios have not been explored. The countersolvent was introduced at the first stage of the York-Scheibel extraction column, and feed was introduced at the sixth stage.

The countersolvents used and their effects on the relative distributions of the hydrocarbons are shown in Table VIII.

EXTRACTIVE DISTILLATION OF LIQUID–LIQUID EXTRACT. The extract portion was fed into a continuous still, 2 inches in diameter, made up of the following sections in ascending order: a 100-ml capacity reboiler, a vacuum-jacketed, silvered column 3 feet long containing helipak, an intermediate feed section, and a vacuum-jacketed silvered 15-plate Oldershaw column, another feed section and a 10-plate Oldershaw column, topped off with a solenoid operated liquid dividing reflux head. The extract entered the still through a preheater at the lower feed section. The reflux ratio and boil-up rate were adjusted to fix the bottoms compo-

Reformate with DCB at 27°C with Countersolvent

Effluent Rate, ml/min		Distribution of Hydrocarbons in Extracts				
Raffinate	Extract	A	B	T	X	C_9
6.3	42	30	12	25	24	8
8.2	39	26	10	26	27	11
13	43					
9.4	42					
		27	9	27	22	15

Distillation of DCB Extract

	Temperature °C			Bottoms: Distillate	Benzene % in Distillate
Extract	DCB	Distillate	Bottoms		
145	89	65	192		19
149	94	83	194	26	30
108	94	84	192	25	35
134	88	73	180	20	3.6
121	88	86	190	30	11
131	89	70	180	35	0

sition to contain benzene and have a boiling point of between 170° and 195°C (*See* Table IX).

The preferred ratio of fresh solvent to feed in the extractive still has not been determined but should be a function of the specific countersolvent used in the initial extraction stage, the number of plates in the distillation section, and the amount of benzene tolerated in the overhead product. We have selected arbitrarily a feed-to-solvent ratio in the E. D.

Table X. Separation of Aromatics from Dicyanobutane by Continuous Distillation of Bottoms from Extractive Still at Atmospheric Pressure

			Temperature, °C		
Countersolvent Used in Extraction	Upper Feed	Lower Feed	Distillate	Bottoms	Bottoms: Distillate
0	204		148	278	
7% Benzene in n-Heptane		230	85–120	270	5.2
Cyclohexane	178	244	143	270	7.5
Isoheptane					10.6
n-Hexane	170	240	75–165	271	8.9

Table XI. Distillation of Aromatic Portion (No Countersolvent in Extraction)

		Composition of Distillate				
Temp., °C	Percent Distilled	A	B	T	C_8	C_9
80	3	1.2	98.8			
80	1.2	5	75	20		
106	.6	14	14	72		
111	15	5	0	95		
111	4	3.3	0	96.7		
112	10	18	0	66	16	
132	15	6	0	0	94	
137	12	3.3	0	0	96.7	
141	3	6	0	0	94	
151	5	6	0	0		94
159	31 (residue)	2	0	0	0	98

column of 4:1 by volume (E.D. = Extractive Distillation). This fresh solvent entered the E.D. column through a preheater attached to the upper feed section. Under these conditions when n-heptane was the countersolvent, almost all the benzene was found in the overhead. When isoheptane or cyclohexane was the countersolvent, part of the benzene was in the distillate and part in the bottoms. When hexane was used, all the benzene was in the still bottoms along with all the other aromatics.

The aromatics were stripped from the solvent in the same continuous still, introducing the mixture through both feed sections. Table X shows some of the conditions in this distillation.

These solvent-free aromatic portions were then distilled through a 1-inch diameter, 3-foot long Podbelniak column to give the results shown in Tables XI to XV. Purities of better than 99% were obtained when countersolvent was used. The bottoms solvent was recycled as fresh DCB.

SOLVENT RECOVERY FROM RAFFINATE. The small amount of solvent left in the raffinate portion can be removed by a water wash. The latter

Table XII. Distillation of Aromatic Portion
(Benzene–n-Heptane as Countersolvent)

Temp., °C	Percent Distilled	Composition of Distillate				
		A	B	T	C_8	C_9
56	.7	82	0	18		
108	.4	19	0	81		
111	.7	2.05	0	97.5		
111	3.7	.2	0	99.8		
111	18	0	0	100		
111	5.8	0	0	100		
111	7.3	0	0	100		
111	6.1	0	0	60	40	
139	2.5	0	0	3.7	96.7	
139	18.7	0	0	0	100	
141	6.4	0	0	0	100	
144	4.2	0	0	0	100	
150	1.1	.3	0	0	99.7	
159	7.4	0	0	0	2	98
163	2.4	0	0	0	0	100
164	14.7 (residue)	0	0	0	0	22

Table XIII. Distillation of Aromatic Portion
(Cyclohexane as Countersolvent)

Temperature, °C	Percent Distilled	Composition of Distillate				
		A	B	T	X	C_9
77	2	50	50			
77	5	50	50			
77	2	32	68			
79	3	31	53	16		
108	1	6.4	.7	93		
109	18	.4	0	99.6		
110	7	.3	0	99.7		
111	1	1.7	0	71	27	
132	2	1	0	10	89	
139	27	0	0	0	100	
150	1.5	0	0	0	58	47
156	30 (residue)	0	0	0	0	

can then be extracted with an aromatics stream to remove most of the dicyanobutane. The number of stages and solvent ratios required are shown in Table XVI. Thus a four-stage extraction of a raffinate stream with 1/20th its volume of water reduces the solvent content from about

Table XIV. Distillation of Aromatic Portion
(Isoheptane as Countersolvent)

Temperature, °C	Percent Distilled	Composition of Distillate				
		A	B	T	X	C_9
77	1.4	7.5	92.5			
82	.6	7.5	92.5			
85	1.7	18.6	72.4	8.9		
110	1.1	7.5	3.2	89.3		
112	22.2	0.1	0	99.9		
111	6.4	0	0	78	22	
139	1.7	0	0	.6	99.4	
	17.3	.2	0	0	99.8	
144	4.8	0	0	0	100	
145	6.9	0	0	0	100	
150	3.3	0	0	0	97	3
158	31 (residue)	0	0	0	0	101

Table XV. Distillation of Aromatic Portion
(n-Hexane as Countersolvent)

Temperature, °C	Percent Distilled	Composition of Distillate				
		A	B	T	X	C_9
76	.3	14	86			
80	.6	1.3	98.7			
80	1.0	.6	99.4			
80	6.4	.1	96.9	3		
106	.6	.1	28.8	71.1		
109	3.5	.1	5.5	94.4		
111	13.3	.1	0	99.9		
111	8.9	0	0	100		
111	3.7	0	0	100		
111	.8	0	0	67	33	
131	.6	.1	0	12.5	87.4	
134	.7	.6	0	3.2	96.2	
135	13.3	.4	0	.3	99.3	
136	10.9	.2	0	0	59	40.8
144	3.2	.4	0	0	42	57.6
145	2.4	.2	0	0	8.5	91.3
147	1.4	0	0	0	5.5	94.5
153	1.1	.4	0	0	5.5	94
161	3.3	.1	0	0	0	1.2
162	.8	0	0	0	0	.3
169	22.9 (residue)					

0.2% to less than .005%. Then 3.3 volumes of toluene per volume of water reduces the dicyanobutane content to .001% from an initial 4.0% in four stages of extraction. Since this water is recycled, its solvent content is significant only with respect to removal of dicyanobutane from the raffinate.

THE COMBINED OPERATION. The entire process of removal of aromatics and recovery of solvent-free raffinate is shown schematically in Figure 9. The numbers on this diagram are illustrative only and probably can be varied widely within limits to be determined for the particular operation. A clean-up still for the solvent is included to work on a slip-stream of unassigned proportions for removal of tar from the solvent. The net loss of hexane or light countersolvent may be compensated by distilling a light cut from the raffinate or by using some other light stream available from the refinery to feed in at F.

Table XVI. Distribution of DCB in Four-Stage Countercurrent Extraction

	HC/Water	DCB Content, %		
		Raffinate	Water	Toluene
Initial	∞	.18	0.0	
	0.0		4.0	0.0
1st Stage	20.	.102	3.64[a]	
	3.3		.0011[a]	.004
2nd Stage	20.	.048	1.60	
	3.3		.0099	.005
3rd Stage	20.	.019	.73	
	3.3		.153	.025
4th Stage	20	<.005[a]	.13	
	3.3		.45	1.29[a]

[a] Effluent streams from extractors. K (water)/(raff) = 36: K (tol)/(water) = 2.9.

Summary

There are several significant practical advantages of dicyanobutane over sulfolane:

(1) Its lower melting point avoids the necessity for steam tracing lines and equipment, especially for outdoor winter operations.

(2) Its higher thermal stability allows for stripping the extract phase at atmospheric pressure (rather than vacuum) resulting in lower capital cost for smaller diameter, thinner walled distillation columns, as well as lower power consumption.

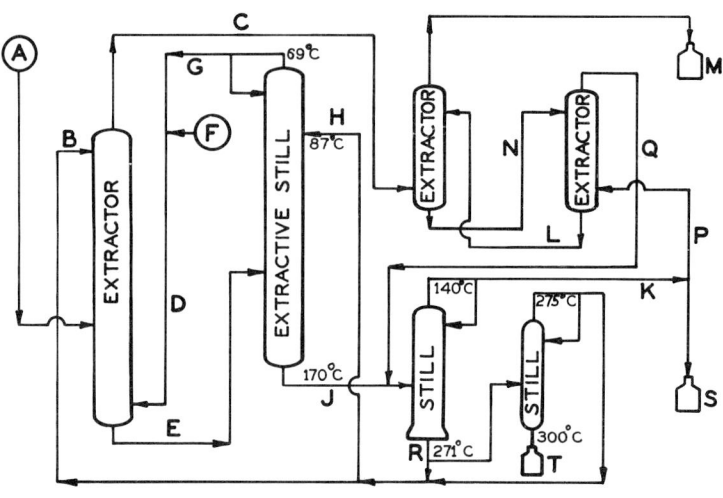

STREAM	A	B	C	D	E	F	G	H	J	K	L	M	N	P	Q	R	S	T
DCB	0	300	0.3	0	299.7	0	0	86	385.7	0	0	0	0.3	0	0.3	386	0	0
ALIPHATICS	65	0	65	0	0	0	0	0	0	0	0	65	0	0	0	0	0	0
HEXANE	0	0	4	17	13	4	13	0	0	0	0	4	0	0	0	0	0	0
AROMATICS	35	0	0	0	35	0	0	0	35	45	0	0	0	10	10	0	35	0
WATER	0	0	0	0	0	0	0	0	0	0	3.1	0.1	3.0	0	0	0	0	0
TAR	0.1	0	0	0	0.1	0	0	0.1	0.1	0	0	0	0	0	0	0.1	0	0.1

Figure 9. Process diagram for aromatics separation

(3) Its lower density, together with its higher capacity for hydrocarbons, requires only 60% of the total solvent on a weight basis. Assuming DCB priced at 42¢/lb and sulfolane at 60¢/lb, on a comparable basis of aromatic removal, DCB will cost less than half the cost of sulfolane. In addition, processing equipment can be designed for smaller solvent streams. Since the solvent stream in the sulfolane process is approximately six times the hydrocarbon feed stream, a sizable capital savings may be realized.

(4) The absence of water in the solvent allows for lower heat requirements to evaporate the water.

Literature Cited

(1) Cinelli, E., Girotti, P. L., Tesei, R., *Hydrocarbon Process. Petrol. Refiner* **1963**, 42 (8), 141.
(2) Colburn, A. P., Schoenborn, E. M., *Trans. Am. Inst. Chem. Engr.* **1945**, 41 (4), 421.
(3) Cousserans, G., *Hydrocarbon Process. Petrol. Refiner* **1964**, 43 (6), 149.

(4) Deal, G. H., Evans, H. D., Oliver, E. D., Papadopoulos, M. N., *Petrol. Refiner* **1959,** 38 (9), 185.
(5) Esso Research & Engineering Co., British Patent **812,114** (April 15, 1959).
(6) Feldman, J., Saffer, B., British Patent **1,067,042** (April 7, 1964).
(7) Feldman, J., Saffer, B., French Patent **1,475,984** (March 1, 1966).
(8) Feldman, J., Thomas, M., U. S. Patent **3,408,397** (October 29, 1968).
(9) *Ibid.*, **3,350,439** (October 31, 1967).
(10) Grote, H. W., *Chem. Eng. Progr.* **1958,** 54 (8), 43.
(11) Hutton, D. G., *Hydrocarbon Process. Petrol. Refiner* **1964,** 43 (8), 139.
(12) Koppers, H. G.m.b.H., Belgian Patent **707,674** (July 10, 1968).
(13) McKinnis, A., U. S. Patent **2,441,827** (May 18, 1948).
(14) Moormeier, L. F., Feldman, J., *Hydrocarbon Process. Petrol. Refiner* **1965,** 44 (12), 151.
(15) Porter, P. E., Deal, C. H., Stross, F. H., *J. Am. Chem. Soc.* **1952,** 78, 2999.
(16) Porter, R. S., Johnson, J. F., *Am. Chem. Soc., Div. Petrol. Chem., Preprints* **1959,** 4 (3), 51.
(17) Von Rock, H., *Chem. Ingr. Tech.* **1956,** 28, 489.
(18) Saunders, K. W., *Ind. Eng. Chem.* **1951,** 43, 121.
(19) Sheets, M. R., Marchello, J. M., *Hydrocarbon Process. Petrol. Refiner* **1963,** 42 (12), 99.
(20) Shoji, K., Yanagida, T., Mizukoshi, S., Inoue, S., Okawa, S., *Koru Taru* **1964,** 16 (6), 239; *Chem. Abstr.* **1964,** 61, 11825e.
(21) Still, C., German Patent **1,269,603** (July 17, 1968).
(22) Voetter, H., Kosters, W. C. G., World Petroleum Congress, 6th, Frankfurt am Main, Germany, June 1963.

RECEIVED January 12, 1970.

16

Carbon Black from Petroleum Oil

T. A. RUBLE

Continental Carbon Co., Houston, Tex.

Over 450 million gal/year of petroleum oil are consumed in the United States for producing carbon black, an amorphous form of carbon used principally in automotive tires. Broadly, the process consists of completely burning a fuel with air, injecting the atomized feedstock oil into the hot products of combustion, and recovering the carbon black by filtration. Reactor configuration and dimensions, residence time, and temperature are the critical factors. Oil feedstock is characterized by high degree of aromaticity and low ash components, particularly, sodium, potassium, lithium, iron, and copper.

Carbon black, a relatively pure form of amorphous carbon, is now being produced in the United States at a rate of almost 3 billion pounds annually. Of this, at least three-fourths is made from petroleum refinery streams, utilizing some 480 million gallons of oil; the remainder is produced from natural gas (4). Four of the seven major producers are affiliated with major oil refining companies and are considered part of their petrochemical complexes. Certainly then, the classification of carbon black as a petrochemical should be justified.

Carbon black finds its way into many products: inks, paints, paper, fertilizer, plastics, and explosives to name a few. By far the major use, however, is in automotive tires which consume 65% of the total production. The present day tire contains roughly 1 pound of carbon black for each 2 pounds of rubber and provides both the bounce and wear characteristics desired by the user. The properties carbon black imparts to rubber compounds are so critical that there are currently more than 20 classified grades of oil blacks. Most distinctions between grades are mainly a function of particle size and structure, although surface chemistry is sometimes a factor for specialty uses. The more important grade designations are illustrated in Table I. Each of those listed are also subdivided according to their structure levels.

Table I. Grade Designations for Carbon Black

Classification	Average Particle Diameter, A
Super Abrasion Furnace (SAF)	180
Intermediate Super Abrasion Furnace (ISAF)	220
Intermediate-Intermediate Super Abrasion Furnace (I-ISAF)	240
High Abrasion Furnace (HAF)	280
Fast Extruding Furnace (FEF)	470
General Purpose Furnace (GPF)	520
Semi-Reinforcing Furnace (SRF)	650

The over-all layout for oil black plants is more or less standard as shown in Figure 1, a typical flow diagram. Fuel gas, preheated combustion air, and preheated oil feedstock are fed continuously into a reactor where the carbon black is formed at temperatures ranging from 2350° to 2750°F. After the reaction is complete, the carbon-laden effluent gases are shock-cooled to 1000°F by direct water quenching prior to entry into the heat exchanger and bag filter. The carbon is removed by glass fabric filtration, then micropulverized to eliminate any hard gritty particles, and conveyed to the pelletizer. Before pelletization the apparent density of the fluffy black is from 3-6 lbs/cu ft. Pelletizing is accomplished by mixing the fluffy black with approximately an equal amount of water in a pug-like mill with extreme agitation. The wet spherical granules containing 30–55% water are dried in a heated rotating drum and discharged over a magnetic separator into storage tanks. The pelleted black is free flowing and relatively dust free, having a bulk density of 20–30 lbs/cu ft. Shipments are made either in covered hopper cars, in 50-lb bags by boxcar or truck-trailer (5, 7).

Needless to say, the focal point of an entire carbon black plant is the reactor. The basic conversion process is universal in that all reactors utilize the concept of burning a fuel with excess air to completion, atomizing oil feedstock into the hot products of combustion for thermal decomposition, and quenching the reaction with direct water sprays.

The earlier oil black reactors were large and unwieldy, containing tons of brick refractory, and several weeks were required to build or repair them. Today, however, by using the heat exchange principle, reactors can be constructed from pieces of standard black iron pipe with castable refractory weighing less than 1 ton (3). Complete repair is accomplished in less than one 8-hour shift.

The necessarily tight control of particle diameter, structure level, and surface area, however, is not quite so simple. Particle size is a function of reaction temperature and time; the higher the temperature, the smaller the particle, and a decrease in residence time will produce a smaller particle. These relationships are shown in Figures 2 and 3. Since the heat for conversion is furnished by burning fuel and a portion of the feedstock, reaction temperature depends on the hydrocarbon-to-combustion air ratio and the speed with which the atomized feedstock is mixed with the hot products of combustion. Rapid mixing is in turn governed by the internal reactor configuration and the dynamic flow pattern as well as the temperature and pressure profile of the elongated reaction chamber. Residence time is controlled by the position of the water quench and total feed rates. The smaller the particle size, the greater the reinforcing effect in rubber as measured by abrasion resistance and tensile strength.

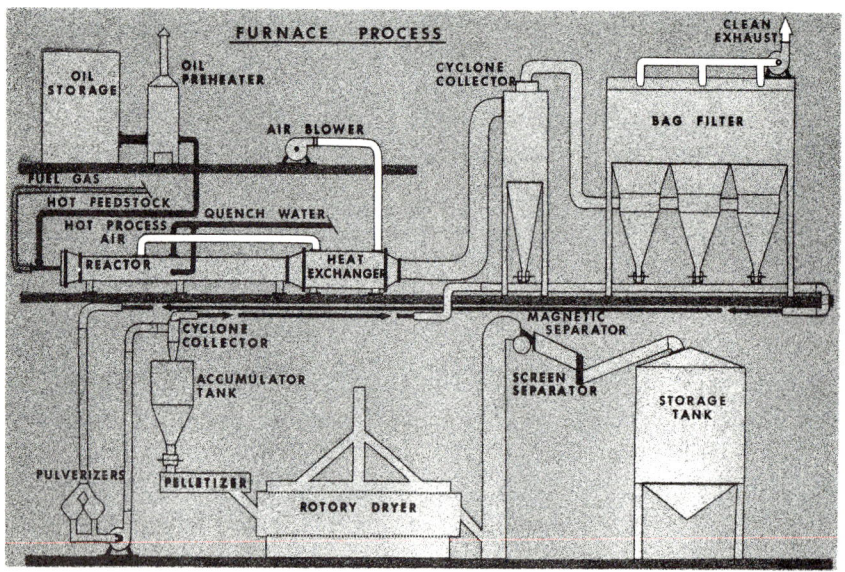

Figure 1. Flow diagram for oil black plants

The chain structure or clustering of particles results from collision and coherence of the carbon particles during their formation. Structure levels can be increased by altering the reactor aerodynamics, changing the feedstock atomization pattern, and/or increasing the feedstock concentration in the reaction zone. The reverse, of course, is the case for decreasing structure within limits. Lower structure may also be obtained by imparting the proper electrostatic charge to the carbon particles during formation, causing them to repel each other (1, 2, 6). Structure contributes to the stiffness and viscosity of the compounded rubber stock which is desirable to varying degrees, depending on end use.

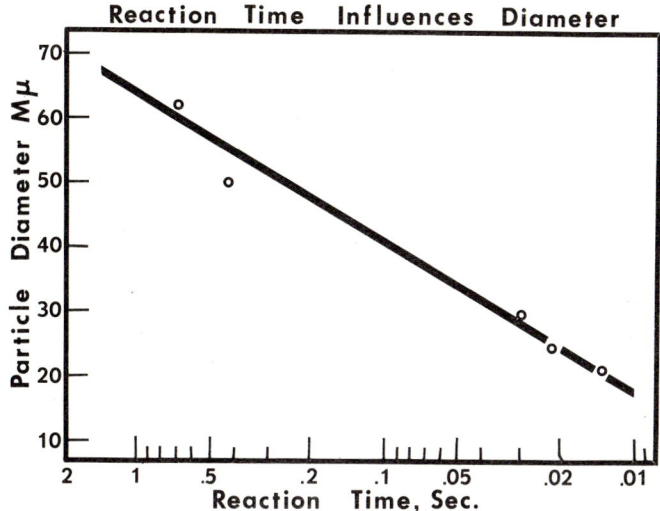

Figure 2.   Effect of reaction time on particle size

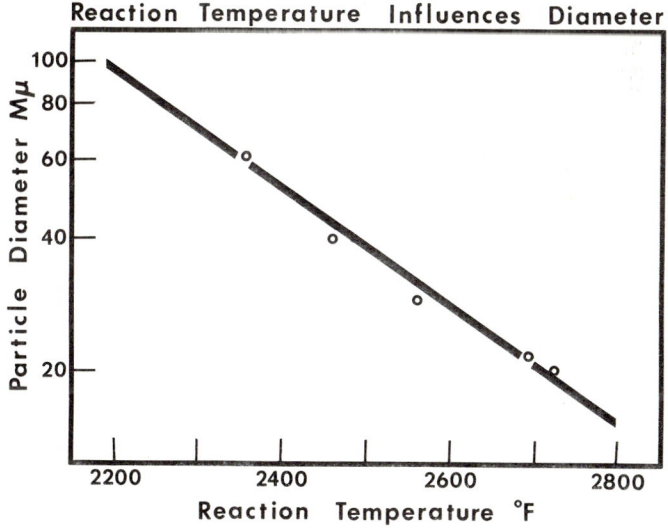

Figure 3.   Effect of reaction temperature on particle size

Surface area is naturally determined by the particle diameter unless an abnormally high value is required for adsorptive purposes. Then a porous particle can be produced by allowing a longer residence time after the reaction zone, causing surface oxidation and pitting. Usually surface oxidation is detrimental to rubber reinforcement.

Exact control of product quality is complicated by the fact that simple tests on blacks alone will not adequately measure their performance in rubber. It is necessary, therefore, to monitor the day-to-day quality by actually incorporating the black into rubber and comparing the compound with a control. Periodically, tread blacks must be compounded and road tested in multisection retreaded tires.

Because of the wide variations in operating conditions, tread and carcass blacks cannot be made in the same reactor; however, all tread grades can be produced in the same reactor as is the case with all carcass grades. Unfortunately, exact reactor construction details and operating details are proprietary information and cannot be revealed; however, exact internal configuration, dimensions, oil spray pattern, and burner design are all extremely critical for optimum quality and yields. One reactor will produce from 20,000–40,000 lbs/day, and a unit consists of from 4 to 5 reactors, discharging into common downstream facilities.

Although carbon black in some form can be made from any hydrocarbon refinery stream, many years of experience and experimentation have narrowed the feedstock specifications greatly from the standpoint of both quality and yield of product. Typical inspection tests are listed in Table II.

Table II. Feedstock Properties

API gravity	−2.0
Viscosity, 122°F, SSF	70.0
BS&W, %	0.08
Ash, %	0.05
Asphaltenes, %	3.0
Carbon, %	90.5
Hydrogen, %	7.5
Sulfur, %	1.5
Sodium, ppm	2.50
Potassium, ppm	0.24
Lithium, ppm	0.30
Iron, ppm	7.30
Copper, ppm	0.10
B.M.C.I.	132

Basically the oil must be highly aromatic and relatively free of impurities, such as sulfur, BS&W, asphaltenes, sodium, potassium, and other ash components.

Aromaticity is indicated by UOP characterization factor and carbon-to-hydrogen ratio; however, the most commonly used indicator is the Bureau of Mines Correlation Index (B.M.C.I.). Carbon black yields in general improve as the aromatic content increases.

Approximately 60% of the sulfur content of the feedstock is converted to gaseous sulfur compounds which are discharged with the effluent stack gases and could cause pollution and corrosion problems. Asphaltenes, or pentane-insoluble materials, increase the grit content by forming undesirable hard coke particles. The presence of sodium and potassium, even in minor amounts, will reduce structure level, thus making it difficult to produce the high structure blacks (*1, 2*). The presence of lithium and sodium above relatively low limits will lower the ignition temperature of the black and will tend to promote oxidation.

Most feedstocks are residuals produced by thermally cracking cat cracker cycle oils and stripping the light ends. Satisfactory feedstocks have also been made by solvent extraction of aromatics from various other refinery streams. With the advent of more active cat cracking catalysts, the heaviest product or decant oil from the cat cracker is suitable without further processing, provided the catalyst is sufficiently removed. Since uniform quality of black depends largely upon the feedstock uniformity, the installation of adequate storage with auxiliary blending equipment is required to eliminate variations.

Although the efficiency of the carbon black process has improved steadily, there is still much that can be done. Depending upon the grade of black and type of feedstock, from 50–65% of the theoretical carbon content is now being recovered from the oil as quality carbon black. A typical stack gas analysis is shown in Table III.

Table III. Stack Gas Composition (Dry Basis)

Gas	Per Cent by Volume
CO_2	4.50
CO	12.10
H_2	12.80
C_2H_2	0.61
CH_4	0.34
N_2	69.65

The composition in Table III will vary according to the grade being produced; the smaller the particle, the leaner the gas. Water vapor content is usually about 45%. These gases can and are being burned under waste heat boilers to generate steam. One European plant utilizes the steam for power generation in sufficient quantity to furnish all the power necessary for plant operation and sells the excess to the local utility company. Some consideration has been given to the utilization of the effluent gaseous components as other chemical raw materials; however, thus far the separation and purification steps pose a most difficult problem, particularly with the large amounts of nitrogen and water present.

The future of oil carbon blacks, at least at the moment, looks bright since it is tied so closely with the rubber industry. Many other materials have been tried as substitutes, but as yet their performance has fallen far short of carbon black.

Literature Cited

(1) Cabot Corp., U. S. Patent **3,010,794**.
(2) Cabot Corp., U. S. Patent **3,010,795**.
(3) "Capacity at Dutch Carbon Black Plant Doubled," *Oil Gas Intern.* **1964**, 4, 12.
(4) "Carbon Black in 1968," *Mineral Industry Surveys*, U. S. Bureau of Mines, Washington, D. C.
(5) "Carbon Black (Oil Black)," *Hydrocarbon Processing* **1967**, 46, 11, 158.
(6) Continental Carbon Co., U. S. Patent **3,223,605**.
(7) "Oil Black," *Ind. Eng. Chem.* **1952**, p. 685.

RECEIVED January 12, 1970.

17

Petrochemical Feedstocks from Residuals

A. R. JOHNSON, S. B. ALPERT, and L. M. LEHMAN

Hydrocarbon Research, Inc., a subsidiary of Dynalectron Corp., 115 Broadway, New York, N. Y. 10016

> *In Japan and Europe there is increasing demand for light distillates for producing petrochemicals and increasing governmental pressure to lower the sulfur content of residuals. The H-Oil process represents a flexible technique to (1) lower the sulfur content of residuals, and (2) maximize the yield of light distillates from low value residuals, or (3) simultaneously and economically satisfy both these objectives.*

The petrochemical refinery has been the subject of considerable discussion at technical meetings and in the literature. The general concept is that of a petrochemical facility, for producing ethylene and its usual co-products, which takes whole crude oil as the primary raw material and produces a minimum of refinery fuel products. This reflects the current situation in which the demand for petrochemical feedstock has grown to where it is now possible to consider the petrochemical plant as a viable operation in its own right, not only as an adjunct to petroleum refining.

In the past, this industry developed on feedstocks which were surplus commodities of petroleum refining and natural gas processing. In the United States this resulted in plants based on ethane and propane since with a large natural gas industry in our gasoline oriented petroleum market these tend to be the surplus materials. Conversely, in fuel oil and distillate oriented markets, such as Europe and Japan, straight-run naphtha has generally been in long supply, and the industry has built itself on that base. Because of this historic dependence on refinery and natural gas by-products, the petrochemical plant has not always been able to plan its own activities independently, being the last point in an economic chain over which it often has little control. Recently, the increasing scale of petrochemical facilities has led to consideration of entire complexes in which production of the usual petroleum products plays little part and in which crude oil would be the principal raw material.

With present technology it is entirely possible to process the bulk of the crude oil into petrochemicals, and several approaches to this problem have been considered. In this study we consider the approach in which ethylene is produced by conventional steam cracking and the heavy residual petroleum fractions are converted into steam cracker charge by means of residue hydrocracking through the H-Oil process.

Aside from the small amount produced by extraction of ethylene as a by-product from petroleum refinery streams, virtually all of this basic chemical is manufactured by the thermal cracking of petroleum fractions. Various methods of accomplishing this operation have been studied, but by far the most commercially successful technique has been thermal cracking of light petroleum fractions mixed with steam in a tubular cracking coil with the requisite heat provided by external firing. This area of technology has been the object of substantial development by several firms, and this process has been improved considerably, both in efficiency of operation and in the range of feedstocks which can be considered for commercial application. Units are now on stream using feeds ranging from ethane to gas oil fractions. Relative economics and the technical aspects of steam cracking these various petroleum fractions were discussed recently (1).

The principal technical problems associated with converting crude oil into ethylene and other light olefins and aromatics stem from conversion of the heavier fractions and in particular the heavy residues which contain asphaltenes, metalloorganic compounds, etc. Several approaches can be taken to solve this problem.

(1) *Direct Contact.* There are several processes at various stages of development for conversion of heavy feedstocks to ethylene. These use various heat carriers and thereby avoid the problem of coke deposition in tubes which can be associated with thermally cracking residues by the more conventional route. This type of process is characterized by the use of moving beds, fluid beds, or gaseous heat carriers. Typical of these processes are the developments of Hoechst, Lurgi, BASF, Unde, etc. (4, 6).

(2) *Residue Conversion.* Instead of converting residue to ethylene directly, these fractions can be converted to lighter fractions which can in turn be steam cracked in a conventional furnace to yield the desired end products. Such conversion can be accomplished in several ways, among which are coking and residue hydrocracking. When the crude available is sweet, and a market exists for coke, coking can be economically attractive. In residue hydrocracking the residual fractions can be converted by hydrocracking into steam cracker charge with a small residue yield which can be consumed as plant fuel. It is this approach which we consider at this time.

Before considering the economics of this approach to ethylene manufacture, we discuss briefly some of the technical aspects of residue hydrocracking.

The H-Oil Process

H-Oil is a new technique for hydrogenating and hydrocracking residues and heavy oils, developed through the joint efforts of Hydrocarbon Research, Inc. and Cities Service Research & Development Co. Several papers have reviewed commercial experience and application studies of this process (2, 3, 5, 7).

The need for better techniques in residue processing has been recognized for a long time. As far back as 1912 Bergius worked on the hydrocracking of Galacian and Roumanian residues. Hydrocarbon Research, Inc. and Cities Service have been active in this area for over 10 years in an effort promoted by the evident need for such a process. Initially, these investigations centered on the fixed bed reactor. In this work, the same several difficulties encountered by other investigators were met— *i.e.,* temperature control, bed plugging and liquid distribution, and maintenance of catalyst activity.

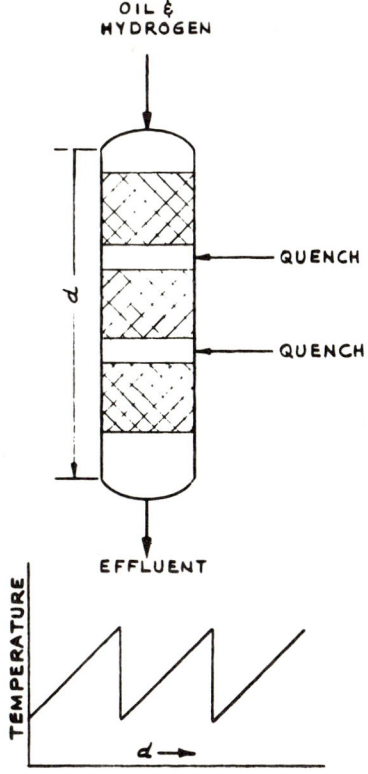

Figure 1. Conventional temperature control in fixed bed reactor

Temperature Control. Hydrogenation reactions are characterized by high exothermic heats of reaction. In fixed bed reactors this is controlled by using multiple bed reactors with quenching between beds (*see* Figure 1). While effective, this has several disadvantages—(a) the temperature is not at the optimum for the desired reaction but is lower or higher through most of the reactor; (b) it is necessary to redistribute the liquid between beds.

Bed Plugging and Liquid Distribution. With residues in high temperature environment, it is important to maintain intimate contact of oil and hydrogen to prevent the formation of coke deposits. Since residues contain sediments, there is a distinct tendency for these materials to deposit in a fixed catalyst bed. Once this occurs, flow distribution is disturbed, and the tendency to form coke deposits is aggravated. This is compounded by the high metals content of these feeds. These metal compounds tend to deposit on catalysts, deactivating the catalyst locally and leading to further coke deposition. This situation requires shutdown and catalyst replacement or regeneration.

Maintenance of Catalyst Activity. All catalysts deactivate with time, and in the presence of residual stocks containing metals, this tendency is aggravated. This requires periodic replacement or regeneration. With a fixed bed reactor this can be achieved only by shutting down the reactor or the entire plant. To solve these problems in the H-Oil process, a different approach is taken.

Instead of passing oil and hydrogen down through the reactor, these materials flow up through the catalyst bed. In addition, liquid is recycled from the top of the reactor to the bottom to attain a velocity sufficient to expand the catalyst bed by 20–40% over its settled volume. This results in a reactor with the following characteristics: (1) the reactor is filled with liquid as the continuous phase; (2) catalyst is maintained in a state of random motion (*see* Figure 2).

This reactor system achieves the basic reactor design characteristic for residuum-hydrogen reactions. This reactor system (the ebullated bed) has these several advantages in this service:

(1) Isothermal. Temperature gradients from top to bottom—even on commercial scale reactors—seldom exceed 5°F and are usually less.

(2) Clean. Since the catalyst is in a state of motion, there is no problem of bed plugging arising from dirty feeds. In fact, on commercial scale operations feedstocks have been processed which contained fine cat cracking catalyst. This catalyst, being finer than the H-Oil catalyst, passed through the reactor without plugging the bed. Since there is excellent oil–hydrogen contact, there is no tendency for coke to form.

(3) Catalyst Can Be Added and Withdrawn. Owing to the ebullated state, the catalyst can be withdrawn from and added to the reactor while the plant is on stream. This permits efficient maintenance at a high

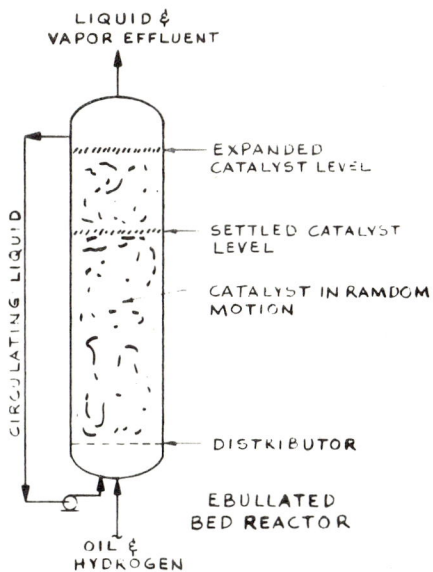

Figure 2. Ebullated bed reactor maintaining catalyst in motion

catalyst activity without shutdown. Thus, there would be no "swing" in conditions to compensate for catalyst activity decline.

Extensive bench scale and pilot plant investigations were conducted on the H-Oil reactor system, using pilot plant equipment ranging in capacity from 5 GPD to 30 BPD. Along with the pilot plant research fundamental studies of the ebullating system were made.

Once the basic technology had been developed in bench scale and pilot plant equipment, it was necessary to demonstrate the novel technical features on a commercial scale. To do this, Cities Service elected to build a 2500-BPD unit at its Lake Charles refinery. This plant has been operating since 1963 and has fully demonstrated the novel technical features of the process.

During the initial period of operation, several areas were identified which required further development for good operation. These problems have been solved, and the unit presently functions as a regular commercial feature of Lake Charles refinery. It is significant to note that the reactor system—the ebullated bed—has functioned in a completely satisfactory fashion from the outset. This has proved to be an extremely effective technique for processing heavy dirty stocks and asphalts to yield lighter materials for further processing. Over this operating history several feedstocks have been processed in the Lake Charles unit. Operating results from some typical operations are shown in Table I. Further com-

Table I. Operating Performance

Feedstock		Sour Vacuum Bottoms		Sour Atmospheric Bottoms	
Material Balance		Feed	Product	Feed	Product
C_1–C_3 lb/day			10,687		8,110
C_4–400°F, BPD			202		193
400°–680°F, BPD			386		330
680°–975°F, BPD			385	690	790
975°F+, BPD		2,030	1,174	810	253
Total		2,030	2,147	1,500	1,566
H_2 Consumption, SCF/B		870		630	
Properties					
C_4 & Heavier	gravity, °API	9.5	21.1	15.0	27.8
	sulfur, wt %	2.68	0.41	2.97	0.55
C_4–400°F	gravity, °API		56.8		61.3
	sulfur, wt %		0.02		0.1
400°–680°F	gravity, °API		32.1		29.2
	sulfur, wt %		0.06		0.21
	aniline pt. °F		158		128
680°–975°F	gravity, °API		22.2		21.6
	sulfur, wt %		0.17		0.49
	aniline pt. °F		175		165
975°F+	gravity, °API		13.3		10.1
	sulfur, wt %		0.65		1.32
	con. carb., wt %		17.4		22.3

Table II. H-Oil Units in Operation and under Construction

Company	Location	Capacity, BPD	Startup Date
Cities Service Oil Co.	Lake Charles, La.	2,500	1963
Kuwait National Petroleum Co.	Shuaiba, Kuwait	28,000	1969
Humble Oil Co.	Bayway, N. J.	16,500	1970
Petroleos Mexicanos	Salamanca, Mexico	18,500	1971

mercial acceptance of the H-Oil Process is shown by the units now in operation or under construction shown in Table II.

H-Oil Process Data

Table III presents several sets of process yields which can be obtained by processing Kuwait atmospheric residue at four different severity levels achieving various yields of light products. As the yield of light products increases, the hydrogen consumption also increases, but at the same time the consumption of hydrogen per unit of steam cracker charge of the Lake Charles H-Oil Unit

	Cutback Asphalt		Heavy Cycle Stock	
	Feed	Product	Feed	Product
		17,600		2,200
		205		76
		1,204	242	637
	1,342	821	2,568	2,338
	1,397	616		
	2,739	2,846	2,810	2,983
	800		1,000	
	11.9	21.0	18.7	24.9
	2.20	0.43	0.88	0.03
		54.6		34.0
		Nil		
		30.1	24.0	26.1
		0.08	0.61	0.02
		104		
		18.5	18.0	24.9
		0.32	0.89	0.03
		143	145	156
		0.6		
		1.17		

Table III. Process Yields—H-Oil Processing of Kuwait Atmospheric Residuum

Case Component—Wt. % on Feed	1	2	3	4
H_2S & NH_3	3.5	3.0	3.6	4.1
C_1	0.4	0.8	1.3	2.0
C_2	0.3	0.9	1.5	2.3
C_3	0.4	1.0	1.7	2.7
C_4	0.4	1.0	0.9	1.9
C_5–180°F	1.0	2.8	3.5	7.2
180°–350°F	3.1	5.0	5.4	12.6
350°–650°F	14.2	24.5	31.7	59.3
650°–975°F	48.6	40.5	38.0	—
975°–1050°F		5.6	3.6	—
1050°F+	29.2	16.3	10.4	10.4
	101.1	101.4	101.6	102.5
H_2 Cons., SCF/B	670	900	1050	1675

actually decreases. This results from the fact that at the lower severities the unconverted residue consumes a significant quantity of hydrogen arising from desulfurization and saturation reactions involving uncon-

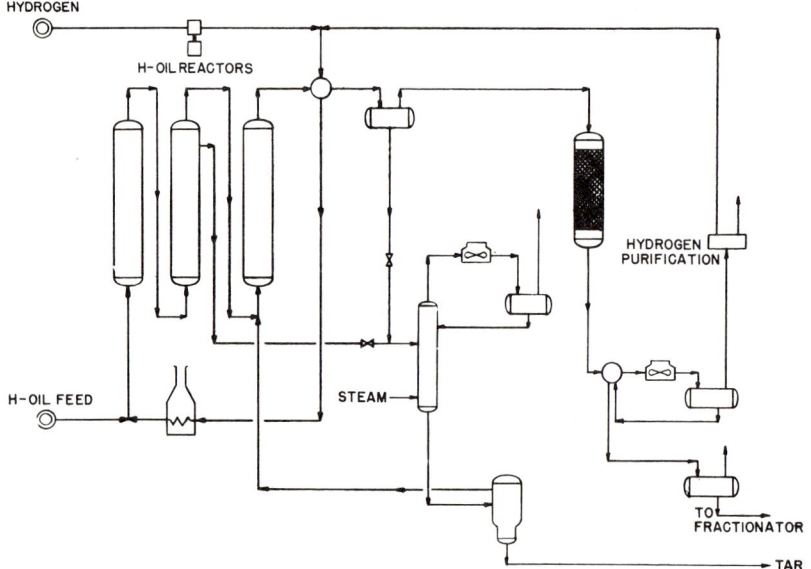

Figure 3.   H-Oil flow diagram

verted molecules. The petrochemical plant would most probably select the most severe operation shown (Case 4) since this results in the maximum yield of steam cracker charge (C_2-650°F).

An H-Oil high conversion process flow diagram is shown in Figure 3. This would be the type employed in Case 4 of Table III. Residual oil is mixed with recycle hydrogen and processed through two H-Oil reactors in series. Liquid effluent from the second reactor is stabilized and vacuum distilled, and the vacuum gas oil from this stream is blended back with the vapor effluent from the second H-Oil reactor and processed in a third reactor for conversion of heavy gas oil. Effluent from this third reactor is cooled and hydrotreated in a fixed bed to saturate olefins produced in the H-Oil operation. Liquid condensed from this vapor effluent is then returned to crude atmospheric distillation, thereby returning a portion of this material boiling above 650°F to the H-Oil unit to achieve extinction of the heavy gas oil fraction. Non-condensible gas from the vapor effluent is passed through a hydrogen purification section, and the hydrogen is recycled back to the first reactor in the processing sequence.

Table IV shows analyses of typical H-Oil distillate products. These are inspections of untreated fractions as they are produced directly from the H-Oil reactor. If these are subjected to a mild treatment (*e.g.*, in the fixed bed reactor described earlier), then these properties would change

Table IV. Analysis of Typical H-Oil Products

Crude Source	North African		Middle East	
Fraction Processed	Gas Oil	Vacuum Resid	Gas Oil	Vacuum Resid
Feedstock Properties				
°API	28.3	18.1	27.6	11.2
% sulfur	0.26	0.6	1.90	4.46
Product Properties				
Lt. nap. (C_5–180°F), °API	80.7	76.2	81.7	76.9
% S	0.02	0.09	0.12	0.34
Oct. No. F–1 cl/3cc	65.6/85.7	73.2/85.3	64.0/84.1	71.8/84.5
Heavy nap. (180–350°F), °API	58.0	56.8	57.9	54.6
% S	0.01	0.13	0.02	0.40
Oct. No. F–1 cl/3cc	30.0/58.7	54.0/70.0	32.3/60.7	48.0/61.3
P	62.6	46.7	63.8	48.7
O	4.7	28.6	3.3	23.8
N	25.5	19.7	21.9	17.3
A	7.2	5.0	11.0	10.0
Kerosene (350°–500°F), °API	41.7	41.9	41.6	39.4
% S	0.06	0.11	0.06	0.50
smoke pt., mm	19	20	17	16
aniline pt. °F	140.1	133.0	132.5	128.5
pour pt. °F	–30	<–35.0	<–35	<–35
Distillate (500°–650°F), °API	30.6	31.8	30.1	28.7
% S	.07	0.13	0.09	0.90
aniline pt. °F	156.3	156.0	148.0	141.0
pour pt., °F	40	20	30	20

somewhat. Principally, the sulfur would be virtually removed, and the olefins would be saturated to paraffins and naphthenes. The two points worthy of note are: first, the treated products are remarkably similar to desulfurized virgin fractions from Middle East crude, and secondly, there is no notable difference between the products produced from North African and Middle East crudes. A further important point is shown in Table V which gives the distribution of iso- and normal paraffins in the light naphtha produced. The proportions of isoparaffins is low, which indicates a good quality steam cracker charge.

Petrochemical Refinery Based on Whole Crude

To illustrate the economics of the approach being considered, we have considered a case in which 1 billion lbs/year of ethylene is produced from whole Kuwait crude using an H-Oil/steam cracking combination. Figure 4 presents the over-all process flow for this operation. Crude oil

Table V. H-Oil Light Naphtha Composition

Charge: West Texas Vacuum Residue, 10.2° API, 2.92 % S

Composition of IBP—200°F Fraction from Pilot Plant Operation

Component	Wt %
C_3	0.62
$i\text{-}C_4$	0.97
$n\text{-}C_4$	5.43
$i\text{-}C_5$	6.73
$n\text{-}C_5$	16.15
2-Methylpentane	12.70
3-Methylpentane	6.85
$n\text{-}C_6$	25.28
Methylcyclopentane	7.85
Benzene	1.93
Cyclohexane	15.49

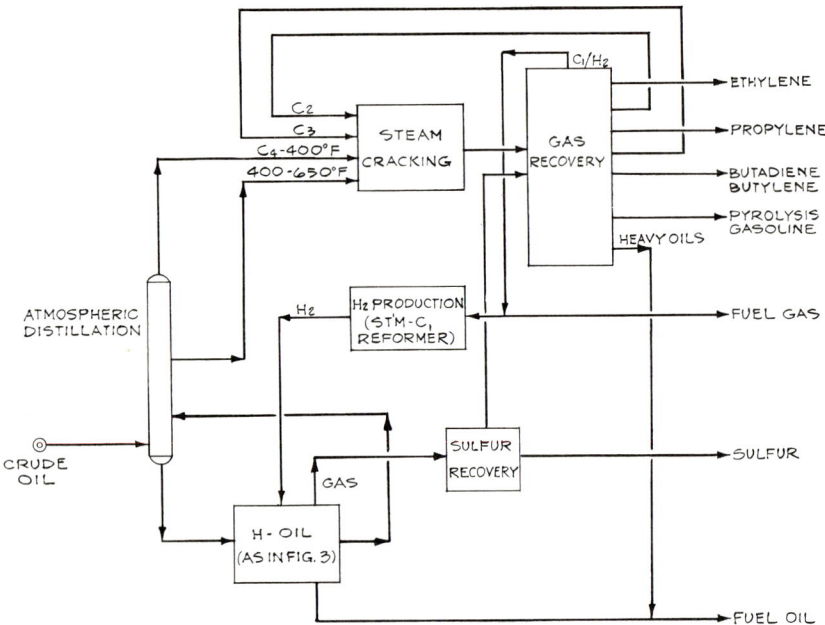

Figure 4. Ethylene production from whole crude

is charged to an atmospheric distillation tower, and the residue boiling above 650°F is processed in an H-Oil unit. The H-Oil severity is set to achieve a high yield of products boiling below 650°F. The distillate products from H-Oil are returned to the atmospheric distillation tower where, together with the virgin materials, they are separated into naphtha and distillate fractions for charge to steam cracking. Gases from the

H-Oil unit are processed for sulfur recovery and then sent for separation through the gas recovery facilities associated with the steam cracker. Remaining unconverted residue from the H-Oil operation is used as a fuel oil component for plant fuel. Ethylene is manufactured by steam cracking of ethane, propane, naphtha, and distillate, and products from these operations are separated in conventional gas recovery facilities. Hydrogen for H-Oil is partially supplied by by-product recovery from steam cracker and H-Oil off-gases supplemented by steam reforming of methane. The heavy oils produced in steam cracking of naphtha and distillate are blended with the H-Oil residue to yield plant fuel.

Table VI presents the over-all material balance on the H-Oil unit used in this case. Conversion is achieved, of all but 9 vol % of the feed, into material boiling below 650°F. The sulfur contents of the distillate fractions produced are quite low, and should it be desirable to control the sulfur content of all of the steam cracker charge, the virgin naphtha and distillate could also be processed in the H-Oil unit to achieve an over-all purification of steam cracker charge.

Table VII presents over-all ethylene plant balances on two bases: (1) a whole crude H-Oil operation (as described above), and (2) an ethylene unit based on naphtha.

In the latter case 34,600 BPCD of naphtha is processed, while in the whole crude case 33,300 BPCD of crude is charged. Process yields have been set for equal production of ethylene, and most of the chemical by-products are roughly equal in quantity for the two cases, with the exception of sulfur. Consequently, comparison of these two alternates is not complicated by involved considerations of by-product marketing. On the fuel yield side, however, as would be expected, there is a considerably

Table VI. H-Oil Processing of Kuwait Atmospheric Residue to Yield Steam Cracking Feedstocks

Charge: Kuwait Atmospheric Residue, 14.5° API, 4.06 % S

Process Yields₃	Wt %	Vol %	°API	% S
H_2S, NH_3	4.1			
C_1	2.0			
C_2	2.3			
C_3	2.7			
C_4	1.9	3.2		
C_5–180°F	7.2	10.4	80	
180°–350°F	12.6	16.4	58	0.01
350°–650°F	59.3	68.2	37	0.05
Tar	10.4	9.0	−5	3.0
	102.5	107.2		

Hydrogen consumption: 1675 SCF/B (chemical)

Table VII. Over-All Plant Balances—Ethylene Production from Naphtha and Kuwait Crude[a]

Basis: Ethylene Production = 1×10^9 lb/yr (454,000 MT/yr)

Charge Stock	Naphtha	Kuwait Crude	Differential Crude Case − Naphtha Case
Charge rate, BPCD	34,600	33,300	
Charge rate, lb/yr × 10^6	3,205	3,780	
Products			
Fuel Gas, BPCD FOE	5,548	4,320	− 1,228
Ethylene, lb/yr × 10^6	1,000	1,000	—
Propylene, "	518	525	+ 7
Butadiene, "	144	141	− 3
Butylenes, "	144	157	+ 13
C_5/400°F, "	705	592	− 113
400°F+ fuel oil, BPCD	1,130	4,210	+ 3,080
Sulfur, MT/yr	—	39,400	+39,400

Differential by-product value[a] of crude case over naphtha case = $1,347,000/yr.[b]

[a] Naphtha steam cracking vs. H-Oil and naphtha and gas oil cracking on crude.
[b] Bases:
    Fuel $ 2.40/B FOE
    Propylene   2.0¢/lb
    Butadiene   4.6¢/lb
    Butylenes   1.0¢/lb
    C_5/400°F   1.3¢/lb
    Sulfur $30.00/MT

larger quantity of fuel oil made in the crude oil operation than when naphtha is processed.

Applying the product values shown in the table, we arrive at a diferential by-product credit for the crude case vs. the naphtha case of approximately $1,300,000/year. Most of this represents sulfur value.

Table VIII summarizes the investment and operating requirements for the H-Oil unit and its attendant facilities, and Table IX compares investment and operating costs for the naphtha and crude oil based facilities. Data on the ethylene plant investment and operating costs were taken from the paper by Freiling, Huson, and Tucker (1). Offsite investments have been taken at 30% of process investment for the naphtha case, as well as for the ethylene plant portion of the crude oil case. For the H-Oil unit and its associated units offsites have been estimated at one-half of this rate—15%. This lower value has been used since the H-Oil system will add little to the storage requirements, and all utilities have been priced to cover the capital requirements for their production.

Comparison of the total annual costs (including capital charges at 15% but excluding feed costs) show that the crude processing route

Table VIII. Investment and Operating Requirements for H-Oil and Associated Facilities

Unit	H-Oil	Others[a]	Total
Investment, $	13,300,000	6,700,000	20,000,000
Utilities			
Fuel, mm Btu/hr	85	229	314
Power, kw	9,100	1,550	10,650
Steam, lb/hr	49,000	39,000	88,000
Cooling water, GPM	6,000	8,600	14,600
Catalyst, $/SD	2,950	240	3,190

[a] Other facilities: crude distillation, hydrogen production and purification, and sulfur recovery.

Table IX. Cost Comparison for Ethylene Production from Naphtha or Crude Oils

Charge Stock	Naphtha	Kuwait Crude
Plant Investments, $ millions		
Process Facilities		
Ethylene	35.0	38.0
H-Oil and associated facilities	—	20.0
Total process	35.0	58.0
Offsites	10.5	14.4
Total plant investment	45.5	72.4
Annual Production Costs, $ million/year		
Operating Costs		
Ethylene	10.7	12.4
H-Oil and associated facilities	—	4.5
Total operating costs	10.7	16.9
Capital charges @ 15%	6.9	10.0
Total costs (without feed costs and product credits)	17.6	27.8

costs $10,200,000 more per year than the naphtha route. After allowing for by-product credit this reduces to $8,900,000. To achieve equal ethylene manufacturing cost, Kuwait crude must be available at a price sufficiently lower than naphtha to cover this difference. Figure 5 shows the relationship between crude and naphtha prices which will result in equal cost ethylene. This relationship is specific for Kuwait crude, and lighter or heavier crudes would show different values relative to naphtha. The most significant item is that at current market prices crude oil can be considered competitive with naphtha in many situations. Consequently, based on presently demonstrated commercial technology, the petrochemical planner can consider whole crude as a possible economic source of his basic raw raterial.

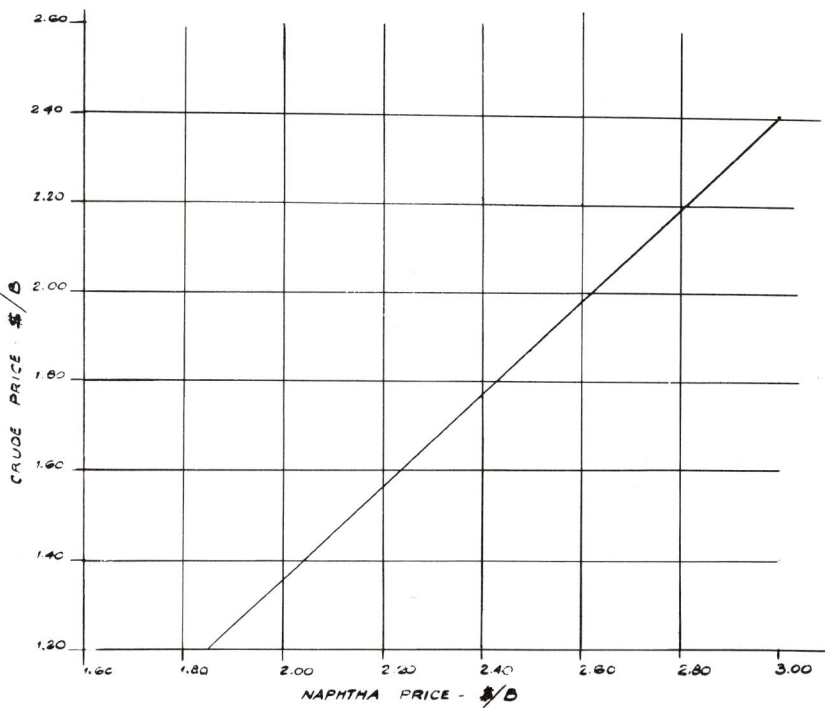

Figure 5. Crude/naphtha price relationship for equal ethylene manufacturing cost

Integration of Fuel Oil Desulfurization and Ethylene Production

In most industrial centers air pollution has become a critical problem, and refiners are under pressure to produce lower sulfur fuel oils to meet local standards. This is true in two areas—the Caribbean and Japan —which are also centers of a burgeoning petrochemical industry. Starting with a high sulfur crude, production of low sulfur fuel oil is a costly operation which adversely affects the refiner's economic position. Our studies have shown that significant savings can be realized by integrating fuel desulfurization with the production of petrochemical feedstocks. To illustrate, let us consider a desulfurization operation which might be located in Japan.

Table X shows the process yields obtained in H-Oil desulfurization of a Kuwait atmospheric residuum to yield 1% sulfur fuel oil. Desulfurization operations are conducted under conditions set to maximize the hydrogenation reactions and to minimize hydrocracking to achieve the requisite degree of sulfur removal at a minimum hydrogen consumption. Inevitably however, the desulfurization reactions are accompanied

Table X. Desulfurization of Kuwait Atmospheric Residue

Feedstock: 17.7°API, 3.71 wt % S.

Process Yields₃	Wt %	Vol %	°API	% S
H₂S, NH₃	3.2			
C₁–C₃	0.5			
C₄–400°F	0.9	1.2	60	<0.07
400°–650°F	16.5	19.0	40	0.2
650°–975°F	49.2	52.3	27	0.55
975°F+	30.6	30.0	15	1.73
	100.9	102.5		

Hydrogen consumption: 540 SCF/B

Table XI. Desulfurization and Hydrocracking of Kuwait Residue

Feedstock: 14.5°API, 4.1 wt % S

Process Yields₃	Wt %	Vol %	°API	% S
H₂S, NH₃	3.7			
C₁	0.6			
C₂	0.6			
C₃	0.6			
C₄–400°F	7.5	10	60	
400°–650°F	16.4	19.3	40	0.1
650°–975°F	35.3	37	22	0.4
975°F+	36.4	35	10	1.6
	101.1	101.3		

Hydrogen consumption: 670 SCF/B

by some carbon–carbon bond cracking, producing some light products. Typically, the fuel oil would be the 250°F+ material, and the lighter products (C_1 through 250°F) would be used to manufacture the hydrogen needed. The fuel oil produced this way is lighter than the feed (23.2°API *vs.* 14.5°API), further illustrating the cracking and hydrogenation which occurs. One method of turning this cracking to a useful purpose would be to utilize the light products for petrochemical feed.

In this study, we have again assumed that the petrochemical manufacturer would purchase whole crude as his basic raw material. As in the previous case, the 650°F and lighter material would be fractionated and used directly as steam cracker charge. Atmospheric residue, as feedstock, would be sent to a desulfurization facility, assumed to be adjacent to the petrochemical plant. This residue would be processed in an H-Oil unit together with the high sulfur fuel oil (also atmospheric residue) which the desulfurization plant was handling. The H-Oil effluent would be separated into low sulfur fuel oil and 650°F and lighter material, which

Table XII. Cost Comparison for Fuel Desulfurization Integration vs. Crude Conversion

	Case 1, Desulf.	Case 2, Desulf. + ethylene Production	Difference, Case 2 − Case 1	Crude Case (From Tables VI–IX)
Fuel oil Charge, BPSD	50,000	50,000	—	—
Crude oil charge, BPSD	—	33,300	33,300	33,300
Ethylene production, lbs/yr		1×10^9	1×10^9	1×10^9
H-Oil investment, $	15,200,000	20,200,000	5,000,000	13,300,000
H-Oil hydrogen consumption, MM SCFD	27	44.6	17.6	31.0

would be returned to the petrochemical plant to be steam cracked together with the virgin fractions.

To achieve the required quantity of steam cracker charge, the severity of this H-Oil operation would be set to provide more light products than shown in the simple desulfurization case in Table X. A typical yield structure might be that shown in Table XI. This severity could be set to match the desired ratio between low sulfur fuel production and ethylene production.

For the case given in Table XI a comparison has been made in Table XII showing the relative attractiveness of this situation with the direct crude utilization case shown earlier. Again, the study is based on production of 1 billion lbs/year of ethylene and the desulfurization of 50,000 BPSD of Kuwait atmospheric residue. The combination case has been set so that the feedstocks processed are the same and the same yield of ethylene is achieved.

The principal effects are:

(1) The net incremental hydrogen consumption to produce steam cracker charge is considerably less in the combination case. This results from the fact that we are not hydrocracking the heaviest fractions of the crude oil, which are considerably more hydrogen deficient than the lighter materials.

(2) Incremental H-Oil investment attributable to steam cracker charge manufacture is much less than in the unintegrated case.

This approach offers an opportunity to reduce the net costs of providing petrochemical feed and desulfurizing fuel oil which could be attractive in certain areas (*i.e.*, Japan and the Caribbean) where facilities of both types are needed.

There are many possible variations on this theme. For example:

(a) High sulfur fuel oil would be the only feedstock (*i.e.*, no whole crude requirement).

(b) The steam cracker charge would be naphtha and lighter only to match an existing facility.

Literature Cited

(1) Freiling, J. C., Huson, B. L., Tucker, W., *Oil Gas Intern.* **1968**, 8 (12), 84.
(2) Griswold, C. R., VanDriesen, R., 31st Midyear Meeting, API, Division of Refining, 1966.
(3) Johnson, A. R., Albert, S. B., Lehman, L. M., 33rd Midyear Meeting, API, Division of Refining, 1968.
(4) Kamptner, H. K., *Am. Chem. Soc., Div. Petrol. Chem., Preprints* **1968**, 13 (4), A23.
(5) McFatter, W., Meaux, E., Mounce, W., VanDriesen, R., 34th Midyear Meeting, API, Division of Refining, 1969.
(6) Suzukawa, Y., Kono, H., *Japan Chem. Quart.* **1969**, 5 (1), 12.
(7) Van Driesen, R., Stewart, N. C., 29th Midyear Meeting, API, Division of Refining, 1964.

RECEIVED June 15, 1970.

INDEX

A

Acetaldehyde	2
Acetone	2
Acetonitrile	217, 224
Acidic catalysts	40
Alkane pyrolysis	2
Alkylaromatics	61
hydrocracking of	56
Alkylbenzenes	140
on acidic catalysts, hydrocracking reactions of	60
cyclization of	58
Alkylcyclohexanes	52
Alkylnaphthalene concentrate	141
Aluminosilicates, crystalline	64
Arrhenius parameters	6
Arrhenius plots	73
Aromatics (see also Petrochemicals)	135
capacity	235
containing ethylbenzene, octafining process for isomerization of C_8	204
demethylation of	32
extraction, solvent for	227
production, chemistry of	21
production via catalytic reforming, chemistry of	20
separation, solvent for	242
Arosolvan	242
Azeotropes	228

B

Bed plugging	274
Benzene (see also Petrochemicals)	137
Bicyclic aromatics of the naphthalene series	141
Biradical reactions	15
Black from petroleum oil, carbon	264
BTX (see also Petrochemicals)	23, 137, 139, 227
Butadiene	135, 137
Diels-Alder reaction of	73
recovery, improved solvents for	215
thermal reaction of	82
thermal reactions of olefins with	86
Butane–butenes system, rate modeling for the	92
Butenes, thermal reactions of	80

C

Carbenes	15
Carbon-alumina	199
Carbon black from petroleum oil	264
Carmex	242
C_8 aromatics containing ethylbenzene, octafining process for isomerization of	204
C_8 aromatics, separation of	206
Catadiene process, Houdry	164
Catalysts (see also Petrochemical processing)	
acidic	40
carbon alumina	199
chromia-aluminosilicates	32, 92, 201
dual functional	22, 38
iron	95
molybdena-alumina	198
nickel	41, 48, 95, 179, 183
platinum-alumina	21, 56, 201
Catalysts, shape selective	64
silica-alumina	41, 18, 208
Catalysts, zeolite-containing	64
Catalytic dehydrogenation of butane	92
Catalytic dehydrogenation of higher normal paraffins to linear olefins	193
Catalytic hydrocracking	38
Catalytic reforming, chemistry of aromatics production via	20
C_5 fraction	138
C_8 fraction	139
C_{10}–C_{13} fraction	141
Chain lengths	11
Chromia-alumina	95, 201
C_8 naphthenes, reforming	29
Cobalt on silica-alumina	45
Coke oven light oils	227
Cope rearrangement	3
Countercurrent extraction	249
Countersolvent	257
Cracked gasolines	227
Cracking mchanism of hydrocarbons	110
C_9 range	140
Crystalline aluminosilicates	64
Cyclic compounds, formation of	87
Cyclic transition state	3
Cyclization	44, 57
of alkylbenzenes	58
Cyclobutanes	16
Cyclododecane	54
Cyclohexane	26, 179
reforming	22, 26
Cyclohexene	86

289

Cycloparaffin hydrocarbons, kinetics of the cracking of vs. olefins .. 120
Cycloparaffins on acidic catalysts, hydrocracking reactions of .. 60
Cycloparaffins, reaction of 48
Cycloparaffin ring, stability of ... 54
Cyclopentadiene 140
Cyclopentanes 26
Cyclopentanes vs. cyclohexanes .. 49
Cyclopropanes 16

D

n-Decane 40, 41
N-Decylbenzene 57
DEG 230
Dehydrogenation
   of butane, catalytic 92
   of higher normal paraffins to linear olefins, catalytic 193
   of propane, oxidative 169
Demethylation of aromatics 32
Demethylation of methylcyclohexane, kinetics of the 179
1,3-Dicyanobutane, properties of .. 243
Dicyanobutane in ternary systems 248
Diels-Alder reaction of olefin and butadiene 73
Diethylene glycol 230, 237
Dimethylcyclohexanes, reforming.. 26
Dimethylformamide 242
Dimethyl sulfoxide 242
Disproportionation 43
Dripolenes, hydrogenated 227
Dual functional catalysts 22, 38

E

Ebullated bed reactor system 274
Economics of petrochemical refinery 38, 127, 130, 132, 137, 142, 211, 277
"Ene" reaction 3
Ethylbenzene 204
Ethylbenzene, octafining process for isomerization of C_8 aromatics containing 204
Ethylcyclohexanes, reforming 26
Ethylene (see also Petrochemicals) 134, 271
   pyrolysis processes for 158
   thermal reaction of 72
Extractive distillation 244
Extraction, solvent for aromatics .. 227

F

Feedstock, petrochemical 123
Feedstocks from residuals, petrochemical 271
Formation of aromatics 90
Formation of cyclic compounds ... 87
Free radical processes 1
Fuels refinery 123
Fuel value 128

Furfural 217, 223

G

Gasoline blending 124
Gasolines, cracked 227
Gas-phase pyrolysis of hydrocarbons 1

H

n-Heptane, reforming 28
Heptane in ternary systems 248
n-Hexadecane 40
Hexamethylbenzene 59
1-Hexene cracking scheme 113
Hoffman-Woodward rules 17
H-Oil process 272-3, 276
Houdry Catadiene process 164
1,5-H shift reaction 3
Hydrocarbons, cracking mechanism of 110
Hydrocarbons, gas-phase pyrolysis 1
Hydrocracking 21
   of alkyl aromatics 56
   chemistry of 38
   of n-decane 41
   paraffin 39
   reactions of cycloparaffins and alkylbenzenes on acidic catalysts 60
Hydrodealkylation 141
Hydrogenated dripolenes 227
Hydrogenation of butenes 95
Hydrogenolysis 44, 56
Hydroisomerization 43

I

Improved solvents for butadiene recovery 215
Indenes 140
Initiation reactions 10
"Intrinsic activation energy" 8
Iron catalyst 95
Isobutane cracking 161
Isomerization 21, 44, 52
   of C_8 aromatics containing ethylbenzene, octafining process for 204
   of radicals 9

K

Kinetics of the demethylation of methylcyclohexane 179
Kuwait crude 24, 276

L

Lead-free gasoline 133
Light catalytic reformate 249
Linear olefins, catalytic dehydrogenation of higher normal paraffins to 193

INDEX

Liquid distribution 274
Liquid-liquid extraction 246

M

β-Methoxypropionitrile 217, 226
Methyl carbonate 242
Methyl Cellosolve 217, 223
Methylcyclohexane 62
  kinetics of the demethylation of 179
4-Methylcyclohexene 86
Methylcyclopentadiene 140, 141
Methylcyclopentene 27, 79
Methyl glutaronitrile 243
Methylindene 141
N-Methylpyrrolidone 239, 242
Middle East crude 279
Modeling of pyrolytic systems 17
Molecular elimination reactions of hydrocarbons 2
Molecular sieve catalysts 155
Molybdena-alumina 198
Morpholine 242

N

Naphthalene series, bicyclic aromatics of the 141
Naphtha reforming 23
Naphtha, straight-run 271
Ni catalyst 40, 45, 48, 95
Nonacidic catalysts 56, 61
n-Nonane, reforming 31
Non-Arrhenius rate constants 13
North African crude 279

O

Octafining material balance 209
Octafining process for isomerization of C_8 aromatics containing ethylbenzene 204
Oils, coke oven light 227
Olefin
  Diels-Alder reaction of 73
  hydrocarbons 110
  catalytic dehydrogenation of high normal paraffins to linear .. 193
  in pyrolysis of petroleum hydrocarbons, secondary reactions of 68
  vs. kinetics of the cracking of paraffin and cycloparaffin hydrocarbons 120
  with butadiene, thermal reactions of 86
Oxidative dehydrogenation of propane 169

P

Palladium mordenite catalysts 48
Paring reaction 48, 59
Paraffin
  hydrocarbons, kinetics of the cracking of vs. olefins 120
  hydrocracking 39
  isomerization 21
Paraffins (see also Petrochemicals) 26
  singly branched vs. unbranched 48
  to linear olefins, catalytic dehydrogenation of higher normal 193
1-Pentene cracking scheme 118
2-Pentene cracking scheme 116
Petrochemical feedstock 123, 271
Petrochemicals
  aromatics ..20, 33, 38, 56, 60, 68, 124, 127, 133–5, 139–40, 204, 228, 242
  benzene (see also BTX) 56–7, 73, 79, 133, 139
  BTX (see also benzene, toluene, xylenes) 23, 137, 139, 146, 228, 242
  butadiene 68, 82, 85, 127, 135, 215, 226
  butenes 80–82, 92, 95, 104
  carbon black 264–270
  cyclic compounds ...57, 60, 62, 79, 81
  cyclohexane 179–192
  cyclohexene 25
  cyclopentane 25
  ethylbenzene 29, 32, 204, 207
  methylbenzene 59
  methylcyclopentane 27, 79
  ethylene 72, 124, 127, 134, 158, 161, 185, 272, 284
  naphtha 241, 282
  naphthenes 29
  olefins 68–90, 110, 124, 127, 193
  paraffins 38, 43, 48, 54, 59, 60, 68, 120, 193
  propylene 74, 124, 127, 133, 135, 153–76
  toluene (see also BTX) ...28, 76, 79, 132, 136, 139
  xylenes (see also BTX) 29, 30, 32, 79, 132, 136, 204, 228
Petrochemical processing
  catalysts in
    acidic 40
    carbon-alumina 199
    chromia-alumina 32, 92, 201
    iron 95
    molybdena-alumina 198
    nickel 179
    nickel-on-kieselguhr 183
    nickel sulfide 41, 48
    nonacidic 61
    platinum-alumina 21, 56, 201
    silica-alumina 41, 48, 208
  catalytic hydrocracking in..38–65, 195
    catalysts used in 38, 40–45, 195
  catalytic reforming
    in 20–33, 124, 137

Petrochemical Processing (Continued)
H-Oil 271, 278
Houdry Catadiene 165
octafining 204, 213
reactions in
   biradical 15
   cracking 124, 138, 146, 153
   cyanoethylation 227
   cyclization 57, 61
   dealkylation 57, 206
   dehydrogenation 21, 26, 41, 92,
                  104, 137–9, 153,
                  164, 169, 193–203
   demethylation 32, 179, 183
   Diels-Alder 87
   disproportionation 43
   hydrocracking 21, 38, 42, 64–5,
                  110, 206, 273
   hydrogenation 92, 206, 273
   hydrogenolysis 62, 185, 189
   initiation 10
   isomerization 2, 9, 21, 43,
                      59, 204–6
   metathesis 5
   paring 48, 52, 59
   pyrolysis 1–17, 68, 158, 162
   termination 12
   transalkylation 32
separation and purification
   in 124, 215, 227, 228, 242
   crystallization 33
   distillation ...32, 215, 228, 242, 244
   fractionation 32
   Parex process 33
   solvent extraction 242
   Udex process 230, 242
thermal pyrolytic cracking
   in 1–17, 68, 138, 158
Petrochemical refinery 20, 123–136,
                   137, 229, 279
   economics of 38, 127, 132, 137,
                  142, 211, 279
   vs. fuels refinery 124–7, 131,
                  137, 142
Petroleum oil, carbon black from 264
Phenanthrene 62
Platinum-alumina 201
Platinum on silica-alumina 46
Polycyclic aromatics 62
Polymethylcyclopentanes, hydro-
   genolysis of 56
Product optimization in the petro-
   chemical refinery 137
Products from pyrolysis 139
Products of thermal cracking 138
Product price influence 132
Profitability 142
Propane, oxidative dehydrogenation
   of 169
Propylene (*see also* Petrochemicals) 135
   manufacture 153
   shortage 154
   thermal reaction of 74

Pyrolysis 1–17, 68, 158, 162
   of hydrocarbons, gas phase 1
   of petroleum hydrocarbons, sec-
      ondary reactions of olefins in 68
   processes for ethylene 158
   products from 139
   very low pressure 13
Pyrolytic systems, modeling 17

R

Radical fission 14
Radical reactions with unsaturates 8
   isomerization of 9
   tetramethylene type 16
   trimethylene type 16
Raffinate, solvent recovery from .. 259
Rate constants, non-Arrhenius ... 13
Rate modeling for the butane–butenes system 92
Rate models 101
Reactions of cycloparaffins 48
Refinery complexity 127
Reformate, light catalytic 249
Reformed gasolines 227
Reforming 22
   catalysts 22
   chemistry of aromatics produc-
      tion *via* catalytic 20
   C_8 naphthenes 29
   cyclohexane 22, 26
   ethyl- and dimethylcyclohexanes 26
   methylcyclopentane 27
   n-heptane 28
   n-nonane 31
   reactions 25
Residuals, petrochemical feedstocks
   from 271
Residue hydrocracking 272
RON 139

S

Secondary reactions of olefins in
   pyrolysis of petroleum hydro-
   carbons 68
Selective solvent 215
Separation of C_8 aromatics 206
Shape selective catalysts 64
Shock tube techniques 13
Silica-alumina catalysts 154
Singly branched *vs.* unbranched
   paraffins 48
Solvent
   for aromatics extraction 227, 242
   recovery from raffinate 259
   for butadiene recovery, improved 215
Steam cracking 272
Straight-run naphtha 271
Styrenes 140
Sulfolane 239, 243, 248

T

Tetraethylene glycol 227, 239

INDEX

Tetralin 62
Tetramethylene type radical 16
Termination reaction 12
Ternary systems 248
Thermal cracking, products of ... 138
Thermal reactions
   of butenes 80
   of butadiene 82
   of ethylene 72
   of olefins with butadiene 86
   of propylene 74
Toluene 28, 137, 248
Transalkylation 32
Transition state, cyclic 3
Triethylene glycol 236
Trimethylene type radicals 16

U

Udex extraction process 229, 242
Unsaturates, radical reactions with 8

V

Vapor-phase catalytic dehydrogenation 99
Very low pressure pyrolysis 13
4-Vinycyclohexene 86
Voevodsky reaction 110

W

Whole crude, petrochemical refinery based on 279

X

Xylenes (see also BTX and Petrochemicals) 137, 204

Z

Zeolitic catalysts 64, 154